BIRKHÄUSER

Trends in Mathematics is a series devoted to the publication of volumes arising from conferences and lecture series focusing on a particular topic from any area of mathematics. Its aim is to make current developments available to the community as rapidly as possible without compromise to quality and to archive these for reference.

Proposals for volumes can be sent to the Mathematics Editor at either

Birkhäuser Verlag
P.O. Box 133
CH-4010 Basel
Switzerland

or

Birkhäuser Boston Inc.
675 Massachusetts Avenue
Cambridge, MA 02139
USA

Material submitted for publication must be screened and prepared as follows:

All contributions should undergo a reviewing process similar to that carried out by journals and be checked for correct use of language which, as a rule, is English. Articles without proofs, or which do not contain any significantly new results, should be rejected. High quality survey papers, however, are welcome.

We expect the organizers to deliver manuscripts in a form that is essentially ready for direct reproduction. Any version of TeX is acceptable, but the entire collection of files must be in one particular dialect of TeX and unified according to simple instructions available from Birkhäuser.

Furthermore, in order to guarantee the timely appearance of the proceedings it is essential that the final version of the entire material be submitted no later than one year after the conference. The total number of pages should not exceed 350. The first-mentioned author of each article will receive 25 free offprints. To the participants of the congress the book will be offered at a special rate.

Positivity

Karim Boulabiar
Gerard Buskes
Abdelmajid Triki
Editors

Birkhäuser
Basel · Boston · Berlin

Editors:

Karim Boulabiar
Institut Préparatoire aux
Etudes Scientifiques et Techniques
Université de Carthage
B.P. 51
2070 La Marsa
Tunisia
e-mail: karim.boulabiar@ipest.rnu.tn

Gerard Buskes
Department of Mathematics
University of Mississippi
Hume Hall 305
P. O. Box 1848
University, MS 38677-1848
USA
e-mail: mmbuskes@sunset.olemiss.edu

Abdelmajid Triki
Département de Mathématiques
Faculté des Sciences de Tunis
Campus Universitaire
1060 Tunis
Tunisia
e-mail: Triki@fst.rnu.tn

2000 Mathematical Subject Classification 06A70, 06D22, 06F15, 28B05, 46A17, 46A40, 46B22, 46E30, 46L05, 46L50, 46L52, 47B34, 47B40, 47B65, 47D40, 47G10, 47L07, 46L10, 46M05, 54C40, 54G05, 54H10

Library of Congress Control Number: 2007933653

Bibliographic information published by Die Deutsche Bibliothek. Die Deutsche Bibliothek lists this publication in the Deutsche Nationalbibliografie; detailed bibliographic data is available in the Internet at http://dnb.ddb.de

ISBN 978-3-7643-8477-7 Birkhäuser Verlag AG, Basel - Boston - Berlin

Basel · Boston · Berlin
P.O. Box 133, CH-4010 Basel, Switzerland
Part of Springer Science+Business Media
Printed on acid-free paper produced from chlorine-free pulp. TCF ∞
Printed in Germany
ISBN 978-3-7643-8477-7 e-ISBN 978-3-7643-8478-4

9 8 7 6 5 4 3 2 1 www.birkhauser.ch

Contents

Preface vii

B. Banerjee and M. Henriksen
Ways in which $C(X)$ mod a Prime Ideal Can be
a Valuation Domain; Something Old and Something New 1

D.P. Blecher
Positivity in Operator Algebras and Operator Spaces 27

K. Boulabiar, G. Buskes, and A. Triki
Results in f-algebras 73

Q. Bu, G. Buskes, and A.G. Kusraev
Bilinear Maps on Products of Vector Lattices: A Survey 97

G.P. Curbera and W.J. Ricker
Vector Measures, Integration and Applications 127

J. Martínez
The Role of Frames in the Development of Lattice-ordered
Groups: A Personal Account 161

B. de Pagter
Non-commutative Banach Function Spaces 197

A.R. Schep
Positive Operators on L^p-spaces 229

A.W. Wickstead
Regular Operators between Banach Lattices 255

Preface

This collection of surveys is an outflow from the 2006 conference Carthapos06 in Tunis (Tunisia). Apart from regular conference talks, five survey talks formed the core of a workshop in Positivity, supported by the National Science Foundation. The conference organizers (Karim Boulabiar, Gerard Buskes, and Abdelmajid Triki) decided to expand on the idea of core surveys and the nine surveys in this book are the harvest from that idea.

Positivity derives from an order relation. Order relations are the mathematical tool for comparison. It is no surprise that seen in such very general light, the history of Positivity is ancient. Archimedes, certainly, had the very essence of positivity in mind when he discovered the law of the lever. His method of exhaustion to calculate areas uses a principle that nowadays carries his name, the Archimedean property. The surveys in this book are slanted into the direction that Archimedes took. Functional analysis is heavily represented. But there is more. Lattice ordered groups appear in the article by Martinez in the modern jacket of frames. Henriksen and Banerjee write their survey on rings of continuous functions. Blecher and de Pagter in each of their papers survey parts of non-commutative functional analysis. Positive operators are the main topic in the papers by Curbera and Ricker, Schep, and Wickstead. And positive bilinear maps are the protagonists in the survey by Bu, Buskes, and Kusraev. The conference organizers (and editors of this volume) write about f-algebras.

Carthapos06 was more than just a conference and workshop in Africa. It brought together researchers in Positivity from many directions of Positivity and form many corners of the world. This book can be seen as a culmination of their paths meeting in Tunisia, Africa.

June 5, 2007

G. Buskes
Oxford, U.S.A.

Positivity
Trends in Mathematics, 1–25

Ways in which $C(X)$ mod a Prime Ideal Can be a Valuation Domain; Something Old and Something New

Bikram Banerjee (Bandyopadhyay) and Melvin Henriksen

Abstract. $C(X)$ denotes the ring of continuous real-valued functions on a Tychonoff space X and P a prime ideal of $C(X)$. We summarize a lot of what is known about the reside class domains $C(X)/P$ and add many new results about this subject with an emphasis on determining when the ordered $C(X)/P$ is a valuation domain (i.e., when given two nonzero elements, one of them must divide the other). The interaction between the space X and the prime ideal P is of great importance in this study. We summarize first what is known when P is a maximal ideal, and then what happens when $C(X)/P$ is a valuation domain for every prime ideal P (in which case X is called an SV-space and $C(X)$ an SV-ring). Two new generalizations are introduced and studied. The first is that of an almost SV-spaces in which each maximal ideal contains a minimal prime ideal P such that $C(X)/P$ is a valuation domain. In the second, we assume that each real maximal ideal that fails to be minimal contains a nonmaximal prime ideal P such that $C(X)/P$ is a valuation domain. Some of our results depend on whether or not $\beta\omega \setminus \omega$ contains a P-point. Some concluding remarks include unsolved problems.

1. Introduction

Throughout, $C(X)$ will denote the ring of real-valued continuous functions on a Tychonoff space X with the usual pointwise ring and lattice operations and $C^*(X)$ will denote its subring of bounded functions, and all topological spaces considered are assumed to be Tychonoff spaces unless the contrary is stated explicitly. (Recall that X is called a Tychonoff space if it is a subspace of a compact (Hausdorff) space. Equivalently if X is a T_1space and whenever K is a closed subspace of X not containing a point x, there is an $f \in C(X)$ such that $f(x) = 0$ and $f[K] = \{1\}$.) An element of $C(X)$ is nonnegative in the usual pointwise sense if and only if it a square. So algebraic operations automatically preserve order. This makes the notion of positivity essential for studying $C(X)$. This simple observation was used with great ingenuity by M.H. Stone in 1937 to make the first thorough study of $C(X)$ as a ring. It was restricted to the case when X is compact. Among the many interesting results in this seminal paper is that $C(X)$ determines X. That is, if X and Y are compact spaces and $C(X)$ and $C(Y)$ are algebraically isomorphic, then X and Y are homeomorphic.

This study was broadened to include unbounded functions in [Hew48] by Stone's student E. Hewitt. While this paper contains a number of serious errors, it set the tone for a lot of the research that led to the book [GJ76]. (It was published originally in 1960 by Van Nostrand). For more background and history of this subject, see [Wa74], [We75], [Hen97], and [Hen02]. Our general sources for general topology are [E89] and [PW88].

Sections 2 and 3 survey some of what has been done in the past about integral domains that are homomorphic images of a $C(X)$ and the prime ideals P that are kernels of such homomorphisms. We concentrate especially on the cases when $C(X)/P$ is a valuation domain. In Section 2, we review some of what is known when P is maximal; i.e., when $C(X)/P$ is a field. Section 3 recalls what is known about spaces X such that at $C(X)/P$ is a valuation domain whenever P is a prime ideal of $C(X)$. They are called SV-spaces. The remainder of the paper focuses on new research beginning with the study in Section 4 of almost SV-spaces; that is, spaces X and rings $C(X)$ in which every maximal ideal of $C(X)$ contains a minimal prime ideal P such that $C(X)/P$ is a valuation domain. Section 5 is devoted to the study of products of almost SV-spaces and logical considerations concerning the validity of some results. The one-point compactification of a countable discrete space is not an SV-space, but the consequences of the assumption that it is an almost SV-space are studied in Section 6. Spaces X and rings $C(X)$ in which every real maximal ideal of $C(X)$ contains a prime ideal such that $C(X)/P$ is a valuation domain are examined in Section 7. In the final Section 8, two related papers and the contents of a book are discussed briefly, some sufficient conditions are given to say more about valuation domains that are homomorphic images of a ring $C(X)$, and some unsolved problems are posed.

2. What happens when the valuation domains are fields?

A commutative ring A such that whenever a and b are nonzero elements of A, it follows that one of them divides the other, is called a *valuation ring*. Below, we are interested only in the case when A is also an integral domain, in which case such a ring A is called a *valuation domain*. We begin with the case when the valuation domain is a field, and recall that the kernel of a homomorphism onto a field is a maximal ideal. Let $\mathcal{M}(A)$ denote the set of maximal ideals of A. This set is nonempty as long as A has an identity element

Recall that a field F is said to be *real-closed* if its smallest algebraic extension is algebraically closed. Equivalently, F is real-closed if it is totally ordered, its set

F^+ of nonnegative elements is exactly the set of all squares of elements of F, and each polynomial of odd degree with coefficients in F vanishes at some point of F. As is shown in Chapter 13 of [GJ76], if $M \in \mathcal{M}(C(X))$, then $C(X)/M$ is a real-closed field. We continue to quote facts from [GJ76].

If $f \in C(X)$, then $Z(f)$ denotes $\{x \in X : f(x) = 0\}$, and we let $coz(f) = X \backslash Z(f)$. If $S \subset C(X)$, we let $\mathcal{Z}[S] = \{Z(f) : f \in S\}$. Thus $\mathcal{Z}[C(X)]$ (which we abbreviate by $\mathcal{Z}[X]$) is the family of all zerosets of functions in $C(X)$. A subfamily $\mathcal{F}$ of $\mathcal{Z}[X]$ that is closed under finite intersection, contains $Z(g)$ whenever if contains some $Z(f) \in \mathcal{F}$, and does not contain the empty set is called a *z-filter*. Note that an element f is in some proper ideal if and only if $Z(f) \neq \varnothing$. It follows that if I is a proper ideal of $C(X)$, Then $\mathcal{Z}[I]$ is a z-filter.

An ideal I is *fixed* or *free* according as $\cap\{Z(f) : f \in I\}$ is nonempty or empty. A maximal ideal M is fixed if and only if $\mathcal{Z}[M] = \{x\}$ for some $x \in X$, in which case M is denoted by M_x. Clearly, $C(X)/M$ always contains a copy of $\mathbb{R}$. The maximal ideal M is called *hyper-real* if $C(X)/M$ contains $\mathbb{R}$ properly and is called *real* otherwise. Every fixed maximal ideal is real, but the converse fails to hold. If every real maximal ideal of $C(X)$ is fixed, then X is called a *realcompact space*. Subsequent to the appearance of [GJ76], hyper-real fields are also called *H-fields*.

Recall that the continuum hypothesis CH is the assumption that the least uncountable cardinal ω_1 is equal to the cardinality 2^ω of the continuum.

2.1 Definition. Suppose that an ordered set L satisfies: If A and B are countable subsets of L such that $a < b$ whenever $a \in A$ and $b \in B$, then there is an $x \in L$ such that $a < x < b$ whenever $a \in A$ and $b \in B$ (Symbolically we write this conclusion as $A < x < B$.) Then L is called an *η_1-set.*

Much of what is known about H-fields of cardinality no larger than 2^ω is summarized next.

2.2 Theorem

(a) Every H-field is both real-closed and an η_1-set.

(b) All real-closed fields that are η_1-sets of cardinality ω_1 are (algebraically) isomorphic.

(c) Every η_1-set has cardinality at least 2^ω.

(d) All H-fields of cardinality 2^ω are isomorphic if and only if CH holds.

All but part of (d) are shown in Chapter 13 of [GJ76]. That there is only one H-field (in the sense of isomorphism) of cardinality 2^ω implies CH is due to A. Dow in [D84]. Some more detail about what may happen if CH fails: see [R82].

There are a large number of results concerning H-fields of large cardinality in [ACCH81] that depend on various set-theoretic hypotheses and use proof techniques involving combinatorial set theory. Most of its contents are beyond the scope of this article. (Some errors in [ACCH81] are pointed out by A. Blass in his review in Math. Sci. Net. None of them affect what is written above.)

3. When every prime ideal of C(X) is a valuation prime; SV rings and spaces

As in [DW96], a prime ideal P of $C(X)$ is called a *valuation prime* if $C(X)/P$ is a valuation domain. Chapter 14 of [GJ76] is devoted to the study of the set of prime ideals of $C(X)$, and a little is said about the order-structure of $C(X)/P$ when P is a prime ideal of $C(X)$, but the first thorough study of valuation primes and their associated valuation domains appears in [CD86]. Inspired by this, a number of authors began to investigate rings $C(X)$ and spaces X such that every prime ideal of $C(X)$ is a valuation prime. Such rings and spaces are called SV-rings and SV-spaces respectively. See [HW92a], [HW92b], and [HLMW94].

The order structure of $C(X)/P$ is described completely when X is the one-point compacification $W(\omega+1)$ of $W(\omega)$ in [M90] and to a lesser extent in Chapter 4 of [DW96] when X is compact. See Section 6 below.

The *Stone-Čech compactification* βX is the compact space that contains X as a dense subspace such that each member of the subring $C^*(X)$ of bounded functions in $C(X)$ has a (unique) extension in $C(\beta X)$. Thus $C^*(X)$ and $C(\beta X)$ are isomorphic.

What follows next will be used often below. See [GJ76].

3.1 Proposition and definitions. Suppose X is a Tychonoff space. Then:

(a) There is a bijection from βX onto the set $\mathcal{M}(\mathcal{C}(X))$ of maximal ideals of $C(X)$ given by $p \to M^p = \{f \in C(X) : p \in cl_{\beta X} Z(f)\}$.

(b) Each prime ideal of $C(X)$ is contained in a unique maximal ideal M^p, and the intersection of all the prime ideals contained in M^p is $O^p = \{f \in C(X) : cl_{\beta X} Z(f)$ is a neighborhood of $p\}$.

(c) The prime ideals containing a prime ideal P form a chain (that is, are totally ordered) under set inclusion, and P contains minimal prime ideals. The number of minimal prime ideals contained in the maximal ideal M is called its *rank* $Rk(M)$ and is denoted by ∞ unless it is finite. If X is compact, the *rank* $Rk(X)$ of X is the supremum of the ranks of the maximal ideals of $C(X)$. It is known that if the rank of each maximal ideal of $C(X)$ is finite, then $Rk(X) < \infty$. (See [HLMW94].)

(d) If $p \in X$, then M^p is denoted by M_p and O^p by O_p. If $O_p = M_p$, then p is called a *P-point.* A space all of whose points are P-points is called a *P-space.* X is a *P-space* if and only if $Z(f)$ is clopen for each $f \in C(X)$ if and only if $C(X)$ is a von Neumann regular ring. Moreover, every compact P-space is finite.

(e) If for all $p \in \beta X$, the prime ideals containing O^p are totally ordered, then X is called an *F-space.* (Equivalently, X is an F-space if O^p is prime for every $p \in \beta X$.) Every P-space is an F-space, every F-space is an SV-space and the implied inclusions are proper. Moreover, no SV-space can contain a nontrivial convergent sequence; that is, a closed copy of the one-point compactification $W(\omega+1)$ of the space $W(\omega)$ of finite ordinals. (See 3.4 below and [HLMW94].)

If $P \subset Q$ are prime ideals of $C(X)$, then $C(X)/Q$ is a homomorphic image of $C(X)/P$. Thus, if P is a valuation prime, then so is Q. Hence:

3.2 Proposition. X is an SV-space if and only if every minimal prime ideal of $C(X)$ is a valuation prime.

Let $mC(X)$ denote the set of minimal prime ideals of the ring $C(X)$. If $f \in C(X)$, let $h(f) = \{P \in mC(X) : f \in P\}$ and let $h^c(f) = mC(X)\backslash h(f)$. Using $\{h^c(f) : f \in C(X)\}$ as a base for a topology, $mC(X)$ becomes a zero-dimensional Hausdorff space called *the space of minimal prime ideals* of $C(X)$. See [HJe65]. This space has been studied extensively, but we recall only those facts known about it that are relevant to this paper.

Recall that if I is an ideal of $C(X)$ such that $f \in I$ and $Z(g) = Z(f)$ imply that $g \in I$, then I is called a *z-ideal.* As is shown in [HJe65], the map $P \to P \cap C^*(X)$ is an order preserving homeomorphism of $mC(X)$ onto $mC^*(X)$. So $C(X)$ and $C(\beta X)$ have homeomorphic spaces of minimal prime ideals. Again a prime z-ideal P of $C(X)$ is valuation prime if and only if $P \cap C^*(X)$ is a valuation prime of $C^*(X)$ by Corollary 2.1.12 of [CD86]. Thus we have:

3.3 Proposition. X is an SV-space if and only if βX is an SV-space.

Thus, to determine the algebraic properties of an SV-ring $C(X)$, there is no loss of generality in assuming that X is compact.

3.4 Remark. It has long been known that every F-space is an SV-space. (See, for example [L86].) If X and Y are two disjoint F-spaces, then the attaching of X and Y at two non P-points respectively of X and Y; serves as an example of an SV-space which fails to be an F-space and consequently the class of SV-spaces contains the class of F-spaces properly. (This is noted in [HW92a] and [HW92b], and implicitly in [CD86].)

Recall that a space Y is said to be C^*-*embedded* (resp. C-*embedded*) in a space X if the map $f \to f|Y$ is a surjection of $C^*(X)$ onto $C^*(Y)$ (resp. of $C(X)$ onto $C(Y)$). We close this section with more known facts about SV-spaces.

3.5 Proposition and remarks

(a) Every C^*-embedded subspace of an SV-space is an SV-space and consequently closed subspaces of compact SV-spaces are SV-spaces. (See [HW92a]).

(b) If a compact space X can be expressed as the union of finitely many closed subspaces such that each of them is an SV-space, then X becomes an SV-space, but not every SV-space can be represented in this way. (See [HW92a] and [L03]. The example in [L03] is the result of a complicated construction.)

(c) If X is compact and each point of X has a closed neighborhood that is an F-space, then X becomes an SV-space. Though the converse need not hold. Let X be the union of two disjoint copies of $\beta\omega$ (ω denotes the set of countable ordinals). and Y be the space by identifying the corresponding points of

$\beta\omega \setminus \omega$. Then every nonisolated point of Y has no F-space neighborhood, though Y becomes a compact SV-space. (See [HW92a].)

(d) If X be an *almost discrete space*, i.e., a Hausdorff space with exactly one nonisolated point, then X is an SV-space if and only if X is of finite rank. (See [HW92b].)

(e) Every compact SV-space has finite rank. It is not known if there is a compact space of finite rank which fails to be an SV-space. (See [HLMW94]).

4. The first generalization; almost SV rings and spaces

Below $\beta\omega$ and $\alpha\omega$ will abbreviate $\beta(W(\omega))$ and $W(\omega+1)$ respectively.

4.1 Definition. $C(X)$ is called an *almost SV-ring* and X an *almost SV-space* if each maximal ideal of $C(X)$ contains a valuation prime that is a minimal prime ideal.

Next, it is shown how to create a large class of almost SV-spaces that are not SV-spaces.

4.2 Theorem. The space obtained by attaching the nonisolated point of $W(\omega+1)$ to a compact F-space at a non P-point is an almost SV-space that is not an SV-space.

Proof. Let Y denote a compact F-space with a point q such that O_q is not maximal. (For example, Y could be $\beta\omega$ and q any point of its nonisolated points.) Let X denote the result of attaching the spaces Y and $\alpha\omega$ at the points $q \in Y$ and $\omega \in \alpha\omega$, and call the resulting point p. By 3.1(e), X is not an SV-space because it contains a sequence of distinct points converging to p. Because Y is an F-space and each point of $\alpha\omega$ other that ω is a P-point, to see that X is an almost SV-space, it suffices to show that there is a minimal valuation prime of $C(X)$ containing O_p.

Let φ and ψ denote respectively the restriction maps of $C(X)$ onto $C(\alpha\omega)$ and $C(Y)$. Clearly each is a surjective homomorphism. Because O_q is not maximal, its inverse image $\psi^{-1}(O_q)$ is a nonmaximal prime ideal of $C(X)$ containing O_p. We show next that $O_p = \varphi^{-1}(O_\omega) \cap \psi^{-1}(O_q)$. For, f is in this intersection if and only if its restriction to $\alpha\omega$ vanishes on a neighborhood of ω, and its restriction to Y vanishes on a neighborhood of q if and only if f vanishes on a neighborhood of p.

Because in any commutative ring a prime ideal that contains the intersection of two ideals must contain one of them, it follows that any prime ideal of $C(X)$ that contains O_p must contain at least one of $\varphi^{-1}(O_\omega)$ or $\psi^{-1}(O_q)$. Moreover, because q is not a P-point of Y, $\varphi^{-1}(O_\omega)$ is not contained in $\psi^{-1}(O_q)$. Hence there cannot be any prime ideal of $C(X)$ containing O_p and properly contained in $\psi^{-1}(O_q)$. Thus the latter is a minimal prime ideal of $C(X)$ containing O_p.

Suppose $\pi : C(Y) \to C(Y)/O_q$ is the canonical homomorphism. Then $\pi \circ \psi : C(X) \to C(Y)/O_q$ is a surjective homomorphism with whose kernel $\ker(\pi \circ \psi) =$

$\psi^{-1}(O_q)$. So $C(X)/\ker(\pi \circ \psi)$ and $C(Y)/O_q$ are isomorphic. Because Y is an F-space, this latter is a valuation domain and we may conclude that $\psi^{-1}(O_q)$ is a minimal valuation prime ideal. This concludes the proof of the theorem. □

We digress to quote some results in [CD86] that will be useful in what follows.

4.3 Concepts and results from [CD86].

(a) Suppose p is a point in a Tychonoff space X and $\Im$ is a z- filter of subsets of X. If for every $f \in C(X \setminus \{p\})$ such that $0 \le f \le 1$, there is $Y \in \Im$ such that $\lim_{x\to p} f|Y$ exists, then $\Im$ is called a $P(p)$ *filter*. See [CD86] for a discussion of the properties of such filters. The authors do not describe $P(p)$ filters as z-filters, but treat them as one whenever they use it.

(b) Suppose X is compact and each of its points is a G_δ point. Substantial use will be made below of a mapping γ introduced in a more general setting by C. Kohls in [K58] and described in more detail in [CD86]. See Section 2.2 of [CD86] for proofs and more details about the assertions made below

  (1) Suppose $p \in X$ is nonisolated and $Y = X \setminus \{p\}$. Then γ is a bijection from the set Q of prime z-ideals of $C(Y)$ such that $\mathcal{Z}[Q]$ converges to some point in $\beta Y \setminus Y$ onto the family $\Im$ of all nonmaximal prime z-ideals of $C(X)$ contained in M_p. Then $\gamma(Q)$ is defined by $\mathcal{Z}[\gamma(Q)] = \{cl_X Y : Y \in \mathcal{Z}[Q]\} = \{Y \cup \{p\} : Y \in \mathcal{Z}[Q]\}$.

  (2) The prime z-ideal Q is maximal if and only if $\gamma(Q)$ is an immediate prime z-ideal predecessor of M_p by 3.2(2) of [CD86].

  (3) $\mathcal{Z}[Q]$ is a $P(p)$ filter if and only if $\mathcal{Z}[\gamma(Q)]$ is a $P(p)$ filter.

  (4) If P is a valuation prime contained properly in M_p and $Q \subset P$ is a minimal prime ideal, then $\mathcal{Z}[Q]$ is a $P(p)$ filter. (See Theorem 2.2.2 of [CD86].)

Recall that $\upsilon X = \{p \in \beta X : M^p \text{ is a real maximal ideal}\}$ and that $C(\upsilon X)$ and $C(X)$ are isomorphic. υX is called the (Hewitt) *realcompactification* of X. The proof of the next result is an exercise.

4.4 Theorem. The following assertions are equivalent.

(1) X is an almost SV-space.
(2) υX is an almost SV-space.
(3) βX is an almost SV-space.

4.5 Corollary. If Y is a dense C^*-embedded subspace of an almost SV-space X, then Y is an almost SV-space.

Proof. For then βY and βX are homeomorphic, so the conclusion follows from Theorem 4.4. □

Recall again that the question of whether a compact space of finite rank is an SV-space was left as an open problem in Section 6 of [HLMW94]. So, the second part of the hypothesis of the next theorem may be redundant, and its conclusion may be too weak.

4.6 Theorem. If X is a compact space of finite rank and each of its points has a compact neighborhood that is an an almost SV-space, then X is an almost SV-space.

Proof. Because X is compact, $\{M_p : p \in X\}$ is the set of all maximal ideals of $C(X)$. We need to show that each M_p contains a minimal prime ideal that is a valuation prime. For each $p \in X$, there is a compact neighborhood $T = T(p)$ of p that is an an almost SV-space. Let $\varphi : C(X) \to C(T)$ denote the map that sends each $f \in C(X)$ to its restriction to T. The map.φ is an epimorphism since T is C-embedded in X. As usual, $O_p = \{f \in C(X) : p \in int_X Z(f)\}$ and letting $O_p^T = \{f \in C(T) : p \in int_T Z(f)\}$, we see that $\varphi^{-1}(O_p^T) = O_p$. Note also that the inverse image under φ of a pair of incomparable prime ideals of $C(T)$ containing O_p^T is a pair of incomparable prime ideals of $C(X)$ containing O_p because φ is an epimorphism.

Let m and n denote respectively the ranks of p with respect to T and X. Because X is compact, it follows from Corollary 1.8.2 of [HLMW94] that $m \leq n. < \infty$. It follows that O_p^T is the intersection of m incomparable minimal prime ideals $\{P_i^T\}_{i=1}^m$ of $C(T)$ and hence that O_p is is the intersection of the minimal prime ideals $\{\varphi^{-1}(P_i^T)\}_{i=1}^m$. In any commutative ring, if a prime ideal contains a finite intersection of ideals, it must contain one of them. So, we may conclude that $\{\varphi^{-1}(P_i^T)\}_{i=1}^m$ is the collection of all minimal prime ideals of $C(X)$ that contain O_p. Because T is an almost SV-space, there is a j such that $1 \leq j \leq m$ and P_j^T is a minimal valuation prime of $C(T)$ containing O_p^T. If $\pi : C(T) \to C(T)/P_j^T$ denotes the canonical homomorphism, then $\pi \circ \varphi$ is an epimorphism from $C(X)$ onto $C(T)/P_j^T$ whose kernel is $\varphi^{-1}(P_j^T)$. Therefore $C(X)/\varphi^{-1}(P_j^T) = C(T)/P_j^T$ and hence $\varphi^{-1}(P_j^T)$ is a minimal valuation prime of $C(X)$ contained in M_p. Since $p \in X$ is arbitrary, this completes the proof. □

Use will be made below of the concept that follows in deriving a sufficient condition for a compact perfectly normal space to be an almost SV-space.

4.7 Definition. For any space X, a point $p \in \beta X$ such that O^p is a prime ideal of $C(X)$ is called a *βF-point*. If O^p is a valuation prime, then p is called a *special βF-point.*

No example is known of a βF-point that is not a special βF-point.

4.8 Lemma. Suppose p is a nonisolated G_δ-point of a compact space X, and let Q denote a prime z-ideal of $C(X \setminus \{p\})$ such that the prime z-filter $\mathcal{Z}[Q]$ converges to some point in $\beta(X \setminus \{p\}) \setminus (X \setminus \{p\})$. Then the following are equivalent:

(1) For every $f \in C(X \setminus \{p\})$ such that $0 \leq f \leq 1$, there exists a $Y \in \mathcal{Z}[Q]$ and a $g \in C(X)$ such that $f|Y = g|Y$.
(2) For every $f \in C(X \setminus \{p\})$ such that $0 \leq f \leq 1$, there exists a $Y \in \mathcal{Z}[Q]$ such that $\lim_{x \to p} f|Y$ exists, i.e.; $\mathcal{Z}[Q]$ is a $P(p)$ filter.

Proof. If (1) holds, then $\lim_{x\to p} g|Y$ exists and consequently $\lim_{x\to p} f|Y$ also exists. So (2) holds.

If (2) holds, define $h : Y \cup \{p\} \to \mathbb{R}$ by letting $h|Y = f|Y$ and $h(p) = \lim_{x\to p} f|Y$. Clearly $h \in C(Y \cup \{p\})$. Since X is compact and $Y \cup \{p\}$ is a closed subset of X, and hence is C-embedded in X. If g is an extension of h over X, then $g|Y = f|Y$. □

Combining Lemma 4.8 and Theorem 2.3.2 of [CD86], we obtain:

4.9 Theorem. Suppose p is a nonisolated G_δ-point of a compact space X, and let Q denote a prime z-ideal of $C(X \setminus \{p\})$ such that the prime z-filter$\mathcal{Z}[Q]$ converges to some point in $\beta(X \setminus \{p\})\setminus (X \setminus \{p\})$. Then the following are equivalent:

(1) $\gamma(Q)$ is a valuation prime z-ideal of $C(X)$ contained in M_p.
(2) Q is a valuation prime z-ideal and $\mathcal{Z}[Q]$ is a $P(p)$ filter.

4.10 Theorem. If X is compact and perfectly normal, and for every nonisolated point p of X, there is a free $P(p)$ z-ultrafilter $\mathcal{Z}[M^q]$ on $X \setminus \{p\}$ such that q is a special βF point of $\beta(X \setminus \{p\})$, then X is an almost SV-space.

Proof. A nonisolated point p of the perfectly normal space X is a G_δ-point. Suppose $\mathcal{Z}[M^q]$ is a free $P(p)$ z-ultrafilter on $X\setminus\{p\}$ such that q is a special βF-point. Thus O^q becomes a valuation prime ideal of $C(X \setminus \{p\})$ and $\mathcal{Z}[O^q]$ clearly converges to q in $\beta(X \setminus \{p\})\setminus (X \setminus \{p\})$. Since γ is a bijection, $\gamma(O^q)$ is a minimal prime ideal of $C(X)$ contained in M_p. Because $\mathcal{Z}[M^q]$ is a $P(p)$ filter, $\gamma(M^q)$ becomes an immediate prime z-ideal predecessor of M_p which is a valuation prime by 2.3.3 of [CD86]. Now $\gamma(O^q)$ is a minimal prime ideal of $C(X)$ contained in $\gamma(M^q)$, which is properly contained in M_p and since $\gamma(M^q)$ is a valuation prime, this implies that $\mathcal{Z}[\gamma(O^q)]$ is a $P(p)$ filter by 2.2.2 of [CD86]. Consequently, by 4.3(b)(3) above, $\mathcal{Z}[O^q]$ becomes a $P(p)$ filter. Finally, because O^q is a valuation prime and $\mathcal{Z}[O^q]$ is a $P(p)$ filter, $\gamma(O^q)$ becomes a valuation prime by Theorem 4.9. But p is an arbitrary nonisolated point, so this completes the proof. □

4.11 Theorem. $\alpha\omega$ is an almost SV-space if and only if there exists a free ultrafilter Ψ on ω such that for every $f \geq 0$ in $C^*(\omega)$, there is a $Y \in \Psi$ such that $\lim_{x\to\omega} f|Y$ exists.

Proof. Let φ be the unique continuous extension of the inclusion map $i : \omega \to \alpha\omega$ over $\beta\omega$. Recall from 4.3(b)(1) that the map γ is a bijection of the family of all prime z-ideals Q of $C(\omega)$ such that $\mathcal{Z}[Q]$ converges to point of $\varphi^{-1}(\omega)$ (where $\varphi^{-1}(\omega) = \beta\omega \setminus \omega$) onto the family of all nonmaximal prime z-ideals of $C(\alpha\omega)$ contained in M_ω. Since ω is a P-space, the set of prime z-ideals Q of $C(\omega)$ such that $\mathcal{Z}[Q]$ converges to a point of $\beta\omega\setminus\omega$ is the set of maximal ideals M^q such that $q \in \beta\omega \setminus \omega$. So by 4.3(b)(1), the set of minimal prime ideals of $C(\alpha\omega)$ contained in M_ω is given by:$\{\gamma(M^q) : q \in \beta\omega \setminus \omega\}$. Thus $\alpha\omega$ is an almost SV-space if and only if $\gamma(M^q)$ is a valuation prime for some $q \in \beta\omega \setminus \omega$ if and only if $\mathcal{Z}[M^q]$ is a $P(\omega)$ z-ultrafilter on ω (by Corollary 3 of [CD 86]) if and only if there exists a free ultrafilter Ψ on ω such that for every $f \geqq 0$ in $C^*(\omega)$, there exists $Y \in \Psi$ such that $\lim_{x\to\omega} f|Y$ exists. □

Henceforth, we will write $A \cong B$ to abbreviate the statement that the rings A and B are isomorphic, and we will write $X \approx Y$ to abbreviate the statement that the topological spaces X and Y are homeomorphic.

The *Stone extension theorem* states that if $f : X \to Y$ is continuous and Y is compact, then f has a continuous extension $f^* : \beta X \to Y$. (See 6.5 of [GJ76].)

An algebra A such that $C^*(X) \subset A \subset C(X)$ is said to be an *intermediate algebra* of $C(X)$ and is said to be a *c-type algebra* if also $A \cong C(Y)$ for some Tychonoff space Y. We let $A^* = \{f \in A : |f|$ is bounded by some positive integral multiple of $1\}$. If A is an intermediate algebra of $C(X)$ then clearly $A^* = C^*(X)$. For more background on intermediate c-type algebras, see [DGM97].

4.12 Theorem. $C(X)$ is an almost SV-ring if and only if every c-type intermediate algebra of $C(X)$ is an almost SV-ring.

Proof. Let A be an intermediate c-type algebra of $C(X)$ and $\upsilon_A X = \{p \in \beta X : f^*(p) \in \mathbb{R}$, for all $f \in A\}$ where f^* is the Stone extension of f over βX to the two-point compactification $\mathbb{R} \cup \{\pm\infty\}$ of $\mathbb{R}$. Clearly $X \subset \upsilon_A X \subset \beta X$. By Corollary 2.8 of [HJo61] , $Max(A) \approx Max(A^*)$ and $Max(C^*(X)) \approx \beta X$. Therefore $Max(A) \approx \beta X \approx \beta(\nu_A X)$. Consequently $A \cong C(\upsilon_A X)$ and the rest of the proof follows by Theorem 4.4. □

5. Product spaces and set-theoretic considerations

In Theorem 3.1 of [CD86], it is shown that if $\alpha\omega$ contains a nonmaximal valuation prime, then the space $\beta\omega \setminus \omega$ contains a P-point. It is noted also in this paper that W. Rudin showed that if CH holds, then $\beta\omega \setminus \omega$ contains a dense set of P-points, and Shelah showed that there are models of ZFC in which $\beta\omega \setminus \omega$ has no P-points. (See [Wi82].) It follows that there are models of ZFC in which $\alpha\omega$ is not an almost SV-space. In Theorem 3.3.4 of that paper it is shown that these results also hold if ω is replaced by any infinite discrete space.

This yields another difference between SV and almost SV-spaces. While closed subspaces of compact SV-spaces are SV-spaces, there are models of ZFC in which the corresponding result for almost SV-spaces need not hold. In particular, by Theorem 4.2, in a model in which $\beta\omega \setminus \omega$ has no P-points, the space obtained by attacing a copy of $W(\omega+1)$ to a point of $\beta\omega \setminus \omega$ is an almost SV-space with a countable closed subspace that is not an almost SV-space.

Observe that a space is the free union $X_1 \sqcup X_2$ of spaces X_1 and X_2 if and only if $C(X)$ is the direct sum $C(X_1) \oplus C(X_2)$. Because every maximal [resp. proper prime] ideal of $C(X_1 \sqcup X_2)$ is the direct sum of a maximal [resp. proper prime] ideal in one coordinate and the whole ring in the other, it follows that $C(X_1 \sqcup X_2)$ is an almost SV-space if and only if $C(X_i)$ is an almost SV-space for $i = 1, 2$.

Recall also from [HMW03] that X is called a *quasi* P-space if each of the prime z-ideals in $C(X)$ is minimal or maximal. The following facts will be used below.

Fact 1. The one-point compactification αD of an infinite discrete space D is a quasi P-space. (See 2.4 of [HMW03].)

Fact 2. Every infinite locally compact quasi P-space T is a free union of one-point compactifications of infinite discrete spaces. This free union is finite if and only if T is compact. (See 4.1 and 6.1 of [HMW03].)

Fact 3. A prime ideal P of $C(X)$ is minimal if and only if $f \in P$ implies there is a $g \notin P$ such that $fg = \mathbf{0}$. The space $mC(X)$ of minimal prime ideals of $C(X)$ (with the topology decribed just after Prop. 3.2) is always a countably compact zero-dimensional Hausdorff space. Moreover, $mC(X)$ is compact if and only if whenever each function in a prime ideal Q of $C(X)$ has nonempty interior, it follows that $Q \in mC(X)$. See [HJe65].

A space X such that whenever $V \in coz(X)$, there is a $W \in coz(X)$ such that $V \cap W = \varnothing$ and $V \cup W$ is dense in X, is said to be *cozero complemented*. It is well known that $m(C(X))$ is compact if and only if X is cozero complemented. See [HW04]. It will be noted in 5.3 below that if D is an uncountable discrete space, then Ωasv implies that αD is an almost SV-space, even though $\alpha\omega$ is cozero complemented while αD is not cozero complemented.

5.1 Theorem.

(a) If $\alpha\omega$ is an almost SV-space and Y is a compact metrizable almost SV-space with a dense set of isolated points, then $\alpha\omega \times Y$ is an almost SV-space.

(b) If $\alpha\omega$ is not an almost SV-space, then neither is $\alpha\omega \times Y$.

Proof. (a) It follows easily from Fact 3 and the observation that $\alpha\omega \times Y$ is compact that it suffices to show for any $y \in Y$ that $M_{(\omega,y)}$ contains a valuation prime that is a minimal prime ideal. Using Fact 3 again yields this result if y is an isolated point of Y, so we may assume it is not isolated. Because $\alpha\omega \times Y$ is metrizable and Y has a dense set of isolated points, there is is a sequence $\{x_n\}$ of isolated points of $\alpha\omega \times Y$ that converges to (ω, y). Clearly this sequence together with (ω, y) is a subspace T of $\alpha\omega \times Y$ homeomorphic to $\alpha\omega$. Because this latter is an almost SV-space, there is a valuation prime P that is a minimal prime ideal of $C(T)$ contained in the maximal ideal $\{f \in C(T) : f(\omega, y) = 0\}$ of $C(T)$. Let ρ denote the map that sends $f \in C(\alpha\omega \times Y)$ to its restriction to T, and let φ be a mapping that sends each member of $C(T)$ to its coset mod P in the valuation domain $D = C(T)/P$. Clearly $\ker(\rho \circ \varphi) = \rho^{-1}(P)$ is a valuation prime ideal contained in $M_{(\omega,y)}$. It remains only to prove that it is a minimal prime ideal. Because $\alpha\omega \times Y$ is metrizable, the space of minimal prime ideals of $C(\alpha\omega \times Y)$ is compact. So $\rho^{-1}(P)$ is minimal provided that each of its elements is a zero divisor. (See [HJe65].) That will be the case if each $f \in P$ has a zeroset with nonempty interior. Now $f \in \rho^{-1}(P)$ implies $\rho(f) \in P$ implies $f|T \in P$ implies $int_T(f|T) \neq \varnothing$ because P is a minimal prime ideal of $C(T)$. Also, because $\{x_n\}$ is a sequence of isolated points of $\alpha\omega \times Y$, it follow that $intZ(f) \neq \varnothing$. So (a) holds.

(b) If p is an isolated point of Y, then $\alpha\omega \times Y$ is a free union of $X_1 = \alpha\omega \times \{p\}$ and $X_2 = \alpha\omega \times (Y \setminus \{p\})$. Therefore $C(\alpha\omega \times Y)$ is an almost SV-ring if and only if each $C(X_i)$ is for $i = 1, 2$. Since $X_1 \approx \alpha\omega$, it follows that $\alpha\omega$ is an almost SV-space. □

With the aid of a routine induction, (b) implies:

5.2 Corollary. $\alpha\omega$ is an almost SV-space if and only if $(\alpha\omega)^n$ is an almost SV-space for any positive integer n.

The proof of the next theorem depends on some results in [HMW03] where it is shown that a compact space is a quasi P-space if and only if it is a finite free union of one-point compactifications of discrete spaces.

5.3 Theorem. Let D be any uncountable discrete space and let αD denote its one-point compactification. If $\alpha\omega$ is an almost SV-space then so is αD.

Proof. By the preceding remarks, αD is a quasi P-space and hence every prime z-ideal of $C(\alpha D)$ is either minimal or maximal. Clearly αD contains a copy of $\alpha\omega$; say T. Since $\alpha\omega$ is an almost SV-space, there is a minimal nonmaximal prime ideal P of $C(T)$ which is a valuation prime. Let φ denote the restriction mapping of $C(\alpha D)$ onto $C(T)$.

P is a z-ideal since it is a minimal nonmaximal prime ideal of $C(T)$. Therefore $\varphi^{-1}(P)$ is a prime z-ideal of $C(\alpha D)$ which cannot be maximal. So, because αD is a quasi P space, $\varphi^{-1}(P)$ is minimal. Since $C(\alpha D)\ /\ \varphi^{-1}(P)$ and $C(T)/P$ are isomorphic, this completes the proof. □

Next, we prove two results on product spaces under the assumption that $\alpha\omega$ is an almost SV-space.

5.4 Lemma. If $\alpha S = S \cup \{s\}$ and $\alpha T = T \cup \{t\}$ are the one-point compactifications of infinite discrete spaces S and T, and $\alpha\omega$ is an almost SV-space, then $X = \alpha S \times \alpha T$ is an almost SV-space.

Proof. It is shown in Theorem 5.3 that if $\alpha\omega$ is an almost SV-space, then the one-point compactification of any infinite discrete space is an almost SV-space. So, we need only show that $M_{(s,t)}$ contains a valuation prime ideal that is minimal prime. It is easy to find a sequence $\{x(n)\}$ of distinct isolated points of X that converges to (s,t). [If $\{s(n)\}$and $\{t(n)\}$ are sequences of distinct isolated points of S and T, let $x(n) = (s(n), t(n))$.] If $Y = \{x(n)\}_{n=1}^{\infty} \cup \{(s,t)\}$, then $Y \approx \alpha\omega$. So there is a minimal prime valuation prime ideal P of $C(Y)$ contained in $\{f \in C(Y) : f(p,q) = 0\}$. If φ: $C(X) \to C(Y)$ is the restriction map, then clearly $\varphi^{-1}(P)$ is valuation prime. So, we need only show that the prime ideal $\varphi^{-1}(P)$ is minimal. By Fact 3, we need only show that if $f \in \varphi^{-1}(P)$, there is a $g \notin \varphi^{-1}(P)$ such that $fg = \mathbf{0}$. Note also that the zeroset of an element of a minimal prime ideal of $C(Y)$ is infinite.

Suppose first that $(s,t) \in int_Y Z(f|Y)$, in which case $coz(f|Y)$ is a finite set of isolated points of Y. If $Z(f|Y) \setminus (s,t) = \{\mathbf{y}(n)\}_{n=1}^{\infty}$, let $(g|Y)(y(n)) = \frac{1}{n}$ for $n \geq 1$, and $(g|Y) = 0$ otherwise. Then $(f|Y)(g|Y) = \mathbf{0}$, while $(g|Y) \notin P$ because its zeroset is finite. It follows that $\varphi^{-1}(P)$ is minimal. □

5.5 Theorem. If $\alpha\omega$ is an almost SV-space, then the product of two infinite compact quasi P-spaces X and Y is an almost SV-space that is not a quasi P-space (or an SV-space).

Proof. By Fact 2, both X and Y are free unions of finitely many one point compactifications of infinite discrete spaces, so $X \times Y$ is a finite free union of spaces of the form $\alpha S_i \times \alpha T_j$ for infinite discrete spaces S_i and T_j. Each of these summands is an almost SV-space by the lemma, as is their free union.

If $X \times Y$ were a quasi P-space, then so would each of the $\alpha S_i \times \alpha T_j$. By 6.1 of [HMW03], this cannot be the case since each of the latter factors are compact and infinite. Finally, because $X \times Y$ contains a convergent sequence, it cannot be an SV-space. □

6. Some consequences of the assumption that $\alpha\omega$ is an almost SV-space

Henceforth the assumption that the one-point compactification of the countable discrete space ω is an almost SV-space will be denoted by Ωasv. This assumption has been used since the beginning of Section 5.

6.1 Theorem. The following assertions are equivalent:

(a) Ωasv holds.

(b) There is a $p \in \beta\omega \setminus \omega$ such that the maximal ideal M_p of $C(\beta\omega)$ is the immediate prime z-ideal successor of O_p in the class of all z-ideals of $C(\beta\omega)$.

Proof. Let i denote the restriction mapping from $C(\beta\omega)$ to $C(\omega)$. If $p \in \beta\omega \setminus \omega$, then since ω is a discrete space, $i^{-1}(M^p) = O_p$ in $C(\beta\omega)$. Because $\beta\omega \setminus \omega$ is a zeroset, the prime z-filter $Z(i^{-1}(M^p))$ on $\beta\omega$ has an immediate successor in the class of z-filters which is exactly the z-filter generated by $Z(i^{-1}(M^p))$ together with $\beta\omega \setminus \omega$ by Theorem 3.5 of [GJ60]. It follows from Theorem 3.10 of the same paper that this successor is $Z(M_p)$ if and only if p is a P-point of $\beta\omega \setminus \omega$, which as noted above is equivalent to Ωasv. □

6.2 Lemma. Suppose:

(a) Ωasv holds,

(b) X and $mC(X)$ are compact (e.g., if X is compact and perfectly normal), and

(c) for each nonisolated point $p \in X$, there exists an infinite set of isolated points D_p of X such that $D_p \cup \{p\}$ and the one point compactification αD_p of D_p are homeomorphic.

Then X is an almost SV-space.

Proof. Since (a) holds, $Y = \alpha D_p$ is an almost SV-space as noted in the first paragraph of this section. So there is a $P \in mY$ that is a valuation prime contained in $\{f \in C(Y) : f(p) = 0\}$. Letting $\varphi : C(X) \to C(Y)$ denote the restriction map, we see that $C(X)/\varphi^{-1}(P) \cong C(Y)/P$. It remains only to show that $\varphi^{-1}(P) \in mX$. Now $g \in \varphi^{-1}(P)$ implies $g|Y \in P$. By (c), $\mathrm{int}_Y Z(g) \neq \emptyset$ since $P \in mC(Y)$. So since each point of $int_Y Z(g)$ is isolated in X, we know that $int_X Z(g)$ is nonempty as well. From Fact 3 of Section 5, we conclude that $\varphi^{-1}(P) \in mC(X)$. □

A topological space X is said to be *scattered* or *dispersed* if each nonempty subspace Y contains an isolated point of Y. A compact scattered space is necessarily zero-dimensional.

If X is a space, let $X^{(0)} = X$, $X^{(1)} = X \setminus Is(X)$, and for any ordinal η, let $X^{(\eta+1)} = (X^{(\eta)})^{(1)}$. If η is a limit ordinal, then $X^{(\eta)}$ denotes intersection of all $X^{(\beta)}$ such that $\beta < \eta$. From cardinality considerations there is an ordinal α such that $X^{(\alpha)} = X^{(\beta)}$, for each $\beta > \alpha$. If there is an α such that $X^{(\alpha)} = \varnothing$, then X is scattered and the least such α is called the *CB-index* of the scattered space X.

These notions abound in general topology. See, for example, [LR81] or [Se71].

6.3 Theorem. If Ωasv holds, then every compact metrizable scattered space X of Cantor-Bendixon index ≤ 3 is an almost SV-space.

Proof. The hypothesis of Lemma 6.2 will be verified. If $CB(X) = 1$, then X is finite and hence is a P-space. If $CB(X) = 2$, then the set of nonisolated points of X is finite and therefore for every nonisolated point there exists a sequence of isolated points converging to it. If $CB(X) = 3$, then since X is compact and metrizable, $X^{(2)}$ is finite and therefore for all but finitely many points of $X^{(1)}$, there is a sequence of isolated points of X converging to the point. Let $p \in X^{(1)}$ and let $\{x_n\}_{n=1}^{\infty}$ be a sequence of nonisolated points converging to p; i.e., $\{x_n\}_{n=1}^{\infty} \subset X^{(1)}$. By our earlier observation, for all but finitely many members of $\{x_n\}_{n=1}^{\infty}$, there is a sequence of isolated points of X converging to it. Thus every neighborhood of p contains some x_n such that there is a sequence of isolated points of X converging to x_n. Hence we get a sequence $\{y_n\}_{n=1}^{\infty}$ of isolated points of X converging to p. Hence by Lemma 6.2, X is an almost SV-space. □

Remark. Recall that a space X such that whenever $x \in cl_X A$ for some $A \subset X$, there is a sequence of elements of A that converges to x is called a *Frechet space* (or a *Frechet–Urysohn space*). In the proof of Theorem 6.3, metrizability is used only to produce sequences that converge to points in the closures of some subspaces. It follows that the hypothesis that X is metrizable can be weakened to assuming only that X is a Frechet space and $mC(X)$ is compact. Because $mC(X)$ fails to be compact if X is the one-point compactification of an uncountable discrete space, while X is a compact scattered almost SV-space if Ωasv holds, this new result does not generalize Theorem 6.3.

It follows immediately from Corollary 4 in Section 2.2 of [CD86] that if X is a metrizable almost SV-space with a nonisolated point, then Ωasv holds. It follows that Ωasv holds if and only if there is an infinite compact metrizable almost SV-space.

6.4 Corollary. If $[0, 1]$ is an almost SV-space, then Ωasv holds.

Whether or not the converse of this corollary holds is the most important unsolved problem of this paper. See Section 8.

We conclude this section with a statement without proof of a result that provides some circumstantial evidence that $[0,1]$ may be an almost SV-space.

6.5 Theorem. Suppose $p \in [0,1]$ and $g \in C([0,1])$ are such that $g(p) \neq 0$ or there is an open set U of $[0,1]$ such that $Z(g) \cap U = \{p\}$. If Ωasv holds, then there is a minimal prime ideal $Q \subset M_p$ such that whenever $0 \leq f \leq g$, the coset mod Q of g divides coset mod Q of f.

Note that it is enough to assume that $0 < f \leq g$ mod Q.

This theorem does not enable us to decide whether its conclusion holds in case both $Z(g) \cap U$ and $coz(g) \cap U$ are infinite. If this latter case could be handled, it would follow that Ωasv implies $C([0,1])$ is an almost SV-ring.

7. The second generalization; quasi SV-spaces and rings

The task of determining if a space X is an almost SV-space divides naturally into two parts. First we have to find a valuation prime ideal P contained in a maximal ideal M of $C(X)$ and then we have to check whether P is minimal. This part of the problem is more difficult since there is no easy way to determine whether a prime ideal contained properly in P is also a valuation prime. This is part of the motivation for the following definition.

7.1 Definition. A space X such that for each real maximal ideal of $C(X)$ that is not a minimal prime ideal contains a nonmaximal prime ideal P such that $C(X)/P$ is a valuation domain is called a *quasi SV-space* (and $C(X)$ is called a *quasi SV-ring*). In other words $C(X)$ is a quasi SV-ring if for all $p \in \upsilon X$, whenever $M^p \neq O^p$, then M^p contains a nonmaximal prime ideal P such that $C(X)/P$ is a valuation domain.

Remark. The reason for the restriction to real maximal ideals is to make it possible to prove Theorem 7.3 below.

Clearly, every almost SV-space is a quasi SV-space. We have been unable to find an example to show that the converse need not hold.

The following lemma will be used in what follows.

7.2 Lemma. If P is a prime ideal of $C(X)$ contained in real maximal ideal, then the trace of P in $C^*(X)$ is a prime ideal of $C^*(X)$ and $C^*(X)/P \cap C^*(X) \cong C(X)/P$.

Proof. Let i be the restriction of $C(\beta X)$ to $C(X)$ and π be the canonical homomorphism from $C(X)$ onto $C(X)/P$. Now as P is contained in a real maximal ideal, no element of $C(X)/P$ is infinitely large and consequently $\pi \circ i$ becomes an epimorphism from $C(\beta X)$ onto $C(X)/P$. Thus $C(\beta X)/i^{-1}(P) \cong C(X)/P$. Since $C(\beta X) \cong C^*(X)$, it follows that $C^*(X)/P \cap C^*(X) \cong C(X)/P$. □

7.3 Theorem. For any Tychonoff space X the following are equivalent:

(a) X is a quasi SV-space.
(b) υX is a quasi SV-space.
(c) βX is a quasi SV-space.

Proof. (a) and (b) are equivalent since $C(X)$ and $C(\upsilon X)$ are isomorphic.

(a) implies(c) Recall that the collection of maximal ideals of $C^*(X)$ is given by $\{M^{*p} : p \in \beta X\}$; where $M^{*p} = \{f \in C^*(X) : f^\beta(p) = 0\}$ and the collection of all maximal ideals of $C(X)$ is given by $\{M^p : p \in \beta X\}$; where $M^p = \{f \in C(X) : p \in cl_{\beta X} Z(f)\}$. If $p \in \beta X \setminus \upsilon X$ then M^p becomes a hyperreal maximal ideal of $C(X)$ and hence M^{*p} properly contains the prime ideal $M^p \cap C^*(X)$ which is clearly a valuation prime. (See Section 2.1 of [CD86].) Now if $p \in \upsilon X$ and $M^{*p} \neq O^{*p}$ then $M^p \neq O^p$ because if $M^p = O^p$, then $M^{*p} = M^p \cap C^*(X) = O^p \cap C^*(X)$, in which case $M^{*p} = O^{*p}$. Since X is a quasi SV-space and $p \in \upsilon X$, there exists a nonmaximal prime ideal P contained in M^p such that $C(X)/P$ is a valuation domain. Let $P^* = P \cap C^*(X)$. Then by Lemma 7.2, it follows that $C^*(X)/P^* \cong C(X)/P$. Because the latter is a valuation domain, P^* is a nonmaximal valuation prime contained in M^{*p}. Hence $C^*(X)$ is a quasi SV-ring. Because $C^*(X) \cong C\beta X)$, the latter becomes a quasi SV-ring and consequently βX becomes a quasi SV-space.

(c) implies (a) If $p \in \upsilon X$ and $M^p \neq O^p$, there is a nonmaximal prime ideal P of $C(X)$ contained in M^p and by Lemma 7.2, $C(X)/P \cong C^*(X)/P \cap C^*(X)$. Hence $P \cap C^*(X)$ is a nonmaximal prime ideal contained in M^{*p} in $C^*(X)$ and evidently $M^{*p} \neq O^{*p}$. Now since βX is a quasi SV-space and $C(\beta X) \cong C^*(X)$, there is a nonmaximal prime ideal Q containing O^{*p} in $C^*(X)$ such that Q is a valuation prime. We claim that there exists a nonmaximal prime ideal W of $C(X)$ containing O^p such that $W \cap C^*(X)$ contains Q.

To see this, suppose $Q_m \subset Q$ is a minimal prime ideal of $C^*(X)$ containing O^{*p}. As is noted in [HJe65], the mapping that sends each minimal prime ideal of $C(X)$ to its trace on $C^*(X)$ is a surjection. So there is a minimal prime ideal T_m containing O^p in $C(X)$ such that $T_m \cap C^*(X) = Q_m$. Let Ω denote the maximal chain of prime ideals containing T_m in $C(X)$ and $\{T_\alpha\}$ the collection of all nonmaximal prime ideals of $C(X)$ which belong to Ω. Their union T is a prime ideal of $C(X)$. Assume that $T \cap C^*(X) \subset Q$. Now M^p is a real maximal ideal since $p \in \upsilon X$. So $M^p \cap C^*(X) = M^{*p}$, while Q is a nonmaximal prime ideal of $C^*(X)$ contained in M^{*p}. This shows that $T \neq M^p$. Thus T becomes a prime ideal predecessor of M^p; which implies that M^p is an upper ideal. But since every maximal ideal is a z-ideal and a z-ideal can never be an upper ideal, this leads to a contradiction. (See Chapter 14 of [GJ76].)

Thus there must exist a (nonmaximal) prime ideal W of Ω such that $W \cap C^*(X)$ is not contained in Q. Now $T_m \subset W$ implies $Q_m \subset W \cap C^*(X)$ and, as we recall $Q_m \subset Q$. Since the set of prime ideals containing a given prime ideal form a chain, we conclude that $Q \subset W \cap C^*(X)$. Since being a valuation prime is preserved under extensions, and Q is a valuation prime of $C^*(X)$, we conclude that $W \cap C^*(X)$ is a valuation prime of $C^*(X)$. Finally, since M^p is a real maximal ideal of $C(X)$ and W is a nonmaximal prime ideal of $C(X)$ contained in M^p, by Lemma 7.2, $C(X)/W \cong C^*(X)/W \cap C^*(X)$. Because the latter is a valuation domain, this completes the proof. □

Recall that a space X is realcompact if and only if $X = \upsilon X$ and that a metrizable space is realcompact if and only if it is of nonmeasurable cardinality. (See Chapters 8 and 12 of [GJ76].)

7.4 Theorem. Every realcompact metrizable space X is a quasi SV-space if and only if Ωasv holds.

Proof. If X is a quasi SV-space, there is a $p \in \upsilon X = X$ such that M_p contains a nonmaximal prime ideal P of $C(X)$ such that $C(X)/P$ is a valuation domain. Hence by Corollary 4 of Section 2 of [CD86], there is a P-point in $\beta\omega \setminus \omega$ and consequently Ωasv holds.

Suppose Ωasv holds, $p \in X$ is a nonisolated point, and $\{x_n\}_1^\infty$ is a sequence of distinct points converging to p in the metrizable space $X = \upsilon X$. Clearly $Y = \{x_n\}_1^\infty \cup \{p\} \approx \alpha\omega$. Since the restriction mapping φ from $C(X)$ to $C(Y)$ a surjective homomorphism, by Ωasv there exists a non maximal valuation prime ideal P of $C(Y)$ and clearly $C(X)/\varphi^{-1}(P) \cong C(Y)/P$. Because the latter is a valuation domain, $\varphi^{-1}(P)$ is a nonmaximal valuation prime ideal of $C(X)$ contained in the real maximal ideal M_p. This completes the proof. □

If X is a quasi SV-space then it certainly it follows from Theorem 7.3 that every C^*-embedded dense subspace of it is again a quasi SV-space. Here is another condition for a subspace of a quasi SV-space to be a quasi SV-space.

7.5 Theorem. Every open realcompact C-embedded subspace U of a quasi SV-space X is a quasi SV-space.

Proof. If $p \in U$ is such that $M_p^U = \{f \in C(U) : f(p) = 0\}$ is not a minimal prime ideal of $C(U)$, then since U is open and C-embedded in X, we will show that $M_p = \{f \in C(X) : f(p) = 0\}$ is not a minimal prime ideal of $C(X)$.

For, by assumption, there is a $f \in M_p^U$ that does not vanish on a neighborhood of p in U. Because U is C-embedded and open, f has a continuous extension that is in $M^p \setminus O_p$. So M^p is not in $mC(X)$.

Since X is a quasi SV-space, M_p contains a nonmaximal valuation prime ideal P of $C(X)$. If i denotes the restriction map of $C(X)$ onto $C(U)$, then since U is open, $\ker(i) \subset O_p \subset P$. Thus $P/\ker(i)$ becomes a prime ideal of $C(X)/\ker(i)$ and clearly $C(X)/P \cong \frac{C(X)/\ker(i)}{P/\ker(i)}$. Since i is an epimorphism, $C(X)/\ker(i) \cong C(U)$ and $P/\ker(i) \cong i(P)$ and moreover, $i(P)$ is a prime ideal of $C(U)$. Thus $C(X)/P \cong C(U)/i(P)$. As the former is a valuation domain, $i(P)$ becomes a nonmaximal valuation prime ideal of $C(U)$ contained in M_p^U, and because U is realcompact, this completes the proof. □

7.6 Theorem. If a realcompact space X can be expressed as an arbitrary union of open C-embedded subspaces such that each of them is a quasi SV-space, then X is a quasi SV-space.

Proof. Suppose X is realcompact and $\{X_\alpha\}$ is a collection of open C-embedded quasi SV-subspaces such $X = \cup X_\alpha$. If $p \in X$, then $p \in X_\alpha$ for some α. As in the proof of 7.5, let $M_p^{X_\alpha} = \{f \in C(X_\alpha) : f(p) = 0\}$ We will show that if M_p is not a minimal prime ideal of $C(X)$. Then the same assertion will hold for $M_p^{X_\alpha}$ in $C(X_\alpha)$.

For if there is an $f \in M_p \setminus O_p$ and $M_p^{X_\alpha} = O_p^{X_\alpha}$, then $f|X_\alpha$ vanishes on a neighborhood of p in the open subset $X_\alpha$in X. Thus M_p is also minimal prime in $C(X)$. If i is the restriction of $C(X)$ onto $C(X_\alpha)$, then since X_α is a quasi SV-space, there exists a nonmaximal valuation prime ideal P of $C(X_\alpha)$ contained in $M_p^{X_\alpha}$ and $C(X)/i^{-1}(P) \cong C(X_\alpha)/P$. Thus $i^{-1}(P)$ is a nonmaximal valuation prime ideal of $C(X)$ contained in M_p. This completes the proof. □

7.7 Theorem. Finite products of compact quasi SV-spaces are quasi SV-spaces.

Proof. It suffices to prove that the product $X \times Y$ of two quasi SV-spaces is a quasi SV-space. Each maximal ideal of $C(X \times Y)$ is in the set $\{M_{(p,q)} : (p,q) \in X \times Y\}$. If $(p,q) \in X \times Y$ is a nonisolated point, then either p or q is a nonisolated point of X and Y respectively. Assume p is a nonisolated point of X. If W denotes the space $X \times \{q\}$, then $W \approx X$. If i is the restriction mapping from $C(X \times Y)$ onto $C(W)$, then since X is quasi SV-space, there exists a nonmaximal valuation prime ideal P of $C(W)$ contained in the maximal ideal $\{f \in C(W) : f(p,q) = 0\}$ of $C(W)$. Since i is an epimorphism, $C(X \times Y)/i^{-1}(P) \cong C(W)/P$. Because the latter is a valuation domain, $i^{-1}(P)$ is a nonmaximal valuation prime ideal of $C(X \times Y)$ contained in $M_{(p,q)}$. Since (p,q) is an arbitrary nonisolated point, this completes the proof. □

7.8 Definition. A point $p \in \beta X$ is a Qsv-point if M^p contains a nonmaximal valuation prime of $C(X)$ that is a z-ideal.

7.9 Examples

(1) If every $p \in \beta X$ such that M^p is a real maximal ideal of $C(X)$ is a Qsv-point, then X is a quasi SV-space.
(2) Every point of a compact F-space that is not a P-point is a Qsv-point.
(3) If Ωasv holds, then every nonisolated point p of a metrizable space X is a Qsv-point.

For, since Ωasv holds, there is a minimal valuation prime P of $C(\alpha\omega)$ contained in M_ω. If i is the restriction of $C(X)$ onto $C(\alpha\omega)$, then $i^{-1}(P)$ is a nonmaximal valuation prime z-ideal of $C(X)$ contained in M_p and consequently p is a Qsv- point.

Our next result is a sufficient condition for a compact, perfectly normal space to be a quasi SV-space.

7.10 Theorem. Suppose X is compact and perfectly normal. If, for every nonisolated point p of X, there exists a free $P(p)$ z-ultrafilter $\mathcal{Z}[M^q]$ on $X \setminus \{p\}$ such that q is a Qsv-point of $\beta(X \setminus \{p\})$, then X is a quasi SV-space.

Proof. Our hypothesis implies that every $p \in X$ is a G_δ-point, every maximal ideal of $C(X)$ is real, and that there a free maximal ideal M^q of $C(Y)$ which contains a nonmaximal prime z-ideal Q of $C(Y)$ which is a valuation prime, where $Y = X \setminus \{p\}$. By 4.3(b)(2), $\gamma(M^q)$ becomes an immediate prime z-ideal predecessor of the maximal ideal M_p of $C(X)$ in the class of all all z-ideals. (Recall that γ satisfies $Z(\gamma(Q)) = \{cl_X Y : Y \in \mathcal{Z}[M^q]\} = \{Y \cup \{p\} : Y \in \mathcal{Z}[M^q]\}$.) By Corollary 2.3.3 of [CD86], $\gamma(M^q)$ becomes a valuation prime. Because the prime z-ideal Q of $C(Y)$ is contained in M^q, it is clear that $\mathcal{Z}[Q]$ converges to q. Therefore by 4.3(b)(1), $\gamma(Q)$ becomes a prime z-ideal of $C(X)$ contained in the maximal ideal M_p. Clearly $\gamma(Q) \subset \gamma(M^q)$. By Theorem 2.2.2 in [CD86], it follows that if P is properly contained in M_p and is a valuation prime, then $Z(T)$ is a $P(p)$ z-filter for every minimal prime ideal T contained in P. Since any z-filter containing a $P(p)$ z-filter is again a $P(p)$ z-filter, it follows that $Z(\gamma(Q))$ is a $P(p)$ z-filter. By the definition of the mapping 'γ', it follows that $Z(\gamma(Q))$ is a $P(p)$ z-filter if and only if $\mathcal{Z}[Q]$ is. Finally, since Q is a valuation prime of $C(Y)$ and $\mathcal{Z}[Q]$ is a $P(p)$ z-filter, it follows from Theorem 4.9 that $\gamma(Q)$ is a valuation prime contained in M_p. Since p is an arbitrary nonisolated point of X, this completes the proof. □

We conclude this section with two results concerning chains of pseudoprime ideals.

P is called a primary ideal of a commutative ring if $ab \in P$ implies either a or some power of b belongs to P and is called *pseudoprime* if $ab = 0$ implies either a or b belongs to P. It is well known that every prime ideal is primary and every primary ideal is pseudoprime. While it need not hold for arbitrary commutative rings, in a ring $C(X)$, an ideal of $C(X)$ is pseudoprime if and only if it contains a prime ideal. (See [GK60] and [G90].)

If I is an ideal of a commutative ring A and $f \in A$, then (I, f) denotes the smallest ideal of A containing I and f, while $\langle I(f) \rangle$ denotes the principal ideal of A/I generated by the coset $f + I$.

7.11 Theorem. If X is a topological space and $p \in X$ is a Qsv-point then there exists a countable chain of pseudoprime ideals of $C(X)$ contained in M_p which are not primary ideals.

Proof. As $p \in X$ is a Qsv-point, there is a prime ideal P of $C(X)$ properly contained in M_p which is a valuation prime ideal. If $f_1 \in M_p \setminus P$, then (P, f_1) is a pseudoprime ideal since it contains the prime ideal P. Because, as is shown in [K58], no proper principal ideal of $C(X)/P$ is primary, the principal ideal $\langle P(f_1) \rangle$ is not a primary ideal of $C(X)/P$ and consequently (P, f_1) is not a primary ideal of $C(X)$ because $(P, f_1)/P = \langle P(f_1) \rangle$.

Now $(P, f_1) \neq M_p$ since (P, f_1) is not primary. So there is an $f_2 \in M_p \setminus (P, f_1)$. Because $C(X)/P$ is a valuation domain. one of $P(f_1)$ or $P(f_2)$ must divide the other. If $P(f_1)|P(f_2)$ then $f_2 \in (P, f_1)$ – which is a contradiction. Therefore $P(f_2)|P(f_1)$. Hence $\langle P(f_1)\rangle \subset \langle P(f_2)\rangle$ and consequently $(P, f_1) \subset (P, f_2)$. Then there is an $f_3 \in M_p \setminus (P, f_2)$. Continuing this process, we get a countably infinite chain of pseudoprime ideals $(P, f_1) \subset (P, f_2) \subset \cdots \subset (P, f_n) \subset \cdots$ which are not primary. □

Recall from Definition 4.7 that a point p of βX such that O^p is a valuation prime is called a special βF-point. An algebraic characterization of such points follows.

7.12 Theorem. A point $p \in \beta X$ is a special βF-point if and only if the pseudoprime ideals of $C(X)$ containing O^p that are not primary form a chain (i.e., are linearly ordered under set inclusion).

Proof. If p is a special βF-point then O^p is valuation prime and therefore the principal ideals of $C(X)/O^p$ form a chain. This is equivalent to the fact that the ideals of $C(X)/O^p$ form a chain and consequently the ideals of $C(X)$ containing O^p form a chain. In particular, the pseudoprime ideals containing O^pthat are not primary form a chain.

Suppose p is not a βF-point. Then consider two distinct maximal chains Φ and Ψ of prime ideals lying between O^p and M^p in $C(X)$. Suppose P and Q are the minimal prime ideals in Φ and Ψ respectively. Now $\bigcap (\Phi \cap \Psi) = P + Q$ is the minimal member of the (intersecting) chain $\Phi \cap \Psi$, which is a prime z-ideal. Since $P \cup Q$ is not an ideal, there is an $f \in P + Q \setminus (P \cup Q)$.

Now (P, f) and (Q, f) are pseudoprime ideals since they contain prime ideals. Neither of them is primary as is noted in the proof of 7.11. Suppose it were the case that $(P, f) \subset (Q, f)$. Then $P + Q = (Q, f)$ since $(Q, f) \subset P + Q$. Because (Q, f) is not a primary ideal while $P + Q$ is prime, we arrive at a contradiction. Hence neither of (P, f) or (Q, f) is contained in the other. □

8. Remarks and problems

In this section, we refer readers to some papers concerned with residue class rings of the form $C(X)/P$ for P a prime ideal of $C(X)$ whose content we have been able to use only to a very limited extent. These papers inspired us to pose some interesting problems and to derive a few results. Our hope is that some of our readers may be able to make better use of them.

In the long and thorough paper [M90], James Moloney examined closely the residue class domains of $C(\alpha\omega), C^\infty(R)$, (and to a lesser extent $C(X)$ for some other classes of topological spaces) modulo prime ideals assuming CH.

His extraordinary and difficult accomplishment is showing that

$$\{C(\alpha\omega)/P : P \text{ a nonmaximal prime ideal}\}$$

can be divided into precisely 9 distinct isomorphism classes. (Any two members of of the same class are isomorphic, and no two distinct classes contain members that are isomorphic.) The descriptions of these isomorphism classes are order theoretic without any direct description of the algebraic properties of these integral domain. Instead, they involve the cardinality and nature of cofinal and coinitial subsets. There does seem to be a way of describing these results succinctly. The curious reader should examine Theorem 3.2.26 of [M90] and each of the theorems referred to in its proof. No attempt is made to determine when $C(\alpha\omega)/P$ is a valuation domain. We do know, however, from Theorem 3.5(d) that not every prime ideal contained in M_ω is a valuation prime, and that (assuming CH) M_ω contains a valuation prime because CH implies Ωasv. In Section 4 of [M90], some of the results referred to above are applied to some more general spaces. Regrettably, Moloney's interesting results are of little help to us because our goals are different from his. For example, we do not know how to tell which of the equivalence classes described above contains an element $C(\alpha\omega)/P$ such that P is a valuation prime and $P \subset M_\omega$. If we could answer the following questions, we might be able to use some of Moloney's results to reach our main goals. In each case we assume CH.

8.1 Problems

(a) Which of Moloney's 9 equivalence classes contains an element $C(\alpha\omega)/P$ such that P is a valuation prime?

(b) If P is a nonmaximal valuation prime of $C(\alpha\omega)$, can the set of strictly positive elements of $C(\alpha\omega)/P$ have a countable coinitial subset?

Assuming CH, two problems less related to [M90] are:

8.2 Problems

(a) Is $[0,1]$ an almost SV-space?

(b) Is every compact metrizable scattered space with finite CB-index an almost SV-space?

Note that by Theorem 7.4 (and CH), the spaces above are quasi SV-spaces. In [GJ60], the authors pose the question:

(∗) If Q is a prime ideal of $C(Y)$, when is there a space X and a maximal ideal M of $C(X)$ such that $C(X)/M$ and the quotient field of $C(Y)/Q$ are isomorphic?

When (∗) has an affirmative answer, they say that $C(Y)/Q$ is *realized by* $C(X)/M$. It is shown in Theorem 2.3 of [GJ60] that if $C(Y)/Q$ has a realization, then Q is a z-ideal.

8.3 Definition. A Tychonoff space Y and the ring $C(Y)$ is said to be *prime z-sparse* if each nonmaximal prime z-ideal has an immediate successor in the set of prime z-ideals.

Note that any space in which any chain of prime z-ideals is finite is prime z-sparse. Spaces with this finiteness property are studied in [HMW03] and [MZ05]. In particular, the one-point compactification of an infinite discrete space is prime z-sparse. In Example 4.3 of [GJ60], it is shown that $[0,1]^\omega$ is not prime z-sparse. By the *z-dimension* of a space X, we mean the supremum of the lengths of chains of prime z-ideals of $C(X)$. In Section 5 of [MZ05], it is shown that if a compact space X is scattered, then $C(X)$ has finite z-dimension if and only its CB index is finite. (For a precise definition of z-dimension and its properties with emphasis on the case when X is compact, see Sections 4 and 5 of [MZ05].)

8.4 Theorem. If Y is a compact space that is prime z-sparse, and Q is a minimal nonmaximal prime ideal of $C(Y)$ such that $C(Y)/Q$ is a valuation domain, then its set of strictly positive elements has no countable coinitial subset.

Proof. Suppose Q is as above. By Theorem 3.4 of [GJ60], there is a subspace X of Y and a maximal ideal M of $C(X)$ such that $C(X)/M$ and the quotient field of $C(Y)/Q$ are isomorphic. Because Q is not maximal, the field $C(X)/M$ is an η_1-set. (See Chapter 13 of [GJ76].) Hence its set of strictly positive elements has no countable coinitial subset. It follows easily that the valuation domain $C(Y)/Q$ has the property as well. □

The proof of the following corollary follows from the last theorem and the remarks preceding it.

8.5 Corollary. If Y is a compact scattered space with finite CB-index, and Q is a minimal valuation prime ideal of $C(Y)$ such that $C(Y)/Q$ is not a field, then its set of strictly positive elements of $C(Y)/Q$ has no countable coinitial subset.

Next, we include with some brief remarks about the contents of [Sc97].

Note first that the term real closed ring is used by Schwartz in an entirely different way than in [CD86]. Because Schwartz's terminology is used in many papers, we will use it in what follows. We will not repeat the definition of real closed ring. It will be enough for the reader to know that a real closed ring is a lattice ordered ring, that any $C(X)$ is real closed, and that $C(X)/P$ is real closed ring whenever P is a prime ideal of $C(X)$. If M is a maximal ideal of $C(X)$, let $\mathcal{P}(M)$ denote set of prime ideal of $C(X)$ that are contained in M. The author explores the relationship between $C(X)$ being an SV-ring and $\{C(X)/Q : Q \in \mathcal{P}(M)\}$ consisting of valuation domains for a collection of maximal ideals M of $C(X)$. We have been unable to adjust this approach to the study of almost SV-spaces, but hope that some readers of this paper may be able to do so.

In Chapter 4 of the Dales-Woodin book [DW 96], these authors study the residue class rings $C(X)/P$ with which we are concerned in case X is compact. This chapter is not self-contained and the notation used in it differs not only from what we use, but also from many of the articles to which the reader is referred.

If P is a prime ideal of $C(X)$, where X is compact and M_P denotes the unique maximal ideal of $C(X)$ containing P, then P is called *strongly convex* if M_P/P is an interval in the quotient field of $C(X)/P$. The notion of a strongly convex prime ideal may play a major role in studying the valuation prime ideals of $C(X)$ because every valuation prime ideal is necessarily a strongly convex prime. Just studying strongly convex primes will not suffice since the converse of this latter assertion need not hold. For, it is shown in the proof of Theorem 4.40 of [DW96] that there is a nonmaximal valuation prime ideal P of $C(\beta(\omega \times \alpha\omega))$ while there exists a prime ideal $Q \subsetneq P$ which is not even strongly convex and hence fails to be valuation prime.

Actually, there is a much simpler example. By Prop. 4.36 of [DW96], every prime ideal of $C(\alpha\omega)$ is strongly convex. But not every prime ideal of $C(\alpha\omega)$ is a valuation prime since $\alpha\omega$ is not an SV-space. In Proposition 4.37 of [DW96], it is shown that if P is a strongly convex prime ideal of $C(X)$ and X is compact, then the quotient field of $C(X)/P$ is a semi-η_1 field. That is, whenever every element of A is less than every element of B, where A is an increasing and B is a decreasing (countable) sequence of elements of the quotient field of $C(X)/P$, there is a x in the quotient field of $C(X)/P$ strictly in between A and B. Note that R is a semi-η_1 field that is not an η_1-set. It follows from Prop 2.20 of [DW96] assuming CH that if in a semi η_1-field the minimum cardinality of a cofinal (or coinitial) subsets of strictly positive elements is $\geq \aleph_1$ then it is an η_1-field.

The discussion above leads us to believe that a more careful study of the properties of strongly convex ideals and related topics in [DW96] may lead to solutions of some of the problems posed above. We hope this is the case despite the fact that the main focus of this book is on the nature of quotient fields of rings $C(X)/P$ where P is a nonmaximal prime ideal of $C(X)$. The latter are the super-real fields of the title.

Added in proof. We can improve Theorem 6.3 by showing that

Every compact metrizable scattered space is an almost SV-space if Ω-asv holds.

References

[ACCH81] M. Antonovskij, D. Chudnovsky, G. Chudnovsky, and E. Hewitt, *Rings of real-valued continuous functions. II*, Math. Zeit. 176 (1981), 151–186.

[CD86] G. Cherlin and M. Dickmann, *Real closed rings I*, Fund. Math. 126 (1986), 147–183.

[D84] A. Dow, *On ultrapowers of Boolean algebras*, Proceedings of the 1984 topology conference (Auburn, Ala., 1984). Topology Proc. 9 (1984), 269–291.

[DGM97] J.M. Domínguez, J.Gómez, and M.A. Mulero,. *Intermediate algebras between $C^*(X)$ and $C(X)$ as rings of fractions of $C^*(X)$*, Topology & Appl. 77 (1997), 115–130.

[DW96] H.G. Dales and W.H. Woodin, *Super-Real Fields*, Claredon Press, Oxford, 1994.

[E89] R. Engelking, *General Topology*, Heldermann Verlag Berlin, 1989.

[G90] L. Gillman, *Convex and pseudoprime ideals in $C(X)$*, General Topology and Applications, Proceedings of the 1988 Northeast Conference, pp. 87–95, Marcel Dekker Inc., New York 1990.

[GJ60] L. Gillman and M. Jerison, *Quotient fields of residue class rings of continuous functions*, Illinois J. Math 4 (1960), 425–436.

[GJ76] L. Gillman and M. Jerison, *Rings of Continuous Functions*, Springer-Verlag, New York, 1976.

[GK60] L. Gillman and C.W. Kohls, *Convex and pseudoprime ideals in rings of continuous functions*, Math. Zeit. 72 (1960), 399–409.

[Hen97] M. Henriksen, *Rings of continuous functions in the 1950s*. Handbook of the history of general topology, Vol. 1, 243–253, Kluwer Acad. Publ., Dordrecht, 1997.

[Hen02] M. Henriksen, *Topology related to rings of real-valued continuous functions*, Recent Progress in General Topology II, eds. M. Husek, J. van Mill, 553–556, Elsevier Science, 2002.

[Hew48] E. Hewitt, *Rings of real-valued continuous functions I*, Trans. Amer. Math. Soc. 64 (1948) 54–99.

[HJe65] M. Henriksen and M. Jerison, *The space of minimal prime ideals of a commutative ring*, Trans. Amer. Math. Soc. 115 (1965), 110–130.

[HJo61] M. Henriksen and D.G. Johnson, *On the structure of a class of archimedean lattice-ordered algebras*. Fund. Math 50 (1961), 73–94.

[HLMW94] M. Henriksen, S. Larson, J. Martinez, and R.G. Woods, *Lattice-ordered algebras that are subdirect products of valuation domains*. Trans. Amer. Math. Soc. 345 (1994), 195–221.

[HMW03] M. Henriksen, J. Martinez, and R.G. Woods, *Spaces X in which every prime z-ideal of $C(X)$ are minimal or maximal*, Comment. Math. Univ. Carolinae 44 (2003), 261–294.

[HW92a] M. Henriksen and R. Wilson, *When is $C(X)/P$ a valuation ring for every prime ideal P?*. Topology & Appl. 44 (1992), 175–180.

[HW92b] M. Henriksen and R. Wilson, *Almost discrete SV-spaces*, Topology & Appl. 46 (1992), 89–97.

[HW04] M. Henriksen and R.G. Woods, *Cozero complemented spaces; when the space of minimal prime ideals of a $C(X)$ is compact*, Topology & Appl. 141 (2004), 147–170.

[HIJ62] M. Henriksen, J.R. Isbell, and D.G. Johnson, *Residue class fields of lattice-ordered algebras*, Fund. Math. 50 1961/1962 107–117.

[K58] C.W. Kohls, *Prime ideals in rings of continuous functions II*, Duke Math. J. 25 (1958), 447–458.

[L86] S. Larson, *Convexity conditions on f-rings*, Canad. J. Math. 38 (1986), 48–64.

[L03] S. Larson, *Constructing rings of continuous functions in which there are many maximal ideals of nontrivial rank*, Comm. Alg 31 (2003), 2183–2206.

[M90] J. Maloney, *Residue class domains of the ring of convergent sequences and of $C([0,1]),R)$ and $C^\infty([0,1]),R)$*, Pacific J. Math. 143 (1990), 79–153.

[MZ05] J. Martinez and E. Zenk, *Dimension in algebraic frames II: Applications to frames of ideals in $C(X)$*, Comment. Math. Univ. Carolinae 46 (2005) 607–636.

[PW88] J. Porter and R.G. Woods, *Extensions and Absolutes of Hausdorff Spaces*, Springer-Verlag, New York 1988.

[R82] J. Roitman, *Nonisomorphic hyper-real fields from nonisomorphic ultrapowers*, Math. Zeit. 181 (1982), 93–96.

[Sc97] N. Schwartz, *Rings of continuous functions as real closed rings*, Ordered algebraic structures (Curaçao, 1995), 277–313, Kluwer Acad. Publ., Dordrecht, 1997.

[Se71] Z. Semadeni, *Banach Spaces of Continuous Functions*, Polish Scientific Publishers, Warsaw 1971.

[St36] M.H. Stone, *Applications of the theory of Boolean rings to general toplogy*, Trans. Amer. Math. Soc. 41 (1937), 375–481.

[Wa74] R.C. Walker, *The Stone-Čech Compactification*, Springer Verlag, New York 1974.

[We75] M. Weir, *Hewitt-Nachbin Spaces*, North-Holland Math. Studies, American Elsevier, New York 1975.

[Wi82] E. Wimmers, *The Shelah P-point indepence theorem*, Israel J. Math. 43 (1982), 28–48.

Bikram Banerjee (Bandyopadhyay)
Department.of Pure Mathematics
University of Calcutta
West Bengal, India
e-mail: `pbikraman@rediffmail.com`

Melvin Henriksen
Department of Mathematics
Harvey Mudd College
Claremont CA 91711, USA
e-mail: `henriksen@hmc.edu`

Positivity
Trends in Mathematics, 27–71

Positivity in Operator Algebras and Operator Spaces

David P. Blecher

Abstract. This article is aimed at a general reader familiar with the basics of functional analysis. It begins with a quick summary of the most basic 'facts of life' of positivity for Hilbert space operators, or for algebras of operators on a Hilbert space. It being impossible to adequately survey the fundamental role of positivity in the field of operator algebras, since this is so extensive and ubiquitous, in the present article we review selectively some of the general principles in the subject, and give some examples of how positivity plays a central role in the field, even in settings where positivity is not at first in evidence. The topics become more progressively more specialized towards our own current interests, ending with some very recent work of ours and of others.

1. Introduction

In the chapter titled 'Fundamental Tools' in his article 'A view of mathematics', Fields medalist and Crafoord prize winner Alain Connes mentions a conversation in the cafeteria of the IHES, in which the participant mathematicians each listed one main tool which they used in their work. The rest of his chapter is divided into subsections, each devoted to one fundamental tool, and interestingly the very first subsection is entitled 'Positivity'. There, the importance of positivity is briefly illustrated by the fundamental importance of probability theory in mathematics and quantum physics, and also by the theory of operator algebra. Connes says, for example, that 'positivity plays a key role in physics under the name of unitarity which rules out any physical theory in which computed probabilities do not fulfill the golden rule $P(X) \in [0, 1]$." Since probabilities and expectations in quantum mechanics are intimately tied to positivity for Hilbert space operators, for lack of a better name, we shall refer to the latter positivity, or positivity in algebraic systems comprised of Hilbert space operators, as 'quantum positivity'. It is indeed absolutely fundamental and pervasive in modern analysis, noncommutative geometry, quantum physics, and related fields.

---

*Blecher was partially supported by grant DMS 0400731 from the National Science Foundation.

'Quantum positivity' behaves like a *noncommutative* variant of the positivity one sees in many classical Banach lattices. Indeed, many common classical ordered vector spaces are spaces of scalar-valued functions; and the fact that scalar-valued functions commute whereas operators do not, has profound implications for the development of, and the divergence between, these theories. It is not our intention here to survey the subject of quantum positivity, and we make no attempt to be comprehensive or even balanced, either in the text or in the bibliography. This would be impossible, given the ubiquity of this kind of positivity in operator algebras. Also, the article is aimed at a very general mathematical audience, who may not be necessarily be familiar with operators on Hilbert space. Thus we have had to avoid technicalities, which unfortunately are characteristic of the subject, and this also shaped the structure and nature of the article. All definitions are given, although the reader may have to spend time looking back to find them. We divide the paper into several sections, which become progressively more specialized, and progressively closer to some of our current interests. We will not try hard to justify our selection of topics, to a certain extent this was random, or simply motivated by what was possible, by what is covered in other sources (such as Ben de Pagter's beautiful talk at this conference), and by what we needed in later sections. We will begin with a quick review of the basic 'facts of life' of 'quantum positivity'. In the next section we discuss some basics of C^*-algebras and related objects, giving illustrations of how positivity is used profoundly in their theory. We also discuss complete positivity. In Section 3 we discuss some aspects of how positivity is used in the study of linear spaces and algebras of operators on Hilbert space, where positivity may not at first be in evidence, and this is further illustrated in some of the examples in Section 4. In particular, since the Choquet and Shilov boundaries play such a fundamental role in the classical variants of the latter fields, we stress in this article the noncommutative versions of these boundaries. Section 4 presents several recent (and somewhat random) examples of the use of positivity in areas close to our current interests, indeed mostly from the authors work. To those who are surprised that I am writing on positivity, given that most of my work has been on operator spaces or operator algebras in which positivity is not at first in evidence, the point is that nonetheless positivity is a main tool in such settings, for example in canonically associated C^*-algebras, as is explained in parts of Sections 3 and 4. One just has to work a bit harder for ones positivity, which perhaps results in a keener appreciation of it.

In the present article, H, K will denote Hilbert spaces over the complex field. We write $B(H)$ for the algebra of bounded linear operators $T : H \to H$. This is just the $n \times n$ matrix algebra M_n if H finite-dimensional. We are interested in $B(H)_+$, the cone of *positive* operators on H. There are several equivalent definitions of 'positive operator' and the associated Löwner order. For example, T is positive iff $\langle Tx, x\rangle \geq 0$ for all vectors $x \in H$; and we write $T \geq 0$. If H is finite-dimensional, so that T may be regarded as an $n \times n$ matrix A in the usual way, then positivity is equivalent to saying that A is *selfadjoint*, and all of its eigenvalues are nonnegative. We recall that an operator T is selfadjoint if $T = T^*$. Here T^* is the usual *involution* (or *adjoint*) on $B(H)$, defined uniquely by the equation

$$\langle Tx, y\rangle = \langle x, T^*y\rangle, \qquad x, y \in H.$$

For the matrix $A = [a_{ij}]$ above, A^* is just the 'conjugate transpose' $[\overline{a_{ji}}]$. If H is not necessarily finite-dimensional, then eigenvalues are no longer so useful. In this case one looks at the *spectrum* $\mathrm{Sp}(T)$, the set of complex numbers λ such that $\lambda I - T$ is not invertible. Then $T \geq 0$ iff T is selfadjoint and $\mathrm{Sp}(T) \subset [0, \infty)$. And of course $S \leq T$ iff S and T are selfadjoint and $T - S \geq 0$.

Positivity may also be rephrased in terms of a simple algebraic equation:

$$T \in B(H)_+ \;\Leftrightarrow\; T = S^*S, \;\; S \in B(H).$$

In fact, $T \in B(H)_+$ iff $T = S^2$ for a selfadjoint $S \in B(H)$. This operator S can be chosen to also be positive, and we write it as $T^{\frac{1}{2}}$ (it is unique). A good example of a positive operator is an *orthogonal projection* on H, namely an operator satisfying $P = P^2 = P^* \in B(H)$. Every positive operator, which we may assume by scaling to have norm ≤ 1, is in some sense an average of these projections, as we shall discuss.

The good news is that positivity is beautifully related to the underlying algebra, to the spectral theory of T, and to the operator norm $\|T\| \stackrel{def}{=} \sup\{\|Tx\| : x \in H, \|x\| \leq 1\}$. We list the basic facts of life:

- If $T \geq 0$ then $\|T\| = \max \mathrm{Sp}(T)$. This is just the largest eigenvalue if H is finite-dimensional.
- For any $T \in B(H)$, $\|T\| \leq 1$ iff the operator on $H \oplus H$ taking (x, y) to $(x + Ty, T^*x + y)$ is positive. If $T = T^*$ then $\|T\| \leq 1$ iff $-I \leq T \leq I$. These allow innumerable issues about norms to be approached in terms of positivity.
- Every $T \in B(H)$ has a *polar decomposition* $T = U|T|$, with $|T| = (T^*T)^{\frac{1}{2}} \geq 0$ and U a *unitary* operator (this means that $U^{-1} = U^*$) or a *partial isometry* (that is, $U = UU^*U$).
- Every $T \in B(H)$ may be written uniquely as $T = R + iS$ with R, S selfadjoint. Furthermore, any selfadjoint operator R may be written uniquely as a difference of two positive operators whose product is 0 and whose sum is $|T|$. Thus T has a Jordan decomposition $T = T_1 - T_2 + i(T_3 - T_4)$ with $T_i \geq 0$. The latter may also be seen from the *polarization identity* $y^*x = \frac{1}{4}\sum_{k=0}^{3} i^k\,(x + i^k y)^*(x + i^k y)$.
- If $S \leq T$ then $D^*SD \leq D^*TD$, for any operator $D \in B(K, H)$.
- If $T = T^*$ and $\alpha, \beta \in \mathbb{R}$ then $\alpha I \leq T \leq \beta I$ if and only if $\mathrm{Sp}(T) \subset [\alpha, \beta]$.
- $0 \leq S^{-1} \leq T^{-1}$ if $0 \leq S \leq T$ and S is invertible (which implies T invertible).
- There is a reduction from 'quantum positivity' to classical positivity. Indeed, operators in the 'locality' of a positive operator T behave like scalar-valued functions, in the following sense. The *functional calculus* $f \mapsto f(T)$ is an isomorphism between the algebra $C(\mathrm{Sp}(T))$ of continuous functions on the spectrum of T, and the closure of the algebra generated by T and the identity I; and this isomorphism takes the function $f(t) = t$ to T. The calculus is

even a $*$-*isomorphism*: that is, $f(T)^* = \bar{f}(T)$. Thus operators in this closure may be treated as scalar functions. For example, if $f(t) = \sqrt{t}$, then $f(T)$ is precisely the square root $T^{\frac{1}{2}}$ mentioned above. This functional calculus is valid more generally for *normal operators*, that is for operators $T \in B(H)$ which commute with their adjoint T^*. The domain of the functional calculus may also be extended from the continuous functions on the spectrum of T, to the Borel measurable functions.

- From time to time one still uses some ordered vector space techniques, for example geometric Hahn-Banach separation of a cone of positive operators from another set.
- The spectral theorem: Any compact $T \in B(H)_+$ (resp. any $T \in B(H)_+$) is a sum $\sum_i t_i P_i$ (resp. integral $\int_0^{\|T\|} t \, dP(t)$) where P_i are orthogonal projections on H, and t_i are positive scalars (resp. $dP(t)$ is an orthogonal projection-valued measure).

Now for the bad news. First, quantum positivity is very far from 'lattice orderings'. The upper and lower bound properties of the ordering are horrible. This is a continual source of frustration. Some good news on this front: the 'extremely positive' elements – by which we mean the extreme points of the set $\{T \in B(H) : 0 \leq T \leq I\}$ – do form a lattice. Indeed these extreme points are exactly the orthogonal projections on H, which constitute a complete lattice with respect to the natural operations (which correspond to closed joint span or intersection of the ranges of the projections). By the Krein-Milman theorem, an arbitrary positive operator (which we can assume by scaling have norm 1) is a limit of convex combinations of these 'extremely positive' elements (this fact also follows from the last 'bullet' above). We remark that Sherman recently showed that there exists an equivalence relation (originally studied by Kadison and Pedersen) on $B(H)_+$ such that the equivalence classes constitute a complete lattice. Indeed, a more general version of this holds essentially with $B(H)$ replaced by any von Neumann algebra [105]. From time to time physicists find other 'fixes' to the 'lattice failure' problem, and some of these are quite interesting for some purposes. But essentially this 'problem' is just a fact of life, like the sad facts in the next paragraph.

Some other nasty facts of life: unless two positive operators S and T commute, one cannot expect ST to be positive. Powers and exponentials of operators do not often behave as they ought. For example, exponentials behave badly (for example, $e^S e^T \neq e^{S+T}$ in general), and if $0 \leq S \leq T$ one cannot even expect $S^2 \leq T^2$, although, curiously, $S^{\frac{1}{2}} \leq T^{\frac{1}{2}}$. That is, the subject of 'operator monotonicity' is delicate (see, e.g., [88, 15]). Indeed, the usual manipulations with inequalities between real numbers which one is familiar with, are likely to be wrong for operators (unless they are on the list above), and this causes endless frustration. Things can quickly get quite scary, for example, when contemplating replacing a classical argument containing a dozen function inequalities, by operator inequalities. This problem is often insurmountable, or has to be overcome with considerable extra work, or completely different methods. A recent example of this may be found in our project with Labuschagne, surveyed in [27], where we completely generalized the theory of generalized H^p spaces from the 1960s, to a von Neumann algebraic context introduced by Arveson. In particular, one has to take the classical arguments, which feature hundreds of tricks with functions which fail for operators, and replace them with noncommutative tools coming from the theory of von Neumann algebras and unbounded operators. It was continually surprising to us that the results successively generalized; and fortunately this pleasant kind of surprise is often characteristic of the field of operator algebras.

We remark too that there is an industry, currently mostly based in Japan, around the subject of operator inequalities (see, for example, [68]). Physicists also have found ingenious ways of circumventing some of the difficulties mentioned above, for their purposes. We will however not discuss these topics here.

2. Positivity in selfadjoint operator algebras and systems

2.1. C^*-algebras

A C^*-algebra is a Banach algebra A which is also a $*$-vector space (that is, it has an antilinear 'involution' $* : A \to A$) satisfying a list of several natural looking criteria (such as $(xy)^* = y^*x^*$ for $x, y \in A$; these make A a *Banach $*$-algebra*), and also satisfying the so-called *C^*-identity*: $\|x^*x\| = \|x\|^2, x \in A$. Often C^*-algebras are called *operator algebras*, although we prefer to use that term for something a little more general. From the section on positivity in the aforementioned article of Connes, we quote: "The following inequality is in fact the cornerstone of the theory of operator algebras

$$x^*x \geq 0 , \qquad \forall x \in A.$$

C^*-algebras are those abstract algebras endowed with an antilinear involution $x \to x^*$, for which the above inequality "makes sense", *i.e.*, defines a cone $A_+ \subset A$ of positive elements which possess the expected properties. Thanks to functional analysis the whole industry of the theory of convexity can then be applied: one uses the Hahn-Banach theorem to get positive forms, and all the powerful properties of operators in Hilbert space can then be used in this seemingly abstract context."

That is, first, the key point is positivity. If you like, C^*-algebras is a 'positivity theory'; there is a sense in which, explicitly or implicitly, positivity underlies almost every proof in C^*-algebra theory (see for example the texts [18, 45, 74, 97, 114]). Note too that Connes is highlighting the miraculous relationship between the positivity and the algebra, and this miracle only gets more remarkable as one gets deeper into the theory, indeed it is a characteristic feature of the subject of operator algebras. Second, Connes is hinting at the fact that C^*-algebras are just the *selfadjoint* algebras of operators on a Hilbert space (that is, the algebras A such that $a \in A$ iff $a^* \in A$), which are closed in the norm topology. This is the *Gelfand-Naimark theorem*. More precisely, every C^*-algebra A has a faithful (that is, one-to-one) representation on a Hilbert space H, where by 'representation' we mean a $*$-linear homomorphism (that is, a *$*$-homomorphism*) $\pi : A \to B(H)$. The proof is not difficult, except for one technical part (namely showing that in an algebra satisfying the usual definition of a C^*-algebra, the last centered and displayed equation holds automatically (of course one could build this into the definition to save time, as Connes does above)). One may assume that A is unital, that is, it has an identity 1, since every C^*-algebra has a unitization. Then the key ingredient in the remainder of the proof is the following fundamental idea in the subject. For this, we recall that a *state* φ of a C^*-algebra A is a norm 1 functional in A^* which is *positive* in the sense that it takes positive elements to positive elements. If A is unital then states are *unital* too (we say that a function is unital if it takes 1 to 1). The convex set $S(A)$ in A^* consisting of the states is called the *state space*, it is weak* compact if A is unital. There is a *Hahn decomposition*: any $\varphi \in A^*$ is a linear combination $\varphi_1 - \varphi_2 + i(\varphi_3 - \varphi_4)$ of states; indeed this can be done so that φ_1 and φ_2 (resp. φ_3, φ_4) are 'mutually singular' [45, p. 272–273].

Theorem 2.1 (GNS construction). *For any state φ of a C^*-algebra A, there exists a Hilbert space H, a representation π of A on H, and a vector of norm 1 in H, such that $\varphi = \langle \pi(\cdot)\,\zeta, \zeta \rangle$.*

Proof. One endows A with the semi-innerproduct $(a, b) \mapsto \varphi(b^*a)$. The Cauchy-Schwarz inequality for this semi-innerproduct ensures that the completion of the quotient A/N of A by the nullspace N of the associated seminorm, is a Hilbert space. Moreover, the canonical representation $\pi : A \to B(H)$, defined by $\pi(a)[b] = [ab]$ for $a \in A, [b] \in A/N$, is a $*$-homomorphism. Assuming that A is unital for simplicity, we set $\zeta = [1]$, and then it is immediate that $\varphi = \langle \pi(\cdot)\,\zeta, \zeta \rangle$. □

One may construct a faithful representation of A by taking a direct sum, indexed by $S(A)$, of the representations π coming from the GNS construction. That this is faithful follows immediately from the fact that for any positive element $a \in A$ there is a state $\varphi \in A^*$ with $\varphi(a) = \|a\|$, which in turn is a simple consequence of the functional calculus and the Hahn-Banach extension theorem. One can also show easily that φ is an extreme point of the set of states of A iff π is *irreducible*, that is, there are no nontrivial closed $\pi(A)$-invariant subspaces of H.

The terminology 'state' comes from quantum mechanics: the states of a physical system. We remark that one may extend the notion of 'state' to possibly non-selfadjoint algebras of operators on Hilbert space – which are what we prefer to call *operator algebras*. Indeed for any space A with an identity 1 of norm 1, the states of A may be defined to be the functionals $\varphi \in A^*$ with $\|\varphi\| = \varphi(1) = 1$. That these are positive if A is a C^*-algebra is easy: one can clearly assume that A is commutative, by the 8th 'bullet' in Section 1, thus $A = C(\Omega)$ for compact Ω, and then one can appeal to basic facts from measure theory.

The earlier characterizations and properties of positivity in $B(H)$ extend easily to the positive cone A_+ of a C^*-algebra. It is easy to see from some of these that A_+ is closed in the norm topology, and if A is a closed $*$-subalgebra of a C^*-algebra B then $A_+ = A \cap B_+$. Also, states characterize A_+: indeed $a \in A_+$

iff $\varphi(a) \geq 0$ for every state φ. This is quite easy: for the hard direction we can assume that $A \subset B(H)$, consider the states $\langle \cdot \zeta, \zeta \rangle$, for unit vectors $\zeta \in H$, and appeal to an earlier characterization of $B(H)_+$ above.

As an application of some of the positivity ideas above, we prove:

Theorem 2.2. *If $\pi : A \to B$ is a $*$-homomorphism between C^*-algebras then π is positive and contractive (that is, $\pi(A_+) \subset B_+$ and $\|\pi\| \leq 1$). If also π is faithful (that is, one-to-one) then π is isometric.*

Proof. That π is positive is immediate: $\pi(x^*x) = \pi(x)^*\pi(x) \in B_+$. Next, suppose that A is unital. Without loss of generality we may assume that $B = \overline{\pi(A)}$ (otherwise replace B by this algebra). Since $\pi(1)\pi(x) = \pi(x)$ for $x \in A$ it follows that $\pi(1)$ is an identity for B. Suppose that $\lambda \notin \mathrm{Sp}(x)$. Thus there exists an $a \in A$ such that $(\lambda 1 - x)a = a(\lambda 1 - x) = 1$. Applying π we see that $(\lambda 1 - \pi(x))\pi(a) = \pi(a)(\lambda 1 - \pi(x)) = 1$. It follows that $\mathrm{Sp}(\pi(x)) \subset \mathrm{Sp}(x)$, and so by the first 'bullet' in Section 1, $\|\pi(x)\| \leq \|x\|$, for any $x \in A_+$. For general $x \in A$ we have by the C^*-identity that

$$\|\pi(x)\|^2 = \|\pi(x^*x)\| \leq \|x^*x\| = \|x\|^2.$$

Thus π is a contraction.

Next, suppose that π one-to-one and A is unital. By the last displayed equation, to show that π is isometric it suffices to show that $\|\pi(x)\| = \|x\|$ for $x \in A_+$. We may then restrict π to a map between the C^*-algebra generated by x and 1, and the C^*-algebra generated by $\pi(x)$ and 1. In the eighth bullet in Section 1, we can assume that $A = C(K_1)$ and $B = C(K_2)$ for compact sets K_1 and K_2 in $\mathbb{R}$. By the basic theory of commutative Banach algebras, there exists a continuous $\tau : K_2 \to K_1$ such that $\pi(f) = f \circ \tau$ for all $f \in C(K_1)$. Now the result has been reduced to topology, we leave it as an exercise in that subject that τ is surjective. Then it is clear that $\|\pi(h)\| = \|h \circ \tau\|_\infty = \|h\|_\infty$.

Finally, suppose that A is nonunital. We can extend π to a function from A^1 to a unitization of B, by defining $\tilde{\pi}(a + \lambda 1) = \pi(a) + \lambda 1$. It is very easy to see that $\tilde{\pi}$ is a $*$-homomorphism too. Thus it is contractive, and therefore so is π. If π is one-to-one, then it is easy to see that so is $\tilde{\pi}$. By the last paragraph, $\tilde{\pi}$ is isometric, and therefore so is π. □

C^*-algebras need not be unital, but at least they all have positive, increasing, approximate identities of norm ≤ 1. In fact this approximate identity may be taken to be the set of all elements in A_+ of norm < 1, with a suitable ordering. For nonunital C^*-algebras it is often better to not work with the state space $S(A)$, since this may not be compact in this case, but with the (compact) *quasistate* space $Q(A)$ of positive functionals of norm ≤ 1. In what follows, $Q(A)$ may be replaced by $S(A)$ if A is unital.

The systematic study of positivity in C^*-algebras was initiated by Kadison (see, e.g., [71, 72] and [74] and references within). We mention in particular Kadison's *function representation* of any C^*-algebra A. This is the restriction to the set of selfadjoint elements, of the canonical map from the second dual A^{**}, which is also a C^*-algebra in a natural way with the so-called *Arens product* (which is the unique separately weak* continuous Banach algebra product on A^{**} which extends the product of A), to the bounded functions on $Q(A)$. This map is an isometric order isomorphism onto the set B_0 of bounded affine real-valued functions on $Q(A)$ which vanish at 0. Perhaps more importantly, it restricts to an order isomorphism from the selfadjoint elements in A onto the set of functions in B_0 which are continuous (with respect to the weak* topology on $Q(A)$). This allows one to treat collections of selfadjoint elements in A, and their order, as scalar-valued functions, and this in a way that is compatible with the usual tools of convexity theory. This is a powerful device. We note that the elements in A which correspond, under the isomorphism above, to the positive functions which are never 0 on $Q(A) \setminus \{0\}$, are called *strictly positive*, and they are quite important. Note that if A is unital then the strictly positive elements are exactly the positive invertible elements. The existence of strictly positive elements is easily seen to be equivalent to A having a *countable* increasing positive approximate identity (see, e.g., 3.10.5 in [97]).

The elements of A^{**} which correspond in Kadison's function representation to upper or lower semicontinuous functions on $Q(A)$ are also important. For example, the orthogonal projections in this class of element of A^{**} are the *closed* and *open projections* which we will discuss further in Section 2.2. We note that there is a bijective order preserving correspondence between the open projections p and the set of closed left ideals J of A, taking p to $J = A^{**}p \cap A$. 'Left' here can be replaced by right, via the correspondence $J \mapsto J^*$ between left and right ideals. Or, if one wishes for a 'symmetrical' object, instead of ideals one can take instead the *hereditary subalgebras* of A. This is a closed subalgebra D of A with the order-theoretic property that if $a \in A$ and $a \leq d \in D$, then $a \in D$. The bijective correspondence between left ideals J and hereditary subalgebras D is: $D = J \cap J^*$, and $J = AD$. We remark that many properties of C^*-algebras pass to their hereditary subalgebras, which is one of the reason why they are so useful.

Thus much of the algebraic structure of a C^*-algebra A may be captured via positivity, since open projections are defined via the order (see also the definition given of open projections in Section 2.2). Alternatively, the same algebraic information (that is, the one-sided ideals or hereditary subalgebras) can be captured using positivity in a different way, by considering *faces* in $Q(A)$ (recall that a face of a convex set C is a nonempty convex subset F with the property that if $tx + (1-t)y \in F$, for $x, y \in C$ and $0 < t < 1$, then $x \in F$ and $y \in F$). Indeed, there is a bijective order-reversing correspondence between the lattice of such faces which are weak* closed and contain 0, and the lattice of closed left ideals of A.

See, e.g., [42, 49, 97, 18] for more details and history of the topics in the last couple of paragraphs. Much more profound aspects of the state space of a C^*-algebra are developed in great detail in the two volume treatise of Alfsen and Schultz [7, 8].

Von Neumann algebras are the selfadjoint algebras of operators on a Hilbert space which are also closed in the *weak* topology* of $B(H)$. We recall that $B(H)$ is a dual Banach space, and its (strongly unique) predual is the so called *trace class operators* on H. Von Neumann algebras are much simpler in some ways than C^*-algebras – and have many powerful tools, mostly relying ultimately on positivity and projections. The facts about extreme points in the third last paragraph of Section 1, remain true with $B(H)$ replaced by any von Neumann algebra. For example the orthogonal projections in M constitute a complete lattice whose closed convex hull is the positive elements of norm ≤ 1. Thus M is densely (in the norm topology, even) spanned by this lattice of 'extremely positive elements'. This fails for C^*-algebras (for example, $C([0,1])$ has no projections). However the principle can still be used to prove deep results about C^*-algebras, because the second dual of a C^*-algebra is a von Neumann algebra, in a natural way. In fact many proofs which could be proved using positivity arguments, may be considerably simplified by using 'extremely positive' elements, namely projections, in the second dual. The following proof (taken from [112]) illustrates this idea admirably. We consider 'conditional expectations'. These are the positive maps in the next result:

Theorem 2.3. (Tomiyama) *If* $\Phi : A \to A$ *is a linear idempotent* (*that is,* $\Phi \circ \Phi = \Phi$) *map onto a* $*$*-subalgebra* B *of the* C^**-algebra* A*, with* $\|\Phi\| \leq 1$*, then* Φ *is positive, and*

$$\Phi(b_1 a b_2) = b_1 \Phi(a) b_2 \,, \qquad b_1, b_2 \in B, a \in A.$$

Proof. We sketch the key steps of proof. We will assume for simplicity that $1 = 1_A \in B$, although if this is not the case the proof is only a couple of lines longer. That Φ is positive follows from the characterization of A_+ in terms of states: for if φ is a state then so is $\varphi \circ \Phi$ by one of the characterizations of states above, and so $\varphi(\Phi(a)) \geq 0$ if $a \in A_+$. For the last assertion of the theorem, it is easy to show, by going to the second dual, that without loss of generality one may suppose that A is a von Neumann algebra. As we said earlier, any element in A is a limit of linear combinations of 'extremely positive' elements, i.e., orthogonal projections. By symmetry, and taking involutions, we see that it is enough to show that $\Phi(ex) = e\Phi(x)$ for an orthogonal projection $e \in B$ and $x \in A$. Set $f = 1 - e$. For $x, y \in A$ we have

$$\|ex + fy\|^2 = \|(ex + fy)^*(ex + fy)\| = \|x^* ex + y^* fy\| \leq \|ex\|^2 + \|fy\|^2.$$

Now $\Phi(f\Phi(ex)) = f\Phi(ex)$ since $f\Phi(ex) \in B$, and if $\lambda \in \mathbb{R}$ then

$$(\lambda + 1)^2 \|f\Phi(ex)\|^2 = \|f\Phi(ex + \lambda f\Phi(ex))\|^2 \leq \|ex + \lambda f\Phi(ex)\|^2.$$

Combining the last two displayed equations, we have

$$(\lambda + 1)^2 \|f\Phi(ex)\|^2 \leq \|ex\|^2 + \|\lambda f\Phi(ex)\|^2 = \|ex\|^2 + \lambda^2 \|f\Phi(ex)\|^2.$$

Thus $(2\lambda + 1)\|f\Phi(ex)\|^2 \leq \|ex\|^2$, and letting $\lambda \to \infty$ we must have that $\|f\Phi(ex)\| = 0$, so that $f\Phi(ex) = 0$. This means that $\Phi(ex) = e\Phi(ex)$. Replacing e with f we have by the above arguments that $e\Phi(x - ex) = e\Phi(fx) = 0$, so that $e\Phi(x) = e\Phi(ex)$. Hence $\Phi(ex) = e\Phi(x)$ as desired. □

Perhaps the very first result that one meets in von Neumann algebra theory, is the following simple but important fact, which play a crucial role in von Neumann algebra theory, as will be illustrated by the next several results.

Lemma 2.4. *Every increasing bounded net (x_t) of positive elements in a von Neumann algebra M has a least upper bound x, say, in M. Also, the net converges in the weak* topology to x, and also converges strongly (that is, $x_t\zeta \to x\zeta$ for all $\zeta \in H$).*

Proof. Let $\zeta \in H$, and consider the bounded increasing net $(\langle x_t\zeta, \zeta\rangle)$ in $\mathbb{R}$, which certainly has a limit there. By the polarization identity,

$$\langle a\zeta, \eta\rangle = \frac{1}{4}\sum_{k=0}^{3} i^k \langle a(\zeta + i^k\eta), \zeta + i^k\eta\rangle, \qquad a \in M_+, \zeta, \eta \in H.$$

It follows from this that the net $(\langle x_t\zeta, \eta\rangle)$ converges.

The function $(\zeta, \eta) \mapsto \lim_t \langle x_t\zeta, \eta\rangle$ is a bounded bilinear functional on H, since

$$|\lim_t \langle x_t\zeta, \eta\rangle| \leq \sup_t |\langle x_t\zeta, \eta\rangle| \leq \sup_t \ \|x_t\zeta\|\|\eta\| \leq \sup_t \ \|x_t\|\|\zeta\|\|\eta\|.$$

Thus by a basic fact from Hilbert space theory (sometimes called the Riesz representation theorem), there exists an $x \in B(H)$, with $\|x\| \leq \sup_t \ \|x_t\|$, such that $\langle x\zeta, \eta\rangle = \lim_t \langle x_t\zeta, \eta\rangle$. Since $(\langle x_t\zeta, \zeta\rangle)$ increases with limit $\langle x\zeta, \zeta\rangle$, it is easy to see that $0 \leq x_t \leq x$ for every t. Any weak* convergent subnet of (x_t) must converge to x too (since the functionals $\langle \cdot \zeta, \eta\rangle$ are weak* continuous). By Alaoglu's theorem, it follows that $x_t \to x$ weak*. Hence $x \in M$ too. To see that $x_t \to x$ strongly, note that $x - x_t \geq 0$, so that $x - x_t = (x - x_t)^{\frac{1}{2}}(x - x_t)^{\frac{1}{2}}$. Thus for any $\zeta \in H$ we have

$$\|(x - x_t)\zeta\|^2 \leq \|(x - x_t)^{\frac{1}{2}}\|^2\|(x - x_t)^{\frac{1}{2}}\zeta\|^2 = \|x - x_t\|\langle(x - x_t)\zeta, \zeta\rangle \to 0.$$

Finally, to see that $x = \sup x_t$, note that if $S \in B(H)$ with $x_t \leq S$ for every t, then $\langle x_t\zeta,\zeta\rangle \leq \langle S\zeta,\zeta\rangle$ for any $\zeta \in H$. In the limit, $\langle x\zeta,\zeta\rangle \leq \langle S\zeta,\zeta\rangle$, so that $x \leq S$. □

As a first application of this result, we show that the weak* closed two-sided ideals in a von Neumann algebra are in a bijective correspondence with the orthogonal projections p in M that commute with all other elements of M. For such p, clearly Mp is a weak* closed ideal. Conversely, if J is such an ideal, then it is a C^*-algebra, so has a positive increasing approximate identity (x_t) bounded above by 1. If $x_t \to x$ weak* as above, then $x \in J$, so that $xM \subset J$. If $y \in J$ then clearly $xy = y$, so that $x^2 = x = x^*$ and $J = xJ \subset xM$. So $J = xM$. Similarly, $J = Mx$, and so we have for any $y \in M$ that $xy = xyx = yx$. (There is a similar correspondence for one-sided ideals and general orthogonal projections that is only slightly harder.)

Another fact concerning increasing nets, which is a little deeper but is still not very hard, says that a state φ on a von Neumann algebra M is weak* continuous iff for every bounded increasing net (x_t) with weak* limit x, we have $\lim_t \varphi(x_t) = \varphi(x)$. Using this fact, and the Hahn decomposition, it is easy to see that the same thing is true for all functionals $\varphi \in A^*$. Thus weak* continuity in von Neumann algebra settings may be replaced by order theoretic considerations.

Corollary 2.5. *Every $*$-isomorphism between von Neumann algebras is a homeomorphism for the weak* topology.*

Proof. Let $\pi : M \to N$ be a $*$-isomorphism between von Neumann algebras. If φ is a weak* continuous state on N, and if (x_t) is an increasing bounded net of positive elements in M with least upper bound x, then since π is an order isomorphism we have that $(\pi(x_t))$ is an increasing bounded net of positive elements in N with least upper bound $\pi(x)$. Thus $\pi(x_t) \to \pi(x)$ weak*, and so $\varphi(\pi(x_t)) \to \varphi(\pi(x))$. It follows by the remark above the theorem that $\varphi \circ \pi$ is weak* continuous. It is quite easy to see that any weak* continuous $\psi \in N^*$ is a linear combination of four weak* continuous states, and thus $\psi \circ \pi$ is weak* continuous. Hence if $y_s \to y$ weak* in M then $\pi(y_s) \to \pi(y)$ weak* in M; thus π is weak* continuous. Similarly for π^{-1}. □

Theorem 2.6. (Sakai) *If A is a C^*-algebra which has a Banach space predual, then A is $*$-isomorphic, via a weak* homeomorphism, to a von Neumann algebra.*

Proof. This proof does not explicitly use positivity, although it does use conditional expectations, which are positive. Also, it is short, given what we have already proved, and we will refer to the result later. We said earlier that A^{**} is a von Neumann algebra. Suppose that $E^* = A$. The canonical map $E \to E^{**} = A^*$ dualizes to give a weak* continuous contractive surjection $\Phi : A^{**} \to A$. Regard A as a C^*-subalgebra of A^{**}. It is easy to check that Φ extends the identity map on A, so that $\Phi \circ \Phi = \Phi$. Thus Φ is a weak* continuous 'conditional expectation' satisfying Tomiyama's theorem 2.3. Applying that result we have xy and yx are in $\mathrm{Ker}(\Phi)$ for any $x \in A, y \in \mathrm{Ker}(\Phi)$. It follows that for $x, y \in A$ we have $\Phi(xy)$ equals

$$\Phi((x - \Phi(x))(y - \Phi(y)) + \Phi(\Phi(x)(y - \Phi(y)) + \Phi(x\Phi(y)) = \Phi(x\Phi(y)) = \Phi(x)\Phi(y).$$

Hence Φ is a weak* continuous $*$-homomorphism. Thus $\mathrm{Ker}(\Phi)$ is a weak* closed two-sided ideal in A^{**}. By a fact proved in the paragraph after Lemma 2.4, there exists such a projection $p \in A^{**}$, with $\mathrm{Ker}(\Phi) = pA^{**}$. If $a \in A$, then $\Phi((1-p)a) = \Phi(a) - \Phi(p)\Phi(a) = a$. Thus Φ restricts to a surjective weak* continuous faithful $*$-homomorphism from the von Neumann algebra $(1 - p)A^{**}$ onto A. □

Combining the last two results, it follows in a couple of easy lines that every von Neumann algebra has a unique Banach space predual.

We mention a related characterization of von Neumann algebras due to Kadison [73]:

Theorem 2.7. *If A is a C^*-subalgebra of $B(H)$ with $[AH] = H$, and if A contains the upper bound in $B(H)$ of every bounded increasing net of selfadjoint elements in A, then A is a von Neumann algebra.*

See also, e.g., [97] for much more on positivity in C^*-algebras and von Neumann algebras, for example for Pedersen's famous 'up-down theorem', and the C^*-algebraic variant of the Riesz decomposition property (which are respectively 2.4.3 and 1.4.11 in that text).

Positivity plays a fundamental role in the von Neumann algebraic subfield of noncommutative integration theory, and noncommutative L^p spaces. Since other talks at this conference addressed some aspects of this in more detail, we just mention here the very basic idea: the lattice of orthogonal projections plays the role of measurable sets; linear combinations of projections play the role of simple functions; and states on the algebra play the role of integration with respect to a probability measure. We remark in passing that there are generalizations to states on C^*-algebras of the Radon-Nikodým and Lebesgue decomposition theorems from classical measure theory (see [45, 67]). Positivity plays a profound role too in related subjects, such as free probability. In the latter theory, as well as elsewhere, positive maps often appear in conjunction with the notion of entropy.

Contained in the class of C^*-algebras, and containing the von Neumann algebras, there are other important classes of algebras which are defined in terms of their 'order' properties, for example the *AW^*-algebras* (see, e.g., [16, 113]) and the *monotone complete* C^*-algebras. The latter term means that the algebra contains a least upper bound for every bounded increasing selfadjoint net. These algebras are very interesting from an order-theoretic viewpoint, but also present quite formidable difficulties. For example, every monotone complete C^*-algebra is an AW^*-algebras, but the converse is a notorious open problem. From our perspective, a good reason to be interested in such C^*-algebras is that *injective* C^*-algebras (these are the subalgebras A of $B(H)$ which are the range of a conditional expectation $\Phi : B(H) \to A$ in the sense of 2.3 on $B(H)$) are monotone complete. Indeed, if (x_t) is a bounded increasing selfadjoint net in such a C^*-algebra, and if x is a least upper bound of the net in $B(H)$, then it is easy to see that $\Phi(x)$ is the least upper bound in A. More generally monotone complete C^*-algebras (and modules) come up naturally when one considers the *injective envelope* of a subspace of a C^*-algebra, and the latter is a powerful and currently popular tool (see, e.g., [60, 63, 102, 61, 62, 31, 96, 28] for more information on these topics).

2.2. Positivity and noncommutative topology

C^*-algebras may be regarded as a noncommutative variant of topology, by another theorem of Gelfand. This result establishes a bijective correspondence between locally compact Hausdorff spaces, and the class of commutative C^*-algebras: $\Omega \mapsto C_0(\Omega)$. Here $C_0(\Omega)$ denotes the continuous functions vanishing at infinity. Moreover this can be made into a 'duality of categories': there is a well-known correspondence between continuous maps between such topological spaces, and certain $*$-homomorphisms between the associated algebras. Also, Ω is compact iff the associated algebra is unital (that is, has an identity 1). There are many ways in which mathematicians have tried to 'get their hands on the topology', for a noncommutative C^*-algebra. In this subsection we will describe one important way that is extremely intimately tied to the positivity (order). If the C^*-algebra is commutative, so equals $C_0(\Omega)$ as above, we first describe how to recover the open sets U in Ω using the ordering on $C_0(\Omega)$. Namely, these open sets correspond bijectively to the orthogonal projections in the second dual $C_0(\Omega)^{**}$, which are the least upper bounds (and, hence by Lemma 2.4, limits in the weak* topology) of increasing nets of positive elements in $C_0(\Omega)$. The above suggests the following definition (due to Akemann [2, 3, 4], and Giles and Kummer [57]): If B is a C^*-algebra then an orthogonal projection $q \in B^{**}$ is *open* if it is the weak* limit (or equivalently, the least upper bound) of an increasing net of positive elements in B. We say that q is *closed* if $1-q$ is open. By the observation above, these collapse to the usual topological notions if $B = C_0(\Omega)$ (that is, if B is commutative). Many of the properties of open and closed sets generalize to this setting. In particular, there is a beautiful 'noncommutative Urysohn lemma', which we state in the case that B is unital: Given closed mutually orthogonal projections p and q in B^{**}, there exists an element $a \in B$ with $0 \leq a \leq 1$, $ap = pa = 0$ and $aq = qa = q$. This collapses to the usual Urysohn lemma for compact spaces if B is commutative. Open and closed projections always exist in abundance, for example the *left support projection* of any nonzero element $x \in B$ is open (this is the infimum of the projections $p \in B^{**}$ with $px = x$), and similarly for the right support projection.

For the readers convenience, we mention in a little more detail how open sets U in a locally compact Hausdorff space Ω correspond to open projections in $C_0(\Omega)^{**}$. Let $Bo(\Omega)$ denote the commutative C^*-algebra of bounded Borel measurable functions on Ω, with supremum norm. Then $B = C_0(\Omega)$ is a closed $*$-subalgebra of $Bo(\Omega)$, and it is not hard to show that $Bo(\Omega)$ is $*$-isomorphic to a closed $*$-subalgebra of B^{**}, the latter equipped with its usual product discussed earlier. Indeed, recall that $C_0(\Omega)^*$ is isometrically isomorphic to $M(\Omega)$, the space of regular Borel measures on Ω, and if we define $\theta : Bo(\Omega) \to M(\Omega)^* = B^{**}$ by $\theta(g)(\mu) = \int_\Omega g \, d\mu$, then it is easy to see that θ is the desired $*$-isomorphism. What needs to be proved is that a projection $p \in C_0(\Omega)^{**}$ is open if and only if $p = \theta(\chi_U)$ for an open subset $U \subset \Omega$, and this is a pleasant exercise in real analysis.

2.3. Complete positivity

Complete positivity is perhaps even more important than positivity in the study of operator algebras and operator spaces. The key point here is that a matrix of operators is an operator: that is, a matrix

$$[x_{ij}] = \begin{bmatrix} x_{11} & x_{12} & \cdots & x_{1n} \\ x_{21} & x_{22} & \cdots & x_{2n} \\ \cdot & \cdot & \cdots & \cdot \\ \cdot & \cdot & \cdots & \cdot \\ x_{n1} & x_{n2} & \cdots & x_{nn} \end{bmatrix}$$

in $M_n(B(H))$, may be viewed as an operator on $H^{(n)}$:

$$\begin{bmatrix} x_{11} & x_{12} & \cdots & x_{1n} \\ x_{21} & x_{22} & \cdots & x_{2n} \\ \cdot & \cdot & \cdots & \cdot \\ \cdot & \cdot & \cdots & \cdot \\ x_{n1} & x_{n2} & \cdots & x_{nn} \end{bmatrix} \begin{bmatrix} \zeta_1 \\ \zeta_2 \\ \cdot \\ \cdot \\ \zeta_n \end{bmatrix} = \begin{bmatrix} \sum_k x_{1k}\zeta_k \\ \sum_k x_{2k}\zeta_k \\ \cdot \\ \cdot \\ \sum_k x_{nk}\zeta_k \end{bmatrix}.$$

If the latter operator is positive, then we write $[x_{ij}] \geq 0$. That is, $M_n(B(H))$ has a natural cone. Similarly, and more generally, if A is a C^*-algebra, then so is $M_n(A)$, and so it has a natural cone $M_n(A)_+$. We say that a map $T : A \to B(H)$ is *completely positive* if

$$[T(x_{ij})] \geq 0 \quad \text{for all} \quad [x_{ij}] \in M_n(A)_+,\ n \in \mathbb{N}\,.$$

The necessity of considering these 'matrix cones' is admirably motivated in Effros' superb 1978 survey 'Aspects of noncommutative order' [50], where he appeals to the fact (already mentioned in the second bullet in Section 1) that for an operator x,

$$\begin{bmatrix} 1 & x \\ x^* & 1 \end{bmatrix} \geq 0 \quad \Longleftrightarrow \quad \|x\| \leq 1. \tag{2.1}$$

After noting that this formula is the solution to the problem of describing the norm of an operator in terms of order, Effros says: "It would thus seem advisable to regard the ordering on $M_2(A)$, and more generally on all of the matrix algebras $M_n(A)$, $n \geq 1$, as part of the natural "baggage" of a C^*-algebra. The resulting category of "matrix ordered spaces" has proved to be of great value." More motivation from historical sources may be found in Paulsen's excellent and influential monograph [96], which is the standard reference for the theory of completely positive maps on operator systems (and also of completely bounded maps). See also [44] for a treatment of the basic theory of completely positive maps.

Indeed, completely positive maps play a profound role in C^*-algebra theory; and have become more and more central in parts of mathematical physics, as the reader can instantly confirm by putting the words 'completely positive' to an internet search. For example, they have recently become quite prominent in modern theories of quantum information theory, quantum computing, etc, where one looks at completely positive maps on matrix algebras which have special properties with respect to the trace (see, e.g., [93]).

Important in the theory of completely positive maps is the notion of a *dilation*. If $T : X \to B(H)$, then a dilation of T is a map $T' : X \to B(K)$, for a Hilbert space K containing H, with

$$T'(x) = \begin{bmatrix} T(x) & * \\ * & * \end{bmatrix}, \quad x \in A. \tag{2.2}$$

More abstractly, $T' : X \to B(K)$ dilates T if there is an isometry $V : H \to K$ such that $T = V^*T'(\cdot)V$ on X.

An example of a completely positive map on a C^*-algebra A is any representation (that is, $*$-homomorphism) $\pi : A \to B(H)$. Indeed, π is clearly positive, and the same argument shows that the 'amplification' of π to $M_n(A)$ is positive for each $n \in \mathbb{N}$. The following result gave birth to the subject of completely positive maps (and most of what follows in this article):

Theorem 2.8. (Stinespring) *A linear map $T : A \to B(H)$ on a C^*-algebra A is completely positive iff T can be dilated to a representation π of A on a Hilbert space containing H.*

Proof. The usual proof of this, which may be found in many places (e.g., [9, 44, 96]), is very similar to the proof we gave of the 'GNS construction' (Theorem 2.1). Given a completely positive T, the idea to construct π, as in the GNS construction proof, is to find an inner product defined on a simple space containing H on which A has a natural algebraic representation. In this case, the space is $A \otimes H$, and we define the representation of A by $\pi(a)(b \otimes \zeta) = ab \otimes \zeta$ for $a, b \in A, \zeta \in H$. We define the inner product on $A \otimes H$ by

$$\langle a \otimes \eta, b \otimes \zeta \rangle = \langle T(b^*a)\eta, \zeta \rangle , \qquad a, b \in A, \eta, \zeta \in H.$$

The rest can be left as an exercise, following the model of Theorem 2.1. □

If $A = M_n$ and H is finite-dimensional, then T is completely positive iff the matrix $[T(e_{ij})]$ is positive, where $\{e_{ij}\}$ is the standard basis of M_n; and Stinespring's result yields the existence of a finite number of matrices γ_k such that $T(x) = \sum_k \gamma_k^* x \gamma_k$ (see [39]. We remark that the latter results are highly utilized currently in quantum computing and related fields).

Because of Stinespring's theorem, completely positive maps have many useful properties. For example, they satisfy a 'Schwarz inequality':

$$T(x)^* \, T(x) \leq T(x^*x) , \quad x \in A.$$

This inequality follows immediately from (2.2), since if one multiplies the matrix on the right of (2.2), with its adjoint $\pi(x)^*$, uses the fact that π is a $*$-homomorphism, and looks at the term in the 1-1 corner, one sees that $T(x^*x)$ equals $T(x)^* \, T(x)$ plus something positive. A similar, and only slightly more complicated, argument shows that if T is a completely positive unital map on a unital C^*-algebra A, and if $T(c)^*T(c) = T(c^*c)$ for some $c \in A$, then $T(ac) = T(a)T(c)$ for all $a \in A$ (see, e.g., [38, 110] for more along these lines).

The 'conditional expectations' in Theorem 2.3 are completely positive. An important and very useful complement to this fact is the following:

Theorem 2.9. (Choi and Effros [41]) *If $\Phi : A \to A$ is any unital completely positive idempotent map on a unital C^*-algebra, then the range of Φ is a C^*-algebra with new product $\Phi(ab)$, for $a, b \in \mathrm{Ran}(\Phi)$. Moreover $\Phi(\Phi(a)\Phi(b)) = \Phi(a\Phi(b)) = \Phi(\Phi(a)b))$ for $a, b \in A$.*

Operator systems are the unital selfadjoint subspaces of a C^*-algebra. They are the natural domain of definition of (unital) completely positive maps, and they were first studied systematically in this context in [9] (which contains many of the results listed below in this subsection).

A *subsystem* of an operator system $\mathcal{S}$ is a selfadjoint linear subspace of $\mathcal{S}$ containing the 'identity' 1 of $\mathcal{S}$. If $\mathcal{S}$ is an operator system, a subsystem of a unital C^*-algebra A, then $\mathcal{S}$ has a distinguished 'positive cone' $\mathcal{S}_+ = \mathcal{S} \cap A_+$. Then $\mathcal{S}$ has an associated ordering $\leq$, namely $x \leq y$ iff x, y are selfadjoint and $y - x \in \mathcal{S}_+$. If $x \in \mathcal{S}$ then $\frac{x+x^*}{2}$ and $\frac{x-x^*}{2i}$ are selfadjoint, and so any $x \in \mathcal{S}$ is of the form $x = h + ik$ for selfadjoint h, k. Also, since $\|h\|1 + h$ and $\|h\|1 - h$ are positive, it follows that h (and k) is a difference of two elements in $\mathcal{S}_+$.

A linear map $u : \mathcal{S} \to \mathcal{S}'$ between operator systems (or more generally, between $*$-vector spaces), is called *$*$-linear* if $u(x^*) = u(x)^*$ for all $x \in \mathcal{S}$. Some authors say that such a map is *selfadjoint*. We say that u is *positive* if $u(\mathcal{S}_+) \subset \mathcal{S}'_+$. By facts at the end of the last paragraph, any $x \in \mathcal{S}$ may be written as $x = x_1 - x_2 + i(x_3 - x_4)$, and from this it is easy to see that a positive map is $*$-linear. The operator system $M_n(\mathcal{S})$, which is a subsystem of $M_n(A)$, has a canonical positive cone too, and thus it makes sense to talk about *completely positive maps* between operator systems. If T is any map, we write T_n for the same map applied entry-wise to a matrix: $T_n([x_{ij}]) = [T(x_{ij})]$. Thus T is completely positive if each T_n is positive. It is easy to show that for such a map, $\|T\| = \|T(1)\|$. An isomorphism between operator systems which is unital and completely positive, and has a completely positive inverse, we will call a *complete order isomorphism*. The range of a completely positive unital map between operator systems is clearly also an operator system; we say that such a map is a *complete order embedding* if it is a complete order isomorphism onto its range.

Operator systems have a beautiful abstract characterization, for which we will need the following notation. A *matrix ordered vector space* is a $*$-vector space $\mathcal{S}$ with a cone $\mathfrak{c}_n \subset M_n(\mathcal{S})$ for each $n \in \mathbb{N}$, which satisfy two conditions: first, $\mathfrak{c}_n \cap (-\mathfrak{c}_n) = (0)$, second, $\alpha^* x \alpha \in \mathfrak{c}_n$ for any $\alpha \in M_{mn}$ and $x \in \mathfrak{c}_m$. We say that $u \in \mathcal{S}$ is a *matrix order unit* if the diagonal matrix u_n with constant diagonal entry u, is an order unit (that is, for every selfadjoint matrix $x \in M_n(\mathcal{S})$ we have $-tu_n \leq x \leq tu_n$ for some $t \in (0, \infty)$. We recall that u is *Archimidean* if $x \geq 0$ iff $x + tu \geq 0$ for all $t \in (0, \infty)$.

Theorem 2.10. (Choi and Effros [41]) *The operator systems are, up to unital complete order isomorphism, exactly the matrix ordered vector spaces which possess a matrix order unit which is Archimidean.*

We say that a linear map T is *completely contractive* (resp. *completely isometric*) if the map T_n defined a few paragraphs above has norm ≤ 1 (resp. is an isometry), for all $n \in \mathbb{N}$. Suppose that $\mathcal{S}$ is a subsystem of a unital C^*-algebra A. By the Hahn–Banach theorem, the set of states of $\mathcal{S}$ (that is, the set of $\varphi \in \mathcal{S}^*$ with $\varphi(1) = \|\varphi\| = 1$) is just the set of restrictions of states on A to $\mathcal{S}$. Using this fact, and a characterization of A_+ which we stated earlier, it follows that $\mathcal{S}_+$ is exactly the set of elements $x \in \mathcal{S}$ such that $\varphi(x) \geq 0$ for all states φ of $\mathcal{S}$. From this it is clear that if $u \colon \mathcal{S}_1 \to \mathcal{S}_2$ is a contractive unital linear map between operator systems, then u is a positive map (for if $x \in \mathcal{S}_{1+}$, and if φ is a state on $\mathcal{S}_2$ then $\varphi \circ u$ is a state of $\mathcal{S}_1$, so that $\varphi(u(x)) \geq 0$; and so $u(x) \geq 0$). Applying this principle to u_n, we see that a completely contractive unital linear map between operator systems is completely positive. It is easy to see from (2.1) that a completely positive unital map u between operator systems is completely contractive. (For example, to see that u is contractive, take $\|x\| \leq 1$, and apply u_2 to the associated positive matrix in (2.1). This is positive, so that using (2.1) again we see that $\|u(x)\| \leq 1$.) Putting these facts together, we see that a unital map between operator systems is completely positive if and only if it is completely contractive. The same idea, but also applied to the inverse, shows that such a map is a complete order embedding if and only if it is completely isometric.

Arveson's extension theorem [9] states that if $\mathcal{S}$ is a subsystem of a unital C^*-algebra A, and if $u\colon \mathcal{S} \to B(H)$ is completely positive, then there exists a completely positive map from A to $B(H)$ extending u. There is a related result due to Wittstock [120], in which the word 'system' above may be replaced by 'space', and 'completely positive' by 'completely contractive'. Wittstock's result follows very quickly from basic properties of the Haagerup tensor product (which we shall not discuss here). Note that Arveson's result in the case that u is unital, follows immediately from Wittstock's result, and a fact towards the end of the last paragraph.

Arveson also showed in [9] that the category of subspaces X of a unital C^*-algebra A containing 1_A, with morphisms the completely contractive unital linear maps, may be studied in terms of the category of operator systems and completely positive unital linear maps, via the functor taking $X \mapsto X + X^*$. For example, a morphism $T : X \to Y$ in the first category extends uniquely to a morphism $X + X^* \to Y + Y^*$ in the second category. If X and Y are isomorphic in the first category, then $X + X^*$ is completely order isomorphic to $Y + Y^*$ via the obvious map (which is not obviously well defined). This allows one to transfer many questions to a setting where one can use positivity. There is a variant of this trick which works even if X is not unital, which we explain in the second paragraph of Section 3.1 below.

Completely positive maps play a crucial role in Kasparov's bivariate generalization KK of K-theory (see, e.g., [17, 18]). For example, for this theory Kasparov invented an important extension of the Stinespring theorem [18, II.7.5.2] which is used fundamentally. Earlier, completely positive maps also played a pivotal role in the important theory of *extensions* of C^*-algebras (which now to a certain extent is subsumed into the KK-theory just alluded to). An extension of a C^*-algebra C by a C^*-algebra A is just an exact sequence

$$0 \longrightarrow A \longrightarrow B \longrightarrow C \longrightarrow 0$$

of C^*-algebras and $*$-homomorphisms. In the Brown-Douglas-Fillmore (BDF) theory (see, e.g., [36]), one associates an important group $Ext(K)$ to a topological space K, coming from the extensions of $C(K)$ by the compact operators. In the generalization to C^*-algebras, the important fact that the semigroup Ext is a group under reasonable hypotheses, is intimately related to the Stinespring theorem and Kasparov's generalization of it [18, II.7.5.2]. Namely, Ext being a group is related to whether $*$-homomorphisms π from C into the quotient C^*-algebra $B(\ell^2)/K(\ell^2)$ by the compact operators, possess a completely positive 'lifting' (that is, a completely positive linear $T : C \to B(\ell^2)$ whose image in the quotient space is π). See, e.g., [12, 18] for details. Completely positive maps play a remarkable role too in the theory of quasidiagonality (see [37] for a very attractive survey of this topic).

We end by listing a somewhat random selection of papers which contain other interesting results on completely positive maps: [11, 59, 64, 71, 72, 106, 108,

110, 120], and references therein. Perhaps the deepest work on completely positive maps in the last 20 years may be found in the astonishing papers of Kirchberg, some with coauthors (see, e.g., [79, 80, 81, 19, 82, 83], or the survey [94]). These in turn have been used to obtain some of the deepest results on C^*-algebras, such as Kirchberg's profound characterization of separable 'exact' C^*-algebras as precisely the $*$-subalgebras of the famous *Cuntz C^*-algebra* O_2 (see, e.g., [82], and a forthcoming book by Kirchberg). Just to mention some of the most basic ideas: exact C^*-algebras, very loosely speaking, are the ones whose finite-dimensional operator subsystems are close, via completely positive maps, to subspaces of finite matrix algebras M_n. Actually, they have many equivalent definitions, and the proofs of some of the equivalences, and some of the properties of exact C^*-algebras, are quite profound. In [75] it is shown that unital, completely positive maps on a unital 'purely infinite' C^*-algebra A, which factor through finite matrix algebras M_n, can be approximated by 'elementary' positive maps of the form $a \mapsto v^*av$, for isometries $v \in A$. This is a variant of Stinespring's theorem above. These are put together with other remarkable ingredients in an astounding way to conclude that separable 'exact' C^*-algebras are subalgebra of O_2. Indeed, Kirchberg links completely positive maps on operator systems to some of the most important open problems in C^*-algebra theory (see, e.g., [79] and the excellent survey by Ozawa [94]). This led to the solution of some of these problems (see, e.g., [70, 101]).

2.4. C^*-modules and TROs

C^*-modules are a simultaneous generalization of C^*-algebras and Hilbert spaces, and their use may be found, whether explicitly or implicitly, in most modern papers concerning C^*-algebras. Given a C^*-algebra B, a (right) C^*-module Z over B is a right B-module Z possessing a B-valued inner product, satisfying the obvious analogues of the conditions defining a Hilbert space, for example: $\langle z, z\rangle \geq 0$ for all $z \in Z$. A *TRO* (or 'ternary ring of operators') is a subspace Z of a C^*-algebra with $ZZ^*Z \subset Z$. If Z is a TRO then ZZ^* and ZZ^* (these notations mean the *closure* of the span of products from the indicated sets) are obviously C^*-algebras, and Z is a bimodule over these algebras. One makes Z into a right C^*-module by defining the inner product $\langle z, w\rangle = z^*w$, for $z, w \in Z$. Conversely, any C^*-module may be represented as a corner $pA(1-p)$ for a projection p in a C^*-algebra A, and this is a TRO. Thus C^*-modules and TROs can be used somewhat interchangeably. See, e.g., [84] or [28, Chapter 8] for more details on this, or for the facts below. A good example of a C^*-module/TRO is $C_n(A)$, the first column of the C^*-algebra $M_n(A)$, which one can view as the TRO of matrices in $M_n(A)$ supported on the first column only. In fact, every C^*-module over A may be built of these $C_n(A)$'s (see Section 4.7 for a generalization of this fact).

Again, a main tool in the theory of C^*-modules or TROs is positivity. For example one extensively uses positivity in the so-called *linking algebra*, which for a TRO Z is:

$$\mathcal{L}(Z) = \begin{bmatrix} ZZ^* & Z \\ Z^* & Z^*Z \end{bmatrix}.$$

This is clearly itself a C^*-algebra (a $*$-subalgebra of $M_2(A)$, if Z is a TRO in a C^*-algebra A), and it is key to almost everything about Z.

As another example of the role of positivity in this theory, consider the order-theoretic characterization due to Paschke of contractive B-module maps between C^*-modules Z and W, as precisely the maps $T : Z \to W$ satisfying $\langle Tz, Tz \rangle \leq \langle z, z \rangle$ for all $z \in Z$. In fact it is certain subclasses of the bounded module maps that are more important, in particular the maps $T : Z \to W$ for which there is an 'adjoint map' $T^* : W \to Z$ with

$$\langle Tx, y \rangle = \langle x, T^* y \rangle, \qquad x, y \in Z.$$

Such maps are called *adjointable*, and the set of adjointable maps $Z \to Z$ is a C^*-algebra to which one also applies positivity techniques to, in order to obtain important results in the theory of C^*-modules, such as Kasparov's generalization of the Stinespring theorem [18, II.7.5.2], which in turn has many famous applications.

The following is a generalization of Theorem 2.9 to TRO's:

Theorem 2.11. (Youngson, [115] or [28, Theorem 4.4.9]) *If $\Phi : A \to A$ is any completely contractive idempotent map on a TRO, then the range of Φ is a TRO with new ternary product $\Phi(ab^*c)$, for $a, b, c \in \mathrm{Ran}(\Phi)$. Moreover, the latter quantity is unchanged if one replaces any one of a, b, c by an $x \in A$ whose image is a (resp. b, c). Thus for example $\Phi(ax^*c) = \Phi(a\Phi(x)^*c)$, if $x \in A$.*

The usual proofs of the last result uses positivity too (for example, by extending the projection to a completely positive projection on the linking C^*-algebra, and then using Theorem 2.9).

The main constructions with C^*-modules are also all usually done using positivity, although we have shown that this can be avoided for many of them (see, e.g., [28, Chapter 8] and references therein). For example, the most important tensor product of C^*-modules W and Z is defined in terms of a completely positive map from the algebra acting on W into the space of maps on Z; positivity is used powerfully and inescapably here.

The natural morphisms between TROs are the *ternary morphisms*, namely the maps $T : Z \to W$ satisfying $T(xy^*z) = T(x)T(y)^*T(z)$ for $x, y, z \in Z$. Such maps have all the usual amazing properties of $*$-homomorphisms on C^*-algebras. This may be seen from the fact that every such map induces a canonical $*$-homomorphism from $\mathcal{L}(Z)$ to $\mathcal{L}(W)$:

$$\begin{bmatrix} a & z \\ y^* & b \end{bmatrix} \mapsto \begin{bmatrix} \theta(a) & T(z) \\ T(y)^* & \pi(b) \end{bmatrix},$$

where θ and π are $*$-homomorphisms from ZZ^* and Z^*Z into WW^* and W^*W respectively. Conversely, every $*$-homomorphism between the linking algebras, which takes the four corners to the matching corners, is a ternary morphism on the 1-2 corner Z.

Summarizing: C^*-modules and TROs are essentially the same objects; and to all intents and purposes they behave just like C^*-algebras, via the linking algebra.

The morphisms though are different; indeed TRO morphisms are in a correspondence with $*$-homomorphisms (between the linking algebras). Thus positivity is as important for C^*-modules and TROs, as it was for C^*-algebras. For references to the TRO literature see the notes to [28, Chapters 4 and 8].

3. Applications of positivity to noncommutative linear analysis

3.1. Noncommutative linear analysis

An *operator space* is a vector space X with a norm $\|\cdot\|_n$ on the set of matrices $M_n(X)$, for all $n \in \mathbb{N}$, satisfying two simple conditions that we will not need to mention here (see p. 20 in [51]). Ruan's theorem states that these 'are the same as' the linear subspaces of $B(H)$ for a Hilbert space H, up to linear complete isometry. Interestingly, the usual proof of Ruan's theorem (on p. 30 of [51]) relies on an order theoretic Hahn-Banach separation argument. Their theory was developed by Effros, Ruan, the author, Paulsen, Pisier, Junge, Le Merdy, and many others. We regard operator spaces as a noncommutative version of Banach spaces: the first indication of this is the fact that every Banach space is linearly isometric to a closed linear subspace of a *commutative* C^*-algebra; whereas operator spaces are linearly completely isometric to subspaces of general C^*-algebras. Because operator spaces are just vector spaces with (matrix) norms, one can develop a noncommutative variant of Banach space theory for them: a 'noncommutative functional analysis'. This is known as 'operator space theory', and it is a subject with many powerful tools which we are not able to discuss here. We refer the reader to the texts [51, 101, 96, 28] for expositions of this theory.

There are several ways to introduce positivity in the study of operator spaces. The most common way is to exploit the principle (2.1). This is the key idea underlying 'Paulsen's trick', which assigns to any operator space X in a C^*-algebra A, a selfadjoint subspace (indeed an operator system) of the C^*-algebra $M_2(A)$:

$$X \rightsquigarrow \left\{ \begin{bmatrix} \lambda 1_A & x \\ y^* & \mu 1_A \end{bmatrix} : x, y \in X, \lambda, \mu \in \mathbb{C} \right\} \subset M_2(A).$$

This selfadjoint subspace has a unique positive cone which is independent of the particular containing C^*-algebra A, essentially because of (2.1) (see [96, Lemma 8.1] or [28, Lemma 1.3.15] for more details). The canonical morphisms between operator spaces are the so-called *completely bounded* linear maps $T : X \to Y$, which satisfy $\|[T(x_{ij})]\|_n \leq C\|[x_{ij}]\|_n$ for all matrices $[x_{ij}] \in M_n(X)$, all $n \in \mathbb{N}$. If $C \leq 1$ then we mentioned earlier that T is said to be completely contractive. A lemma of Paulsen states that T is completely contractive iff the following map is completely positive:

$$\Phi_T : \begin{bmatrix} \lambda 1_A & x \\ y^* & \mu 1_A \end{bmatrix} \mapsto \begin{bmatrix} \lambda 1_A & T(x) \\ T(y)^* & \mu 1_A \end{bmatrix}.$$

This again relies essentially on (2.1) (see [96, Lemma 8.1] for details); and it has the startling consequence that many facts about maps between operator spaces are best proved via positivity; namely by exploiting the fact that the map in the last displayed equation is completely positive. For example, consider Paulsen's proof from [95] of the following beautiful fact, first proved in unpublished work of Haagerup [58]:

Theorem 3.1. *Any completely contractive linear map $T : X \to B(H)$ on a subspace X of a unital C^*-algebra A may be written as $T = R\pi(\cdot)S$, for a representation π of A on a Hilbert space K which is unital (that is, $\pi(1) = I$), and contractive operators R and S between H and K.*

Proof. In the notation above, Φ_T is completely positive. By the extension theorems of Arveson or Wittstock mentioned towards the end of Section 2.3, one can extend Φ_T to a completely positive map $M_2(A) \to B(H \oplus H)$. This in turn, by Stinespring's theorem mentioned earlier, is the '1-1 corner' of a unital representation of $M_2(A)$ on another Hilbert space. It is quite easy algebra to see that any unital representation of $M_2(A)$ on a Hilbert space gives rise to a unitary operator U from that Hilbert space onto $K \oplus K$, for a subspace K of H, and a unital representation π of A on K, such that via this unitary the first representation becomes simply $[a_{ij}] \mapsto U^*[\pi(a_{ij})]U$, for $[a_{ij}] \in M_2(A)$. In our case, we obtain

$$\begin{bmatrix} 0 & T(x) \\ 0 & 0 \end{bmatrix} = \Theta\left(\begin{bmatrix} 0 & x \\ 0 & 0 \end{bmatrix}\right) = PU^* \begin{bmatrix} 0 & \pi(x) \\ 0 & 0 \end{bmatrix} U_{|H\oplus H} = W'\pi(x)W,$$

for a projection P, where $W = [0 \;\; I]U_{|H\oplus H}$, with a similar formula defining W'. Pre- and post-multiplying by the projection from $H \oplus H$ onto H, and the inclusion from H into $H \oplus H$, gives $T = R\pi(\cdot)S$, for appropriate contractions S, R. □

It is only fair to say that in the last decade or two, there has been a move away from using complete positivity in the study of operator spaces. However it always will play an absolutely crucial role in parts of the subject. Moreover one may expect some of the deepest results to come via complete positivity, as the reader will see by consulting the papers of Kirchberg, some of which are referenced here, or his forthcoming book, and the recent work of Arveson and Junge discussed in Section 4.3.

3.2. Noncommutative function theory

In classical linear analysis one often solves a problem by working in a normed vector space of functions, using topology, measure and integration, positivity, and functional analytic tools. An example of this is the use of Choquet theory in the classical theory of function algebras and function spaces. For example, consider the study of algebras of analytic functions on the disk, or more generally, of vector spaces or algebras of functions on a compact set K. These are usually *nonselfadjoint*: that is $f \in A$ need not imply that $\bar{f} \in A$. Thus there is no 'positivity' immediately in evidence. However positivity quickly appears when one studies such spaces with the usual tools of convexity, such as Choquet theory. For example, if A contains constants (that is, if $1 \in A$) then there exists a smallest closed subset $E \subset K$ such that $\sup_E |f| = \sup_K |f|$, for $f \in A$. This set E is called the Shilov boundary. One solves many problems about A by topological/measure theoretic arguments on E, using the ordering of functions on E, and by the techniques of functional analysis.

When going noncommutative, one might hope (in noncommutative problems which are somehow analogous to classical function theoretic situations) that:

- C^*-algebra theory replaces topology.
- Von Neumann algebra techniques replace arguments using measure and integrals.
- 'Operator space theory' replaces Banach space techniques.

Of course in practice, life is not quite so simple, and the strategy above often fails or is quite difficult. Fortunately, the noncommutative is frequently a land of miracles. In any case, in this endeavor positivity is often the key tool. Unfortunately, in generalizing 'classical analysis arguments', the 'usual tricks' with positivity can fail (as discussed briefly at the end of Section 1), and so one has to find more subtle approaches. Or there might be no natural positivity in evidence at first – it might be deeply hidden. Or we may have to find substitutes for positivity (which often rely on positivity *somewhere*).

When 'going noncommutative', one hopes to replace the compact set K above by a C^*-algebra B, and replace A by a subspace or nonselfadjoint subalgebra of B. These are just the operator spaces (discussed in Section 3.1), and the (possibly nonselfadjoint) operator algebras. Although A has no apparent positivity, positivity is hidden or encoded in many ways (which disappear if one moves to more general categories such as Banach algebras, for example), or positivity arguments can still be applied in an auxiliary space. We give some examples of this. First, if $1 \in A$, then as we said a few paragraphs after Theorem 2.10, we can replace A by the operator system $A + A^*$, and use positivity there. Thus for many purposes one may assume that A is an operator system. If A is nonunital, but is a (possibly nonselfadjoint) operator algebra, then one may easily adjoin a unit. It turns out that the uniqueness of this unitization is a deep and striking application of positivity due to Ralf Meyer [90]. Meyer's result says that if A and B are nonunital subalgebras of $B(H)$ and $B(K)$, and if $\theta : A \to B$ is a surjective isometric (resp. completely isometric) homomorphism, then the homomorphism $a + \lambda I_H \to \theta(a) + \lambda I_K$ for $a \in A$ and $\lambda \in \mathbb{C}$ is also isometric (resp. completely isometric). A little thought shows that this is saying that there is a unique unitization procedure in the category of operator algebras, where we think of two objects as the same if there is a surjective isometric (resp. completely isometric) homomorphism between them. For the purposes of this survey, the interesting part of Meyer's proof is the following idea: the Cayley transform, in the form of a functional calculus applied to a standard conformal map between the unit disk and the right-hand half-plane, transforms the unit ball of A into a set whose elements have a strictly positive real part (exactly as points in the right-hand half-plane have positive real part). This positivity is then used, and then one applies the inverse transform to get the desired result.

As another example of the use of positivity, we mention the very characterization of possibly nonselfadjoint) operator algebras [32], which matches the Gelfand-Naimark characterization of C^*-algebras which we discussed in Section 2.1. This theorem states that operator algebras with an identity of norm 1 are precisely the operator spaces which are also algebras with an identity of norm 1, such that $\|xy\|_n \leq \|x\|_n \|y\|_n$ for all $x, y \in M_n(A), n \in \mathbb{N}$. Examination of all of the ingredients in the usual proofs of this result reveals a crucial use of positivity.

Frequently, one studies an operator space or algebra A using positivity in canonically associated C^*-algebras. In particular, there exists a smallest C^*-algebra $C^*_e(A)$ containing A, at least if $1 \in A$. This is known as the *noncommutative Shilov boundary* or *C^*-envelope*, and its existence is due to Arveson and Hamana (see Section 4.3 for more on this). Its universal property is as follows: for every unital complete isometry $i : A \to B$ into a C^*-algebra B, such that $i(X)$ generates B as a C^*-algebra, there exists a (necessarily unique and necessarily surjective) $*$-homomorphism $\pi : B \to C^*_e(A)$ with $\pi \circ i$ equal to the embedding of X in $C^*_e(A)$. If B is commutative, then $C^*_e(A) = C(E)$, where E is the classical Shilov boundary discussed at the start of this subsection. The C^*-envelope is quite rigidly attached to A; for example any surjective complete isometry between two such spaces A_1 and A_2 extends uniquely to a surjective complete isometry from $C^*_e(A_1)$ to $C^*_e(A_2)$. Moreover, this complete isometry (as with any complete isometry between unital C^*-algebras) is just a unitary operator in the second C^*-algebra multiplied by a $*$-isomorphism between the C^*-algebras.

If X is an operator space not containing the identity operator, then the noncommutative Shilov boundary is not a C^*-algebra in general, it is a C^*-module or TRO (see Section 2.4) containing X, which we will write here as ∂X. One must adapt the universal property in the last paragraph by replacing the words 'C^*-algebra' by 'TRO', '$*$-homomorphism' by the ternary morphisms we met in Section 2.4, and dropping the 'unital' throughout. Note that X inherits a C^*-algebra-valued inner product from ∂X. Positivity in the associated linking algebra is a powerful tool to study the structure of X. This is another way of introducing positivity to study the operator space X.

The noncommutative Shilov boundary is a subobject of the injective envelope, which we have mentioned earlier. For example, if $1 \in A$ then we mentioned that the injective envelope of A is a monotone complete C^*-algebra. The injective envelope is useful because of its abstract properties (see, e.g., [60, 63, 102, 96] and [28, Chapter 4]), and because it is again an object to which one can apply C^*-algebra techniques, and in particular positivity. For example, this approach easily yields a variant, valid for possibly nonselfadjoint operator algebras, of Theorem 2.3. The main idea is to take a projection P on an operator algebra, then extend it to a similar map on the injective envelope of A, which is a C^*-algebra or TRO. Then one applies Theorem 2.3, or 2.9 or [115], to this extension to show that P is a 'conditional expectation'. More specifically, consider the following generalization of [28, Corollary 4.2.9]:

Proposition 3.2. *Let $P : A \to A$ be a completely contractive linear idempotent map on a general operator algebra A, whose range is a subalgebra B of A. If B has a contractive right approximate identity, then $P(ba) = bP(a)$ for all $a \in A, b \in B$.*

Proof. Although positivity is not explicitly visible in this proof, it appears explicitly in the proof of Youngson's theorem 2.11, which we are using (for example, it relies on Theorem 2.9).

First suppose B has a right identity e of norm 1. Suppose that A is a subalgebra of $B(H)$. Let Φ be a minimal A-projection in the sense of, e.g., [60, 96] or [28, Section 4.2], so that $I(A) = \text{Ran}(\Phi)$ is an injective envelope for A inside $B(H)$. By injectivity, extend P and the inclusion map $i : B \to A$ to the injective envelopes, giving maps $P' : I(A) \to I(B)$ and $i' : I(B) \to I(A)$. Since $P' \circ i'$ is the identity map on B, by the 'rigidity' property of the injective envelope $P' \circ i'$ is the identity map on $I(B)$. Thus $Q = i' \circ P'$ is a completely contractive projection on $I(A)$, and $Q_{|A} = P$. We have

$$bP(a) = Q(\Phi(bP(a))) = Q(\Phi(Q(b)eQ(a))) = Q(\Phi(Q(b)Q(e)^*Q(a))),$$

for $a \in A, b \in B$. By the last assertion in Theorem 2.11, the latter equals

$$Q(\Phi(Q(b)Q(e)^*a)) = Q(bea) = P(ba).$$

In the general case, consider $P^{**} : A^{**} \to B^{**}$. This easily satisfies the hypotheses in the previous paragraph, so that $bP(a) = bP^{**}(a) = P^{**}(ba) = P(ba)$. □

Remark. The result does not hold if B has no kind of identity, as one can see by looking at the projection of M_n onto its first row or column. However, a similar proof shows that if P is any completely contractive projection on an operator algebra A, then the range $P(A)$ is completely isometrically isomorphic to an operator algebra with product $P(P(a)P(b))$. For experts, we give a proof: let Φ be as in the proof of Proposition 3.2, its range is $I(A)$. By the main theorem in [75], there exists $u \in I(A)$ such that $ab = \Phi(au^*b)$ for all $a, b \in A$. If $B = P(A)$ we may repeat the proof of Proposition 3.2 to obtain a completely contractive projection on $I(A)$, with $Q_{|A} = P$. By Theorem 2.11, $Q(I(A))$ is a TRO Y whose 'ternary product xy^*z' is $Q(\Phi(xy^*z))$. Define a product on Y by $x \cdot y = Q(\Phi(xQ(u)^*y)) = Q(\Phi(xu^*y))$, the latter by the last assertion in Theorem [115]. There is some suitable representation of Y on a Hilbert space so that the latter product is just $xQ(u)^*y$, and $YQ(u)^*Y \subset Y$. By [32, Remark 2, p. 194] or [75], this makes Y into an operator algebra C. This product restricted to B is $Q(\Phi(aQ(u)^*b)) = Q(\Phi(au^*b)) = Q(ab) = P(ab)$ for $a, b \in P(B)$. Thus B is a subalgebra of C. The above was discovered during conversations with M. Neal around 2003.

We also remark that extremely recently there have been efforts to extend these kinds of results to some other Banach algebras [85].

In turn, this result can be used, almost exactly how we used it in the proof we provided above of Sakai's characterization of von Neumann algebras (Theorem

2.6), to obtain a characterization of unital weak* closed algebras of operators on a Hilbert space (see [22]). Later we were able to drop the unital hypothesis:

Theorem 3.3. [86, 29] *The weak* closed algebras of operators on a Hilbert space are precisely, up to completely isometrically isomorphism which is also a weak* homeomorphic homomorphism, just the (possibly nonselfadjoint) operator algebras which are also a* dual operator space.

The latter term, 'dual operator space', means a little more than being a dual Banach space. Note that the dual Banach space Y^* of an operator space Y has special matrix norms on $M_n(Y^*)$ for $n \geq 2$: the norm of $[\varphi_{ij}] \in M_n(Y^*)$ is ≤ 1 iff the map $y \mapsto [\varphi_{ij}(y)]$ is completely contractive. With these matrix norms Y^* is an operator space, and this is what we mean by a dual operator space.

4. Some recent applications and progress

In this section we present very briefly several extremely recent examples, mostly from work of the author, of how positivity can be used, both directly and indirectly, and sometimes in surprising ways! We also present some recent progress in the subject of positivity in operator spaces.

4.1. Multipliers of operator spaces and noncommutative M-ideal theory

This subsection will be short, as we have amply surveyed this theory elsewhere (e.g., [28, 23]). The main idea which we wish to convey here, is another interesting way to introduce positivity into the study of general operator spaces. Namely, we saw at the end of Section 3.2 that any operator space X has a C^*-algebra-valued inner product. With respect to this inner product, consider the set $\mathcal{A}_\ell(X)$ of maps $T : X \to X$ with

$$\langle\, Tx \,,\, y \,\rangle \;=\; \langle\, x \,,\, Sy \,\rangle\,, \qquad x, y \in X,$$

for some $S : X \to X$. We showed that $\mathcal{A}_\ell(X)$ is a C^*-algebra, and together with Effros and Zarikian proved that it is a von Neumann algebra if X is a dual operator space. Positivity in this C^*-algebra plays a crucial role in understanding certain kinds of structure in X, as is surveyed in [23]. A space with no structure of this type will have $\mathcal{A}_\ell(X)$ trivial.

For example, $\mathcal{A}_\ell(X)$ is key to understanding the *noncommutative M-ideals* in X. Projections in $\mathcal{A}_\ell(X)$ are called *left M-projections* on X. A *right M-ideal of X* is a subspace J with $J^{\perp\perp}$ equal to the range of a projection in $\mathcal{A}_\ell(X^{**})$. We generalized the basics of the classical theory of M-ideals (see [65]) to operator spaces using the latter von Neumann algebra, and using von Neumann algebra techniques, and, in particular, positivity. See [28, Section 4.8] for a short introduction to this topic, or, e.g., [24, 34] for more detail.

For any operator space X, there is an important operator algebra $\mathcal{M}_\ell(X)$ containing $\mathcal{A}_\ell(X)$, the *left multiplier algebra* of X, whose elements consist of maps on X called *left multipliers*. To describe the general left multiplier, we take any (and every) C^*-algebra A containing (a completely isometric copy of) X, and suppose that $a \in A$ satisfies $aX \subset X$. Then the map $x \mapsto ax$ from X to X is the (generic) element of $\mathcal{M}_\ell(X)$. In fact if A is chosen carefully (and this is done in terms of the noncommutative Shilov boundary or the injective envelope discussed above), then one only needs one such C^*-algebra. The importance and role of these multipliers and their theory is discussed in detail in [28, Chapter 4] or [23], and so we will not relate this again here. We just mention, for example, that they yield deep results about dual operator spaces that seem unobtainable by other techniques, such as the fact that any map in $\mathcal{M}_\ell(X)$ is automatically weak* continuous if X is a dual operator space [29]. In turn, this fact is used crucially in the proof of Theorem 3.3, the characterization of dual operator algebras, for example. These multipliers have several interesting characterizations (see [28, Theorem 4.5.2]). For example, there is an order theoretic characterization stating that a linear map $u : X \to X$ is a left multiplier of norm ≤ 1 iff with respect to the inner product discussed above, $[\langle Tx_i, Tx_j \rangle] \leq [\langle x_i, x_j \rangle]$ for all $x_1, \ldots, x_n \in X$ and $n \in \mathbb{N}$. Here the matrices are indexed on rows by i and on columns by j. One nice question which remains open here concerning positivity, is whether $n = 1$ will always suffice in the last characterization. Others of these characterizations of left multipliers require considering the injective envelope of X (discussed earlier), and then applying positivity results surveyed in Section 2, or require C^*-module techniques (which as we have emphasized in Section 2.4, are based heavily on positivity) [23].

4.2. Noncommutative convexity

From early on in the subject of noncommutative convexity, the desire was to replace classical convex combinations of operators $T_1, \ldots, T_n \in B(H)$, with combinations in which the scalars are replaced by operators: namely, $\sum_{k=1}^n \gamma_k^* T_k \gamma_k$, where $\gamma_k \in B(H)$ and $\sum_{k=1}^n \gamma_k^* \gamma_k = I_H$. This has become known as *C^*-convexity*, and notice how positivity is key here. The reader will be able to guess what is a 'C^*-convex set' S. There are a couple of variants of the appropriate notion of 'extreme point' of S in this setting, all being along the lines of $T = \sum_{k=1}^n \gamma_k^* T_k \gamma_k$ as above, with $T, T_k \in S$, perhaps with constraints on the γ_k, implying that T_k or $\gamma_k^* T_k \gamma_k$ is some obvious 'multiple' of T (see, e.g., [87, 91, 69] and references therein). An alternative known as *matrix convexity* has also been studied, which seems more suitable for certain applications to operator spaces. Here if $x \in M_m(X)$ for a vector space X, one considers sums $x = \sum_{k=1}^n \gamma_k^* x_k \gamma_k$, where the γ_k are $m_k \times m$ matrices with scalar entries, with $\sum_{k=1}^n \gamma_k^* \gamma_k = I_m$, and $x_k \in M_{m_k}(X)$. A sequence $\kappa = (K_n)_{n \in \mathbb{N}}$ of sets, with $K_n \subset M_n(X)$, is called *matrix convex* if every such 'matrix convex combination' is in K_m when $x_k \in K_{m_k}$ for $k = 1, \ldots, n$ (see [121]). We say that κ is compact if each K_n is compact. Because space is limited, we will just mention a couple of results in this theory. A *matrix extreme point* of κ is an element $x \in K_m$ such that if x can be written as such a 'matrix convex combination' with $x_k \in K_{m_k}$ for $k = 1, \ldots, n$ (and sometimes there are further conditions on the combinations), then every x_k equals x up to conjugation by a unitary matrix. One of the main results in this theory to date is Webster and Winkler's Krein-Milman theorem [116], which states that the closure of the set of 'matrix convex combinations' of the matrix extreme points of a compact matrix convex sequence of sets κ, equals κ. A quick proof of this result appears in [54] (the main idea appears in the lemma proved below). Recall the usual classical correspondence between compact convex sets C and function systems A, which takes C to the space $A(C)$ of affine continuous scalar functions on K, and takes A to the state space (the positive unital scalar-valued functionals on A). Webster and Winkler extend this correspondence to the compact matrix convex sets, and the operator systems discussed in Section 2.3. Given an operator system X, let $K_n = CPU(X, M_n)$ be the set of completely positive unital maps from X into M_n, which we can view as a subset of $M_n(X^*)$. Then (K_n) is a compact matrix convex set in X^*. Conversely, given any compact matrix convex set $\kappa = (K_n)$, let $A(\kappa)$ be the set of 'matrix affine' (we will not take the time to define this term here) continuous scalar functions on κ. Similarly, $M_n(A(\kappa))$ denotes the 'matrix affine' continuous M_n-valued functions on κ. These have natural positive cones, and one can check that $A(\kappa)$ is an operator system. With a natural qualification, these correspondences are mutual inverses. Thus instead of working with general compact matrix convex sets, one may just work with convex sets consisting of completely positive unital maps.

We say that a completely positive map T is *pure* if any completely positive map S that is dominated by T is a scalar multiple of T. The following sample 'Krein-Milman' result will be quoted later, and is essentially from [54]:

Lemma 4.1. *If X is a finite-dimensional operator system and $n \in \mathbb{N}$ then the set $CPU(X, M_n)$ is the matrix convex hull of the pure CPU maps $T : X \to M_m$, $m \leq n$.*

Proof. Clearly $CP(X, M_n)$ is a closed cone in the completely bounded maps $CB(X, M_n)$. Let $C_m = \{T \in CP(X, M_n) : tr(T(1)) \leq m\}$. Then C_m and $CP(X, M_n) \setminus C_m$ are closed and convex, the C_m are increasing, and $\cup_m C_m = CP(X, M_n)$. By the ordinary finite-dimensional version of the Krein-Milman theorem, C_m is the (classical) convex hull of its extreme points. Thus any element in $CP(X, M_n)$ is a (classical) convex combination of extreme points of C_m for some m. We claim that any extreme point of C_m is of the form tT, for $t \geq 0$ and T pure. This is classical convexity theory: Proposition 13.1 of [99] shows that any extreme point of C_m is of the form tT, for T in an 'extreme ray', and the elements of an 'extreme ray' are exactly the pure elements (top of page 80 in [99]). Next, if $T \in CP(X, M_n)$ is pure, then by Theorem 2.2 in [54], we have $T = v^* R(\cdot) v$ for a pure $R \in CPU(X, M_k)$ and a matrix v of appropriate size.

Putting the above together, any $T \in CP(X, M_n)$ may be written as $\sum_k v_k^* T_k(\cdot) v_k$, for matrices v_k and pure T_k in $CPU(X, M_{n_k})$. If $T(1) = 1$ then $\sum_k v_k^* v_k = 1$, and so T is a matrix convex combination of appropriate pure maps. $\square$

Judging from recent developments (some described in the next section), there will be many exciting developments in this field in the near future. For a few more references on noncommutative convexity and 'matricial extreme points', see for example [5, 52, 53, 55, 89, 91, 119], and references therein.

4.3. The Shilov boundary and completely positive maps

As we have said earlier, the study of operator systems, at least in the category with completely positive maps as morphisms, began in Arveson's papers [9, 10]. The oldest problem in the subject dates to those papers, and it has just recently been solved by Arveson [13], at least in the separable case. Since this a historical and major development in the subject of completely positive maps, which will influence some of the direction of the field for years to come, in this section we would like to advertise this breakthrough by describing the problem, briefly surveying some of the ideas in the proof and in some new work of Junge, and by mentioning some of the things that need to be done in the future.

The problem concerns the noncommutative Shilov boundary, and its relation to what one might call the noncommutative Choquet boundary. As we said in earlier sections, although the noncommutative Shilov boundary is defined and used for general operator spaces, for many purposes (including the problem which we are discussing), one may as well assume that X is an operator system. For example, if X is a unital but nonselfadjoint operator space, then one simply considers the operator system $X + X^*$, and for general X one can play the trick in the second paragraph of 3.1. In the classical case, we have $1 \in X \subset C(\Omega)$, for a compact set Ω which X separates points of, and the Shilov boundary is usually defined to be the closure in Ω of the Choquet boundary. In turn, the Choquet boundary may be defined to be the points $w \in \Omega$ such that 'evaluation at w' is the only state on $C(\Omega)$ (that is, the point mass at w is the only probability measure on Ω) which on X agrees with 'evaluation at w'. In the noncommutative situation, we have a subsystem X of a unital C^*-algebra B, such that X generates B as a C^*-algebra. Arveson's noncommutative variant of the Choquet boundary consists of the irreducible representations π of B on a Hilbert space H (these correspond to evaluation at points of Ω if $B = C(\Omega)$ as above), such that π is the only completely positive unital map $B \to B(H)$ which on X agrees with π. These are called *boundary representations* of X. The open question alluded to above asks if there are sufficiently many boundary representations, in the sense that for any $x \in X$ the norm $\|x\|$ equals the supremum of $\|\pi(x)\|$ over all boundary representations π, and a similar formula holding if $x \in M_n(X)$. Actually, it was unknown in general if there existed any boundary representations at all. If there do exist sufficiently many boundary representations for X, then the C^*-envelope referred to in Section 3.2, may defined to be the C^*-algebra generated by $\{\oplus_\pi \, \pi(x) : x \in X\}$, where the π here are the boundary representations of X (or rather, to avoid set theoretic complications, the π are representatives of the unitary equivalence classes of the boundary representations). An affirmative solution to this problem constitutes a major advance in 'noncommutative Choquet theory', and in particular it goes a long way towards giving boundary representations the primacy that the usual Choquet boundary points play in classical function theory or the theory of uniform algebras. It also gives a more direct (and closer to the classical) construction of the C^*-envelope (or noncommutative Shilov boundary – recall that if B is commutative then the C^*-envelope equals the continuous functions on the classical Shilov boundary). The usual route to the C^*-envelope proceeds via the injective envelope, which we have mentioned earlier. This is a powerful construction but it is not useful for some purposes because it is difficult to 'get ones hands on', and is 'very large'. Even in the classical case it will generally be much larger than, say, a construction such as the 'Dedekind completion' which those in Banach lattice theory will be familiar with. Thus the injective envelope is mostly useful as an abstract tool because of the properties it possesses; one cannot hope to concretely be able to say what it is.

Arveson's recent solution to the open problem above in the case that X is separable, is built from Dritschel and McCullough's solution to the problem if one drops the irreducibility requirement [46], which in turn was influenced by some older ideas already in the literature [1, 92, 122].

The first point is a characterization, essentially due to Muhly and Solel [92], of boundary representations, or rather of representations that satisfy the definition of a boundary representation above except for the irreducibility requirement. Indeed, the restriction of such maps to X are precisely the completely positive unital maps $T : X \to B(H)$ which are *maximal* in the following sense: every completely positive unital map T' dilating T, that is, which satisfies (2.2), is *reducing*, which means that the two 'off-diagonal' entries in the matrix in (2.2) are zero.

The second point is that one can prove that every completely positive unital map from a separable operator system X into $B(\ell^2)$ may be dilated to a map into the bounded operators on a separable Hilbert space H_0 which is maximal in the above sense. This is a refinement, appropriate to the separable case, of a result of Dritschel and McCullough [46], and uses an intricate induction argument.

Third, since X is separable, there exists an embedding map $X \hookrightarrow B(\ell^2)$, and we apply the last fact to this map. Since a dilation of a complete isometry is clearly also a complete isometry, we now have a completely isometric map $T : X \to B(H_0)$ such that every completely positive unital dilation of T is reducing.

Next, one decomposes H_0 as a 'direct integral' $\int_\Omega^\oplus H_w \, d\mu(w)$, for a probability measure μ on an appropriate compact metric space Ω and a family of separable Hilbert spaces H_w. Also, T has a corresponding decomposition $T(x) = \int_\Omega^\oplus T_w(x) \, d\mu(w)$, for completely positive unital maps $T_w : X \to B(H_w)$. It is easy to see that, by construction, there are 'sufficiently many' (in the sense of the last paragraph) of the T_w. The remainder of the proof is essentially just the technical (and lengthy) result that since T is maximal in the sense above, so are almost all of the completely positive maps T_w. Hence by the characterization earlier in this paragraph, almost all of the T_w are restrictions to X of boundary representations, and we are done. Arveson also explains the correspondence between pure states of the operator system X (that is, extreme points of the set of states of X) and boundary representations (recall from Section 1 that the GNS construction gives a correspondence between pure states of a C^*-algebra, and its irreducible representations).

Some natural questions arise in view of this progress, like whether the separability condition can be removed, or whether boundary representations can be constructed more explicitly and canonically (without the measure theory and induction used above)? Marius Junge has some advances in these directions in some soon to be released work, which we will describe (a caricature of) some basic ideas from below. We thank him for many discussions, particularly around the time that this work was starting. Amongst many other remarkable results and new ideas, he has established the existence of sufficiently many boundary representations in the case of subsystems of $\ell^\infty(I, M_n)$, for a set I. We will not describe the details of his proof here, or much of the extraordinary program from [69], since this has not yet been circulated, but it also involves direct integrals of completely positive maps and certain bundles. It seems clear that these ideas will be combined in the near future to completely solve the open problem above.

Since the technical difficulties in the most general case are seemingly quite formidable, for simplicity we consider some of Junge's ideas in the case of an operator subsystem X of a finite-dimensional C^*-algebra $A \subset M_n$. The first point is that there is a useful correspondence, originating in [116], between the extreme points of $\mathrm{Ball}(M_n(X)^*)_+$, and completely positive unital $T : X \to M_m$ which are *extreme maps* in the sense of Section 4.2 (i.e., matrix extreme, or one variant of C^*-extreme). The latter class of maps also coincide with the pure completely positive unital maps $T : X \to M_m$. We recall that pure means that any completely positive map S that is dominated by T is a scalar multiple of T. (To see the more difficult direction of this, suppose that T is extreme, dominating S as above. By Stinespring's theorem 2.8, we may write $S = v^*\pi(\cdot)v$ and $T - S = w^*\theta(\cdot)w$, with π, θ completely positive unital, and then $v^*v + w^*w = I_m$, and the rest is clear.) The above pure maps satisfy a Krein-Milman theorem, their appropriate convex hull is total, as we proved in Lemma 4.1. There is a natural ordering on the pure maps, namely dilation, and one key point is that the maximal ones among the pure maps in this ordering, the 'peaks' or 'tops', constitute a family of irreducible boundary representations, and there are 'sufficiently many' in the sense at the start of this subsection. Thus they give the noncommutative Shilov boundary, as explained there. One can also use the sense of 'maximal' described earlier in the section: it is indeed an easy exercise from, e.g., [9, Corollary 1.4.3], and the earlier described result due essentially to Muhly and Solel, that a completely positive unital map is an irreducible boundary representation iff it is maximal in this sense, and pure.

Suppose that we have an operator subsystem $X \subset M_n$. We will find all the boundary representations of X (thereby constructing the noncommutative Shilov boundary), and moreover we will not use any theory of the injective or C^*-envelope (the task is rather trivial if one is prepared to accept some facts from the latter theories). First we note that if $T : X \to B(H)$ is a pure CPU map, then

$\dim(H) \leq n$. To see this, extend T to M_n and use Stinespring's theorem 2.8 to write T as $T(x) = \sum_{k=1}^{\infty} v_k^* x v_k$ (this is proved similarly to the remark after 2.8). Using the fact that T is pure gives $T(x) = v^* x v$ for some v, which must be an isometry $H \to \mathbb{C}^n$. So $\dim(H) \leq n$.

The following result is due to Junge from 2005 [69]. The proof which we have given is a variant of his; and certainly the main ideas are all his. We reiterate that this is by no means the shortest proof, but it may be the one that really shows what is going on, in an important sense.

Theorem 4.2. *Let X be an operator subsystem of $A = M_{n_1} \oplus \cdots \oplus M_{n_m}$ (of course any finite-dimensional C^*-algebra A can be written this way), which generates A as a C^*-algebra, and let $\pi_j : A \to M_{n_j}$ be the canonical projection.*

- (a) *If $T : X \to M_m$ is a pure CPU map which is maximal among the pure CPU maps, then $T = u^* \pi_j(\cdot)_{|X} u$ to X, for some j and some unitary matrix u.*
- (b) *The π_j which arise as in* (a) *are boundary representations for X, and there are sufficiently many in the sense of the second paragraph of* 4.3.
- (c) *The boundary representations of X are precisely the maps T in* (a).

Proof. (a) If $\tilde{T} : A \to M_m$ is any CPU extension of T, then by the remark after Theorem 2.8 we have $\tilde{T} = \sum_{i,k} v_{ik}^* \pi_k(\cdot) v_{ik}$ for some matrices v_{ik}. Restricting to X, and using the fact that T is pure, we have $T = v^* \pi_j(\cdot) v$ for some j, and a matrix v which is a scalar multiple of v_{ij}, some i. Evaluating at 1 shows that v is an isometry. By Lemma 4.1 we have $\pi_j = \sum_k w_k^* T_k(\cdot) w_k$ for pure CPU maps T_k and some matrices w_k. Then $T = \sum_k (w_k v)^* T_k(\cdot) w_k v$. Again, since T is pure, we have $T = t(w_k v)^* T_k(\cdot) w_k v$, for some k and scalar $t > 0$. Evaluating at 1 shows that $\sqrt{t} w_k v$ is an isometry. By the maximality, $w_k v$ and hence v are invertible, and so v is unitary.

(b) Supposing that $\tilde{T}$ is as in (a), the proof of (a) shows that

$$\tilde{T} = \sum_{i,k} v_{ik}^* \pi_k(\cdot) v_{ik},$$

and if one of these $v_{ik}^* \pi_k(\cdot) v_{ik}$ is nonzero, then v_{ik} is a scalar multiple of a unitary u, and $T = u^* \pi_k(\cdot) u$. It follows that $u^* \pi_k(\cdot) u = v^* \pi_j(\cdot) v$ for v, j as in (a), so that $k = j$. Now it is easy to see that $u = v$. It follows that $\tilde{T} = v^* \pi_j(\cdot) v$. Thus we have shown that $v^* \pi_j(\cdot) v$ is a boundary representation, and hence so is π_j.

To see that there are sufficient many, write the identity map on X, by Lemma 4.1, as a matrix convex combination of a finite collection of pure CPU maps T_k on X. These T_k are 'sufficient many', except that they may not yet be boundary representations yet. If T_k is not maximal among the pure CPU maps, then we replace it by a proper (strict) dilation which is a pure CPU map. Repeating the process, we must eventually stop (since by the comment above the theorem, pure CPU maps must map into a subalgebra of M_n), and now we have 'sufficient many' pure CPU maps which by the last paragraph and (a) are boundary representations, and are unitarily equivalent to restrictions of some of the π_j.

(c) If T were a boundary representation, then its restriction to X would be pure (see, e.g., [9, Corollary 1.4.3]), and maximal in the sense of the Muhly-Solel result stated earlier. Hence it is maximal among the pure. The converse follows from (a) and (b). □

Remarks

1) Once we know that these π_j are boundary representations, it is easy to deduce directly that the associated C^*-envelope $\oplus_j M_{n_j}$ has the important 'rigid' and 'essential' properties of the C^*-envelope (see, e.g., [60, 28] for the definition of these). It is also easy to deduce the 'boundary theorem' of Arveson [10] in this (finite-dimensional) setting.
2) Zarikian had an earlier variant of part of the above, however it used a non-trivial property of the C^*-envelope (rigidity).
3) The second paragraph of the proof of (b) is another useful strategy to construct the noncommutative Shilov boundary for subsystems of M_n.

It is well known that in the case considered by the theorem, the noncommutative Shilov boundary of X may be constructed from the 'blocks' M_{n_k} of A above, by simply removing blocks M_{n_k} from A whose associated minimal projection $p_k \in A$ satisfies $\|x\| = \|x(1-p_k)\|$ for all $x \in X$, and a similar formula for $x \in M_n(X)$. However, again the above may be the most 'elementary' or 'revealing' proof of this fact.

Finally, we see the main task in the next few years as the development of these new tools and ideas to the point where they can be applied to solve concrete problems in 'noncommutative function theory'.

4.4. Positivity in TROs and C^*-modules

We motivate what follows in two main ways. Reflecting on the fundamental importance of algebra to positivity in the subject of C^*-algebras, and the incredibly beautiful and powerful way in which algebra and positivity interact throughout this subject, it is natural to ask the following question: Is there is a class of spaces containing the C^*-algebras, which also have positive cones, in which there is a similarly intricate relationship between the positivity and the algebra? We will describe such a class of spaces here, which we call *ordered C^*-modules*, or what is the same thing, *ordered TROs*, introduced in papers of the author with W. Werner, and with M. Neal.

Our second motivation for looking at ordered TROs is as follows (the reader shall not need the contents of this paragraph for what follows in this subsection, but it is needed at the end of Subsection 4.5). As we said in Section 3.2, one main tool for studying (unital) operator systems is the noncommutative Shilov boundary. This is a powerful tool, and one would like to have an order theoretic variant of it which is valid for nonunital selfadjoint subspaces $\mathcal{S}$ of C^*-algebras, if only because of the grave scarcity of tools in that setting. However, as we have already noted in Section 3.2, the noncommutative Shilov boundary $\partial\mathcal{S}$ of a nonunital space $\mathcal{S}$ is a TRO generated by $\mathcal{S}$. Note that a TRO generated by an ordered operator space will of course also be ordered, with positive cone containing the positive cone $\mathcal{S}_+$ of $\mathcal{S}$. From this one sees that ordered TROs are going to occur naturally if one studies generic ordered operator spaces and their Shilov boundaries. We will discuss this 'ordered boundary' further in Section 4.5, although for simplicity we will focus there mostly on a special case in which the TRO is forced to be a C^*-algebra.

Thus we consider positivity in C^*-modules, or what is the same, in TROs Z inside a C^*-algebra A. One sets $Z_+ = Z \cap A_+$, but we will want to get rid of the dependence on A in the description of the cone of Z (we want to treat Z *intrinsically*). There are two setting one may consider, when Z is selfadjoint [33], or the general case [30]. The algebra begins immediately: in the selfadjoint case it is not hard to check that $Z^2, J(Z) = Z \cap Z^2$, and the closure of $Z + Z^2$, are all C^*-subalgebras of A, in which one may apply the tools of positivity. Moreover, $J(Z)$ is an 'ideal'. Another auxiliary space, in the selfadjoint case, is the *center* of Z, which may be defined to be the set of $z \in Z$ with $yz = zy$ for all $y \in Z$. It is easy to see that $J(Z)_+ = Z_+$, but this is far from being the end of the story. The problem is that $J(Z)$ was defined in terms of A, and one really wants a description of this space, or equivalently of Z_+, which is independent of the particular containing C^*-algebra A. One of the main theorems is that there is a bijective order-preserving correspondence between such cones Z_+ and *open tripotents* (defined below) in the second dual of Z (indeed, in the selfadjoint TRO variant of the theory, these should also be in the selfadjoint part of the center of Z^{**}). A tripotent is an element u with $uu^*u = u$, also known as a partial isometry. We will not give our technical definition of a tripotent being open, but note that it a variant appropriate to tripotents of the notion of 'open projection' that we discussed in Section 2.2. Indeed in [30] we prove that a tripotent u is open iff the projection $\frac{1}{2}\begin{bmatrix} uu^* & u \\ u^* & u^*u \end{bmatrix}$ is an open projection in the sense of Section 2.2, in the second dual of the linking algebra of Z (see Section 2.4 for the definition of this algebra). Given such an open tripotent u, the corresponding cone in Z is $\mathfrak{d}_u = \{z \in Z : u^*z \in (Z^*Z)_+, z = uz^*u\}$. We observe that this link with tripotents puts one in the very algebraic field of 'JB^*-triples' (see, e.g., [103, 14, 47, 48]), and one can use some of the methodology of that subject, and the tripotent variant of the 'noncommutative topology' discussed in Section 2.2 which we develop in [33, 30], to obtain the theory of positive cones in TROs. One idea from the JB^*-triple theory is that of 'local order': in particular any tripotent $u \in Z$ defines a subspace of Z, called the *Peirce 2-space* of u, which is a C^*-algebra in the product xu^*y and has identity u. This C^*-algebra has a positive cone, which in our case is given by the formula for $\mathfrak{d}_u$ above. There are lots of nice algebraic tricks here, and the moral again is that the positivity is beautifully connected to the underlying algebra. See [33, 30] for details.

Example. Let S^2 be the unit sphere, and let Z be the TRO $\{f \in C(S^2) : f(-x) = -f(x)\}$. In this case, the open tripotents u in Z^{**} mentioned above, correspond precisely to open subsets U of the sphere (called blue), which do not intersect $-U$

(called red). Suppose that $S^2 \setminus (U \cup (-U))$ is colored black. Thus the possible associated cones Z_+, may be labeled by such sets U, indeed $Z_+ = \{f \in Z : f(x) \geq 0 \text{ iff } x \in U\}$. These pictures give one a very clear understanding of the behavior of the associated positive cones. For example, the cone is maximal exactly when the black region is the topological boundary of the red region (and hence also of the blue region). Thus, for example, a sphere whose top hemisphere is red and whose bottom hemisphere is blue, with a black equator line, corresponds to a maximal positive cone; but if you thicken the equator to a black band one loses maximality.

4.5. Positivity in nonunital operator spaces, and the Shilov boundary

Until recently, the largest class of operator spaces for which positivity was defined, is the class of (unital) operator systems, which we have already discussed at some length. The case of ordered operator spaces with no unit was ignored, although such spaces occur very naturally: for example, consider the linear span of three generic positive matrices in M_4. It seems interesting to develop some theory for such spaces.

In the late 1990s decade, W. Schreiner began to study 'matrix ordered spaces' which were not necessarily unital [104]. This direction was continued by Wend Werner in Germany [117, 118], Anil Karn and coauthors in India (who have many papers on this topic, e.g., [76, 77, 78] and references therein, which we shall not discuss here except to note note that this work has focused on matrix ordered spaces whose cones satisfy a stronger additional condition than any which we consider here, for example variants of base normed or order unit spaces), and quite recently by the author and coauthors. Although this study is still in a rather preliminary state, being much more subtle and refractory than the unital case, some primary tasks that come to mind to us in this setting are 1) to characterize the cones on an operator space X which correspond to complete order embeddings of X in a C^*-algebra A, 2) to study the process of adjoining a unit to X to obtain a (unital) operator system, and 3) to study the noncommutative Shilov boundary of these spaces (as was done from the beginning in the case of unital operator systems, and as is fundamental in the theory of classical function spaces, as discussed above). One wants all the maps in the universal property of this Shilov boundary to respect the order. Much of what follows consists of a discussion of these three questions. We shall restrict our attention here to the selfadjoint operator space case, the general case may be found in [30].

Let X be an operator space which is also a $*$-vector space such that

$$\|[x_{ji}^*]\|_n = \|[x_{ij}]\|_n, \qquad n \in \mathbb{N},\ [x_{ij}] \in M_n(X).$$

We also assume that we have a cone $\mathfrak{c}_n \subset M_n(X)$ for each $n \in \mathbb{N}$, which are closed in the norm topology, and which make X a matrix ordered vector space in the sense defined above Theorem 2.10. We shall call an operator space satisfying all of the conditions above in this paragraph, a *matrix ordered operator space* (this notation is not quite standard, but does no harm). Certainly any selfadjoint subspace V of a C^*-algebra is a matrix ordered operator space, and it is not hard to see that the dual V^* is one too, and more generally so are spaces $CB(V, B(H))$ of completely bounded maps (see, e.g., [104]). The first of the three questions at the start of the present subsection, asks for the additional conditions on the cones of a matrix ordered operator space which are necessary and sufficient for there to exist a completely isometric complete order embedding of X into a C^*-algebra. If these conditions all hold, we will call X, together with these cones, a *fully ordered operator space*; these being of course the abstract description of selfadjoint subspaces X of a C^*-algebra A say, with cones $M_n(X) \cap M_n(A)_+$. At the present time, the only known answer to this question is Werner's result from [117] that a necessary and sufficient condition is that $\|x\|_n = \sup\{\, |\varphi(\tilde{x})|\, \}$ for all $n \in \mathbb{N}$, and $x \in M_n(X)$, where the supremum is taken over all positive norm 1 functionals φ on $M_{2n}(X)$, and

$$\tilde{x} = \begin{bmatrix} 0 & x \\ x^* & 0 \end{bmatrix}.$$

Turning to the second question, in [117] Werner defines a unitization X^+ of a matrix ordered operator space X which has the following universal property: for any completely contractive completely positive map T from X into a (unital) operator system Y, the unique extension of T to a unital map from X^+ into Y is completely positive. This is the unitization of X which has the smallest possible matrix cones (we remark that Karn gives a nice alternative description of this in [77, 78], which is valid under extra hypotheses on X). We will discuss a different unitization later in this subsection which has the 'biggest cones'.

We turn next to the noncommutative Shilov boundary for matrix ordered operator spaces. There is a simple and reasonable criterion which forces this 'boundary' to be a C^*-algebra (as opposed to being a TRO), namely, that X has a positive cone which densely spans X. For simplicity, we will focus on this case, and briefly discuss the more complicated general case later. In this case, we will again this 'boundary C^*-algebra' as $C^*_e(X)$, and refer to it as the *C^*-envelope* of X. Its universal property is contained in the following result, which is a special case of a more complicated result from [26].

Theorem 4.3. *Suppose that X is a matrix ordered operator space whose positive cone densely spans X, and let $T : X \to B$ be a positive complete isometry into a C^*-algebra. Then the TRO generated by $T(X)$ inside B is a C^*-algebra, indeed equals the C^*-algebra generated by $T(X)$. Moreover, there exists a C^*-algebra $C^*_e(X)$, and a completely positive complete isometry $j : X \to C^*_e(X)$ whose range generates $C^*_e(X)$ as a C^*-algebra, which has the following universal property: for any positive complete isometry $i : X \to A$ into a C^*-algebra, such that $i(X)$ generates A as a C^*-algebra, there exists a (completely positive) $*$-homomorphism $\pi : A \to C^*_e(X)$ such that $\pi \circ i = j$.*

This enveloping C^*-algebra $C^*_e(X)$ has all of the usual properties of the noncommutative Shilov boundary (see, e.g., [63] and [28, Sections 4.3 and 8.3] for a cataloging of these). For example, it gives one a handle on the completely positive surjective complete isometries $T : X \to Y$ between spaces of the type which we are considering. Indeed, such maps extend uniquely to $*$-isomorphisms between the C^*-envelopes, where often they can be classified.

When restricted to the classical case, Theorem 4.3 shows that a Shilov boundary (in the classical sense) exists for the very natural class of function spaces not containing constants, but which is densely spanned by the positive functions it contains. See [26] for details.

One nice application Theorem 4.3, is that it can be used to show that there is a 'biggest' unitization for operator spaces X of the type we are considering. Namely, define X^1 to be the span of X (or rather of its copy inside $C^*_e(X)$) and the identity of the C^*-algebra unitization of $C^*_e(X)$. It is easy to see, from the universal property in the theorem, that amongst all unitizations of X, this one has the biggest positive cone, while Werner's unitization has the smallest cone. We also remark that the new unitization is often easier to describe. For example, for concrete subspaces of M_n, it is usually easy to compute the noncommutative Shilov boundary, and hence the unitization. See [26] for details.

There is one slightly unpleasant feature of the above setup, namely that the canonical completely positive embedding $j : X \to C^*_e(X)$ may not be a complete order embedding. In fact it is easy to see that j is a complete order embedding iff the given cone (or rather, the sequence of matrix cones) on X is maximal, that is there is no strictly larger 'fully ordered operator space' cone structure for X. Thus such maximal cones will play a special role.

Next, we make some remarks on the general case, when X is not necessarily densely spanned by X_+. In this case the noncommutative Shilov boundary is not necessarily a C^*-algebra. It is a selfadjoint TRO in the sense of Section 4.4 if X is selfadjoint, and is just a TRO otherwise, and it has a canonical positive cone. One then uses the theory discussed in Section 4.4. One obtains analogues of the results in the rest of the current subsection, but they are a little more cumbersome to state, and not quite so nice, which is why we chose to focus above on the special case of a densely spanning cone. For details, see [30, 26].

4.6. Peak interpolation and and ideals of operator algebras

The famous Urysohn lemma is perhaps the best result in point set topology, and it certainly is the fundamental tool to study the commutative C^*-algebra $C(\Omega)$. Importantly for this survey, it is an order theoretic statement about (the order interval $[0, 1]$ in) the latter algebra. In the theory of function spaces or uniform algebras, there are variations and refinements of Urysohn lemma known as *peak interpolation* (see, e.g., [56, 111]). An example of the kind of question addressed by peak interpolation: if A is an algebra of functions on a compact set Ω, and if E is a closed subset of Ω, then when is every continuous scalar-valued function h defined on E the restriction of a function f in A? Or the restriction of a function $f \in A$ with $|f| \leq g$, for a given 'control function' g? This question is of course partly order-theoretic, in that the solution f to the 'interpolation problem' has to satisfy a certain order relation. It is also obviously intimately related to the important question of finding, for an open set U containing E, a function f in A which agrees with h on E, and for which $|f|$ is very small outside of U. The 'good' closed sets E from the above perspective are the so-called *p-sets*. Although these sets have cleaner characterizations due to Glicksberg and others, they are *defined* to be the intersections of collections of *peak sets*. In turn, a peak set is a set E for which there exists an $f \in A$, with $f = 1$ on E, and $|f| < 1$ outside E. Peak interpolation theory proceeds by first establishing facts about such sets. These sets are also intimately connected to the closed ideals of A; namely, an important class of ideals of A is in a bijective order-reversing correspondence with the p-sets. Indeed, a combination of results of Hirsberg and Smith yields that the closed ideals J in a uniform algebra $A \subset C(\Omega)$, such that J has a bounded approximate identity, are exactly the sets of functions in A which are zero on a fixed p-set.

The main idea in this section is to describe a noncommutative generalization of parts of the theory of classical peak interpolation developed by our recent student Damon Hay [66], and also by Hay, the author and Neal [25]. This generalization is perhaps not quite in final form yet, but it is already good enough to give new results which seem deep, for example about ideals in operator algebras. The conceptual starting point is the noncommutative Urysohn lemma which we mentioned in Section 2.2. In the light of the classical situation mentioned in the last paragraph, this result suggests the following 'peak interpolation setup': we are given a unital subspace or subalgebra A of a C^*-algebra B, and a closed projection q in B^{**}. We say that q is a *peak projection* if there exists an $a \in A$ with $aq = q$ and $(1-q)|a|^2(1-q)$ is 'strictly less' than 1 (in the sense of Kadison's function representation, for example). This is the noncommutative analog of a 'peak set', and there are a dozen equivalent restatements which the reader may prefer [66]. A *p-projection* is an infimum of a collection of peak projections (and we point out that positivity is used here). We have not been able to establish the full noncommutative version of a theorem of Glicksberg yet, which should say that if A is a unital subalgebra of B then the p-projections are exactly the closed projections in B^{**} which are also in $A^{\perp\perp}$. Nonetheless, we have been able to show that both the latter class of projections, and the (possibly) smaller class of p-projections, do satisfy analogues of the classical 'peak interpolation' results mentioned in the first paragraph of this section. Moreover the proofs use the noncommutative Urysohn lemma, and arguments involving positivity. For example, Hay showed that if $q \in B^{**}$ is a closed projection, then the following are equivalent:

(i) $q \in A^{\perp\perp}$,
(ii) given $\epsilon > 0$, for each open projection $u \geq q$, there exists an element $a \in A$ such that $\|a\| \leq 1 + \epsilon$, $qa = q$ and $(1-u)a^*a(1-u) < \epsilon 1$,
(iii) given $\epsilon > 0$, for every strictly positive $p \in B$ with $p \geq q$, there exists $a \in A$ such that $qa = q$ and $a^*a \leq p + \epsilon$.

If q satisfies a stronger condition than (i) above, if b is a strictly positive element of B, and if $x \in B$ with $x^*qx \leq b$, then for any $\epsilon > 0$ there exists an element $y \in A$ with $qx = qy$ and $y^*y \leq b(1+\epsilon)$. Or, as another example, we proved in [25] that q

is a p-projection for A iff for any open projection $u \geq q$, and for any $\epsilon > 0$, there exists an $a \in A$ of norm ≤ 1, with $aq = q$ and $(1-u)a^*a(1-u) < \epsilon$, and similarly with a^*a replaced by aa^*. If one takes B to be commutative, that is $B = C(\Omega)$ for a compact set Ω, it is easy to see that these results are 'peak interpolation' results of the sort described in the first paragraph of this section. See [35] for some other refinements of the noncommutative Urysohn lemma in the selfadjoint case.

These ideas have interesting applications to the ideal structure of a (not necessarily selfadjoint) operator algebra A, as one might expect from the 'commutative case' (namely, the aforementioned result of Hirsberg and Smith that the closed ideals J in a uniform algebra A, such that J has a bounded approximate identity, are exactly the sets of functions in A which are zero on a fixed p-set), and from what happens in the C^*-algebraic theory of left or right ideals, or hereditary subalgebras (see the discussion in Section 1 several paragraphs above Theorem 2.3). Indeed, we can generalize to our setting key aspects of these theories. For example, Hay showed that closed left ideals J in A, such that J has a contractive right approximate identity, correspond bijectively to the closed projections $q \in A^{\perp\perp}$ that we have discussed above, via the correspondence $J = pA^{**} \cap A$. While this seems natural in view of the C^*-algebra case, the proof is deep, using for example the noncommutative Urysohn lemma and nonselfadjoint variants of it (the peak interpolation discussed above), and positivity in the second dual of a containing C^*-algebra. In [25] we extended Hay's result a little, and used it to develop a theory of hereditary subalgebras of nonselfadjoint algebras paralleling the C^*-algebra case. Thus, we showed that there was a bijective correspondence between such left ideals, the matching class of right ideals, hereditary subalgebras of A, and certain weak* closed faces of the quasistate space of A. Note that if A is nonselfadjoint then there is no obvious correspondence between the above classes of left and right ideals (the trick $J \mapsto J^*$ that works in the C^*-algebra case fails for nonselfadjoint algebras). In fact at the present time this seems to be deep, relying on the other deep results mentioned above. We repeat again that positivity plays a critical but subtle role in this ideal theory, we mentioned this for example in the use of the noncommutative Urysohn lemma, or the peak interpolation results, above.

4.7. A generalization of C^*-modules

Finally, we end with an example which again illustrates the three themes that: (a) with a bit of work, C^*-algebraic notions involving positivity can be generalized to settings where no positivity is at first in evidence, (b) the generalization sits inside an enveloping 'C^*-algebraic object' where positivity can be applied, and which it generates and is tightly connected to, and (c) even without going up to the 'C^*-algebraic object' just referred to, positivity can play an important but deeply hidden role in such settings.

Recall from Section 2.4 that a C^*-module Z is a right B-module Z over a C^*-algebra B possessing a B-valued inner product satisfying the axioms for a Hilbert space, for example, $\langle z, z\rangle \geq 0$ for all $z \in Z$. As we indicated, these objects and their theory are an extremely powerful tool, and it seemed desirable to generalize the theory to the case that B is replaced by a possibly nonselfadjoint operator algebra A. This is theme (a) mentioned in the last paragraph. We use the term *rigged module* for this generalization of C^*-modules. An example of an object that one would wish to be a rigged module is the right A-module $C_n(A)$, the 'first column' of $M_n(A)$. In this example, an enveloping 'C^*-algebraic object' (mentioned in theme (b) above), may be taken to be the C^*-module $C_n(C_e^*(A))$, where $C_e^*(A)$ is the C^*-envelope or noncommutative Shilov boundary mentioned in Section 3.2 and in other sections. Notice how $Y = C_n(A)$ generates $C_n(C_e^*(A))$; indeed $YC_e^*(A)$ is dense in $C_n(C_e^*(A))$.

There are many equivalent definitions of a rigged module Y over A, some involving a superspace which is a C^*-module, others involving another left module X and a pairing $X \times Y \to A$. In fact, the simplest to state of these definitions, which we give momentarily, was only recently proved in [25] to be equivalent. This equivalence solves a 15 year old problem attributable to Paulsen, and seems to be deep. Indeed the proof uses a deep fact about one-sided ideals in operator algebras described in the last subsection, which in turn needs the full force of the results in that subsection on 'peak interpolation', which in turn used positivity as we saw there. This illustrates theme (c) above. This 'simplest definition' of a rigged module, is that these are the operator spaces Y which are also right A-modules, such that Y 'asymptotically factors' through spaces of the type $C_n(A)$ in the last paragraph. That is, there are nets of completely contractive module maps φ_t, ψ_t between Y and $C_{n_t}(A)$, with $\psi_t(\varphi_t(y)) \to y$ for all $y \in Y$. Here $n_t \in \mathbb{N}$. With earlier definitions of 'rigged module' we had shown that the theory of C^*-modules will generalize (see [20] and references therein). Moreover, we had shown that any rigged module Y over A could be represented as an A-submodule of a C^*-module Z over $C_e^*(A)$ with $YC_e^*(A)$ dense in Z. In fact Z is a certain tensor product of Y and $C_e^*(A)$, and also equals the noncommutative Shilov boundary of Y under a reasonable extra condition (see [23]). Thus many facts about Y can be proved by applying the theory of C^*-modules to Z, a setting in which positivity is a natural tool, and then pulling the results down to Y. This again illustrates theme (b) above.

References

[1] J. Agler, An abstract approach to model theory, pp. 1–23 in *Surveys of some recent results in operator theory,* Vol. II, Longman Sci. Tech., Harlow 1988.

[2] C.A. Akemann, The general Stone-Weierstrass problem, *J. Funct. Anal.* **4** (1969), 277–294.

[3] C.A. Akemann, Left ideal structure of C^*-algebras, *J. Funct. Anal.* **6** (1970), 305–317.

[4] C.A. Akemann, A Gelfand representation theory for C^*-algebras, *Pacific J. Math.* **39** (1971), 1–11.

[5] C.A. Akemann and G.K. Pedersen, Facial structure in operator algebra theory, *Proc. London Math. Soc.* **64** (1992), 418–448.

[6] E.M. Alfsen, *Compact convex sets and boundary integrals,* Springer-Verlag, New York-Heidelberg, 1971.

[7] E.M. Alfsen and F.W. Schultz, *Geometry of state spaces of operator algebras,* Birkhäuser Boston, Inc., Boston, MA, 2003.

[8] E.M. Alfsen and F.W. Shultz, *State spaces of operator algebras. Basic theory, orientations, and C^*-products,* Birkhäuser Boston, Inc., Boston, MA, 2001.

[9] W.B. Arveson, Subalgebras of C^*-algebras, *Acta Math.* **123** (1969), 141–224.

[10] W.B. Arveson, Subalgebras of C^*-algebras II, *Acta Math.* **128** (1972), 271–308.

[11] W.B. Arveson, *Noncommutative dynamics and E-semigroups*, Springer Monographs in Mathematics, Springer-Verlag, New York, 2003.

[12] W.B. Arveson, Notes on extensions of C^*-algebras, *Duke Math. J.* **44** (1977), 329–355.

[13] W.B. Arveson, The noncommutative Choquet boundary, Preprint 2007, math.OA/0701329, to appear in *Journal of Amer. Math Soc.*

[14] M. Battaglia, Order theoretic type decomposition of JBW*-triples, *Quart. J. Math. Oxford,* **42** (1991), 129–147.

[15] J. Bendat and S. Sherman, Monotone and convex operator functions, *Trans. Amer. Math. Soc.* **79** (1955), 58–71.

[16] S.K. Berberian, *Baer *-rings,* Springer-Verlag, New York-Berlin, 1972.

[17] B. Blackadar, *K-theory for operator algebras,* Second edition, Math. Sci. Res. Inst. Pub, 5, Cambridge University Press, Cambridge, 1998.

[18] B. Blackadar, Operator algebras. Theory of C^*-algebras and von Neumann algebras, Encyclopaedia of Mathematical Sciences, 122, Springer-Verlag, Berlin, 2006.

[19] B. Blackadar and E. Kirchberg, Generalized inductive limits of finite-dimensional C^*-algebras, *Math. Ann.* **307** (1997), 343–380.

[20] D.P. Blecher, A generalization of Hilbert modules, *J. Funct. Anal.* **136** (1996), 365–421.

[21] D.P. Blecher, The Shilov boundary of an operator space and the characterization theorems, *J. Funct. Anal.* **182** (2001), 280–343.

[22] D.P. Blecher, Multipliers and dual operator algebras, *J. Funct. Anal.* **183** (2001), 498-525.

[23] D.P. Blecher, Multipliers, C^*-modules, and algebraic structure in spaces of Hilbert space operators, pp. 85–128 in *Operator algebras, quantization, and noncommutative geometry,* Contemp. Math., 365, Amer. Math. Soc., Providence, RI, 2004.

[24] D.P. Blecher, E.G. Effros, and V. Zarikian, One-sided M-ideals and multipliers in operator spaces, I, *Pacific J. Math.* **206** (2002), 287–319.

[25] D.P. Blecher, D.M. Hay, and M. Neal, Hereditary subalgebras of operator algebras, to appear *J. Operator Theory,* math.OA/0512417

[26] D.P. Blecher, K. Kirkpatrick, M. Neal, and W. Werner, Ordered involutive operator spaces, to appear in *Positivity.*

[27] D.P. Blecher and L.E. Labuschagne, Von Neumann algebraic H^p theory, To appear in *Proceedings of fifth conference on function spaces,* Contemp. Math., math.OA/0611879

[28] D.P. Blecher and C. Le Merdy, *Operator algebras and their modules – an operator space approach,* Oxford Univ. Press, Oxford (2004).

[29] D.P. Blecher and B. Magajna, Duality and operator algebras: automatic weak* continuity and applications, *J. Funct. Anal.* **224** (2005), 386–407.

[30] D.P. Blecher and M. Neal, Open partial isometries and positivity in operator spaces, Preprint 2006, math.OA/0606661

[31] D.P. Blecher and V.I. Paulsen, Multipliers of operator spaces, and the injective envelope, *Pacific J. Math.* **200** (2001), 1–17.

[32] D.P. Blecher, Z.-J. Ruan, and A.M. Sinclair, A characterization of operator algebras, *J. Funct. Anal.* **89** (1990), 188–201.

[33] D.P. Blecher and W. Werner, Ordered C^*-modules, *Proc. London Math. Soc.* **92** (2006), 682–712.

[34] D.P. Blecher and V. Zarikian, *The calculus of one-sided M-ideals and multipliers in operator spaces,* Mem. Amer. Math. Soc. **842** (2006).

[35] L.G. Brown, Semicontinuity and multipliers of C^*-algebras, *Canad. J. Math.* **40** (1988), 865–988.

[36] L.G. Brown, R.G. Douglas, P.A. Fillmore, Extensions of C^*-algebras and K-homology, *Ann. of Math.* **105** (1977), 265–324.

[37] N.P. Brown, On quasidiagonal C*-algebras, *Operator algebras and applications,* 19–64, Adv. Stud. Pure Math. **38** Math. Soc. Japan, Tokyo, 2004.

[38] M.-D. Choi, A Schwarz inequality for positive linear maps on C^*-algebras, *Illinois J. Math.* **18** (1974), 565–574.

[39] M.-D. Choi, Completely positive linear maps on complex matrices, *Linear Alg. and Applns,* **10** (1975), 285–290.

[40] M.-D. Choi and E. G. Effros, The completely positive lifting problem for C^*-algebras, *Ann. of Math.* **104** (1976), 585–609.

[41] M.-D. Choi and E. G. Effros, Injectivity and operator spaces, *J. Funct. Anal.* **24** (1977), 156–209.

[42] F. Combes, Sur le faces d'une C^*-algébre, *Bull. Sci. Math.* **93** (1969), 57–100.

[43] A. Connes, A view of mathematics, Preprint (2005), available from www.alainconnes.org

[44] J.B. Conway, *A Course in Operator Theory,* Graduate Studies in Mathematics, 21, Amer. Math. Soc. Providence, RI, 2000.

[45] J. Dixmier, *C^*-algebras,* North-Holland Publ. Co., Amsterdam, 1977.

[46] M. Dritschel and S. McCullouch, Boundary representations for families of representations of operator algebras and spaces, *J. Operator Theory,* **53** (2005), 159–167.

[47] C.M. Edwards and G.T. Rüttimann, *On the facial structure of the unit balls in a JBW*-triple and its predual,* J. London Math. Soc. **38** (1988), 317–332.

[48] C.M. Edwards and G.T. Rüttimann, *Inner ideals in C*-algebras,* Math. Ann. **290** (1991), 621–628.

[49] E.G. Effros, Order ideals in a C^*-algebra and its dual, *Duke Math. J.* **30** (1963), 391–411.

[50] E.G. Effros, Aspects of noncommutative order, pp. 1–40 in *C^*-algebras and applications to physics*, Lecture Notes in Math., 650, Springer, Berlin, 1978.

[51] E.G. Effros and Z.-J. Ruan, *Operator Spaces*, London Mathematical Society Monographs, New Series, 23, The Clarendon Press, Oxford University Press, New York, 2000.

[52] E.G. Effros and S. Winkler, Matrix convexity: operator analogues of the bipolar and Hahn-Banach theorems, *J. Funct. Anal.* **144** (1997), 117–152.

[53] D.R. Farenick, Extremal matrix states on operator systems, *J. London Math. Soc.* **61** (2000), 885–892.

[54] D.R. Farenick, Pure matrix states on operator systems, *Linear Algebra Appl.* **393** (2004), 149–173.

[55] D.R. Farenick and P.B. Morentz, C^*-extreme points in the generalized state spaces of a C^*-algebra, *Trans. Amer. Math. Soc.* **349** (1997), 1725–1748.

[56] T.W. Gamelin, *Uniform Algebras,* Second edition, Chelsea, New York, 1984.

[57] R. Giles and H. Kummer, A non-commutative generalization of topology, *Indiana Univ. Math. J.*, **21** (1971/72), 91–102

[58] U. Haagerup, Decompositions of completely bounded maps on operator algebras, Unpublished manuscript (1980).

[59] D. Hadwin, Completely positive maps and approximate equivalence, *Indiana Univ. Math J.* **36** (1987), 211–228.

[60] M. Hamana, Injective envelopes of operator systems, *Publ. R.I.M.S. Kyoto Univ.* **15** (1979), 773–785.

[61] M. Hamana, Regular embeddings of C^*-algebras in monotone complete C^*-algebras, *J. Math. Soc. Japan* **33** (1981), 159–183.

[62] M. Hamana, Modules over monotone complete C^*-algebras, *Internat. J. Math.* **3** (1992), 185–204.

[63] M. Hamana, Triple envelopes and Silov boundaries of operator spaces, *Math. J. Toyama University* **22** (1999), 77–93.

[64] F. Hansen, An operator inequality, *Math. Ann.* **246** (1979/80), 249–250.

[65] P. Harmand, D. Werner, and W. Werner, *M-ideals in Banach spaces and Banach algebras,* Lecture Notes in Math., 1547, Springer-Verlag, Berlin–New York, 1993.

[66] D.M. Hay, Closed projections and peak interpolation for operator algebras, to appear *J. Int. Eq. Oper. Th.*, math.OA/0512353

[67] M. Henle, A Lebesgue decomposition theorem for C^*-algebras, *Canad. Math. Bull.* **15** (1972), 87–91.

[68] F. Hiai and H. Kosaki, Means of Hilbert space oeprators. Lectures Notes in Mathematics, 1820. Springer-Verlag, Berlin, 2003.

[69] M. Junge, Minimal sets of complete contractions on operator systems (tentative title), Draft (2005) and revision (March 2007).

[70] M. Junge and G. Pisier, Bilinear forms on exact operator spaces and $B(H) \otimes B(H)$, *Geom. Funct. Anal.* **5** (1995), 329–363.

[71] R.V. Kadison, *A representation theory for commutative topological algebras*, Mem. Amer. Math. Soc. **7** (1951).

[72] R.V. Kadison, A generalized Schwarz inequality, *Ann. of Math.* **56** (1952), 494–503.

[73] R.V. Kadison, Operator algebras with a faithful weakly-closed representation, *Ann. of Math.* **64** (1956), 175–181.

[74] R.V. Kadison and J.R. Ringrose, *Fundamentals of the theory of operator algebras*, Graduate Studies in Mathematics, 15, Amer. Math. Soc. Providence, RI, 1997.

[75] M. Kaneda and V.I. Paulsen, Quasi-multipliers of operator spaces, *J. Funct. Anal.* **217** (2004), 347–365.

[76] A.K. Karn and R. Vasudevan, Characterization of matricially Riesz normed spaces, *Yokohama Math. J.* **47** (2000), 143–153.

[77] A.K. Karn, Adjoining an order unit to a matrix ordered space, *Positivity* **9** (2005), 207–223.

[78] A.K. Karn, Corrigedem to the paper "Adjoining an order unit to a matrix ordered space", Draft, 2006.

[79] E. Kirchberg, On nonsemisplit extensions, tensor products and exactness of group C^*-algebras, *Invent. Math.* **112** (1993), 449–489.

[80] E. Kirchberg, On subalgebras of the CAR-algebra, *J. Funct. Anal.* **129** (1995), 35–63.

[81] E. Kirchberg, On restricted peturbations in inverse images and a description of normalizer algebras in C^*-algebras, *J. Funct. Anal.* **129** (1995), 1–34.

[82] E. Kirchberg and N.C. Phillips, Embedding of exact C^*-algebras in the Cuntz algebra O_2, *J. Reine Angew. Math.* **525** (2000), 17–53.

[83] E. Kirchberg and S. Wassermann, C^*-algebras generated by operator systems, *J. Funct. Anal.* **155** (1998), 324–351.

[84] E.C. Lance, *Hilbert C^*-modules – A toolkit for operator algebraists,* London Math. Soc. Lecture Notes, 210, Cambridge University Press, Cambridge, 1995.

[85] A.T.-M. Lau and R.J. Loy, Contractive projections on Banach algebras, Preprint 2006.

[86] C. Le Merdy, An operator space characterization of dual operator algebras, *Amer. J. Math.* **121** (1999), 55–63.

[87] R.I. Loebl and V.I. Paulsen, Some remarks on C^*-convexity, *Linear Algebra Appl.* **35** (1981), 63–78.

[88] K. Löwner, Über monotone matrixfunktionen, *Math. Z.* **38** (1934), 177–216.

[89] B. Magajna, C^*-convex sets and completely bounded bimodule homomorphisms, *Proc. Roy. Soc. Edinburgh Sect. A* **130** (2000), 375–387.

[90] R. Meyer, Adjoining a unit to an operator algebra, *J. Operator Theory* **46** (2001), 281–288.

[91] P.B. Morentz, The structure of C^*-convex sets, *Canad. J. Math.* **46** (1994), 1007–1026.

[92] P.S. Muhly and B. Solel, An algebraic characterization of boundary representations, pp. 189–196 in *Nonselfadjoint operator algebras, operator theory, and related topics*, Oper. Th. Adv. Appl., 104, Birkhäuser, Basel, 1998.

[93] M. Nielsen and I. Chuang, *Quantum computation and quantum information*, Cambridge University Press, 2000.

[94] N. Ozawa, About the QWEP conjecture, *Internat. J. Math.* **15** (2004), 501–530.

[95] V.I. Paulsen, Every completely polynomially bounded operator is similar to a contraction, *J. Funct. Anal.* **55** (1984), 1–17.

[96] V.I. Paulsen, *Completely bounded maps and operator algebras,* Cambridge Studies in Advanced Math., 78, Cambridge University Press, Cambridge, 2002.

[97] G.K. Pedersen, *C^*-algebras and their automorphism groups,* Academic Press, London (1979).

[98] G.K. Pedersen, *Analysis now,* Graduate Texts in Mathematics, 118, Springer-Verlag, New York, 1989.

[99] R.R. Phelps, *Lectures on Choquet's theorem*, 2nd Edition, Lecture Notes in Mathematics, Vol. 1757, Springer-Verlag, Berlin, 2001.

[100] G. Pisier, *Similarity problems and completely bounded maps,* Second, expanded edition, Lecture Notes in Math., 1618, Springer-Verlag, Berlin, 2001.

[101] G. Pisier, *Introduction to operator space theory,* London Math. Soc. Lecture Note Series, 294, Cambridge University Press, Cambridge, 2003.

[102] Z.-J. Ruan, Injectivity of operator spaces, *Trans. Amer. Math. Soc.* **315** (1989), 89–104.

[103] B. Russo, *Structure of JB^*-triples, Jordan algebras (Oberwolfach, 1992),* 209–280, de Gruyter, Berlin, 1994.

[104] W.J. Schreiner, Matrix regular operator spaces, *J. Funct. Anal.* **152** (1998), 136–175.

[105] D. Sherman, On the dimension theory of von Neumann algebras, to appear, Math. Scand., math.OA/0503747

[106] R.R. Smith and J.D. Ward, Matrix ranges for Hilbert space operators, *Amer. J. Math.* **102** (1980), 1031–1081.

[107] W.F. Stinespring, Positive functions on C^*-algebras, *Proc. Amer. Math. Soc.* **6** (1955), 211–216.

[108] E. Størmer, Positive linear maps of operator algebras, *Acta Math.* **110** (1963), 233–278.

[109] E. Størmer, Positive linear maps of C^*-algebras, pp. 85–106 in *Foundations of quantum mechanics and ordered linear spaces,* Lecture Notes in Phys., Vol. 29, Springer, Berlin, 1974.

[110] E. Størmer, Multiplicative properties of positive maps, *Math. Scand.* **100** (2007), 184–192.

[111] E.L. Stout, *The theory of uniform algebras,* Bogden and Quigley, 1971.

[112] S. Stratila, *Modular theory in operator algebras,* Editura Acedemiei and Abacus Press, 1981.

[113] S. Stratila and L. Zsidó, *Operator algebras, the general Banach algebra background (Tentative title)*, Theta Series in Advanced Mathematics, Bucharest, to appear.

[114] M. Takesaki, *Theory of Operator Algebras I*, Springer, New York, 1979.

[115] M.A. Youngson, Completely contractive projections on C^*-algebras, *Quart. J. Math. Oxford* **34** (1983), 507–511.

[116] C. Webster and S. Winkler, The Krein-Milman theorem in operator convexity, *Trans. Amer. Math. Soc.* **351** (1999), 307–322.

[117] W. Werner, Subspaces of $L(H)$ that are $*$-invariant, *J. Funct. Anal.* **193** (2002), 207–223.

[118] W. Werner, Multipliers on matrix ordered operator spaces and some K-groups, *J. Funct. Anal.* **206** (2004), 356-378.

[119] S. Winkler, *Matrix convexity,* Ph.D. thesis, U.C.L.A., 1996.

[120] G. Wittstock, Extensions of completely bounded C^*-module homomorphisms, pp. 238–250 in *Operator algebras and group representations, Vol. II (Neptun, 1980),* Monogr. Stud. Math., 18, Pitman, Boston, MA, 1984.

[121] G. Wittstock, On matrix order and convexity, pp. 175–188 in *Functional analysis: surveys and recent results,* North-Holland Math. Stud., 90, North-Holland, Amsterdam, 1984.

[122] S.L. Woronowicz, Nonextendible positive maps, *Commun. Math. Phys.* **51** (1976), 243–282.

David P. Blecher
Department of Mathematics
University of Houston
Houston, TX 77204-3008, USA
e-mail: `dblecher@math.uh.edu`

Positivity
Trends in Mathematics, 73–96

Results in f-algebras

K. Boulabiar, G. Buskes, and A. Triki

Contents

1.	Introduction	73
2.	Averaging operators	74
3.	Square-mean closed and geometric-mean closed f-algebras	80
4.	Maximal rings of quotients and (extended) orthomorphisms	85
5.	Order bounded disjointness preserving operators	88
6.	A new representation theorem	91
References		94

1. Introduction

We wrote a survey [18] on lattice ordered algebras five years ago. Why do we return to f-algebras once more? We hasten to say that there is only little overlap between the current paper and that previous survey. We have three purposes for the present paper. In our previous survey we remarked that one aspect that we did not discuss, while of some historical importance to the topic, is the theory of averaging operators. That theory has its roots in the nineteenth century and predates the rise of vector lattices. Positivity is a crucial tool in averaging, and positivity has been a fertile ground for the study of averaging-like operators. The fruits of positivity in averaging have recently (see [24]) started to appear in probability theory (to which averaging operators are close kin) and statistics. In the first section of our paper, we survey the literature for our selection of old theorems on averaging operators, at the same time providing some new perspectives and results as well.

Our second goal is to update the information from our previous survey on representation of disjointness preserving operators. Substantial new results have been obtained since and we intend to show that many of them can be understood from a generalized point of view, i.e., the structure theory of f-algebras. Indeed,

The material in this paper is based upon work supported by the National Science Foundation under Grant No. INT 0423522.

in Section 6 we will prove the following new theorem that summarizes a rather large portion of the literature (e.g., [2, 3, 12, 13, 14, 32, 33]) on representation of order bounded disjointness preserving operators.

Theorem 1.1. *Let A be an nth-root closed semiprime f-algebra and let B be a semiprime f-algebra. If $T : A \to B$ is an order bounded disjointness preserving operator then there exist an algebra and lattice homomorphism $S : \mathrm{Orth}\,(A) \to \mathrm{Orth}^{\infty}\,(\mathrm{Orth}\,(B^{\mathrm{ru}}))$ and an element $w \in \mathrm{Orth}^{\infty}\,(\mathrm{Orth}\,(B^{\mathrm{ru}}))$ such that*

$$T(f) = wS(f) \qquad \textit{for all} \quad f \in A.$$

Finally, to be able to present a proof of the latter theorem, we felt the need to lead the reader through the theory of various extended orthomorphisms and rings of quotients as available in the literature.

Last but not least, we have been involved in a study of the so-called square of a vector lattice [17], which in effect enables a systematic translation from the theory of order bounded bilinear maps that are separately disjointness preserving into the theory of order bounded disjointness preserving operators. The glue needed to achieve that translation is provided by so-called orthosymmetric bilinear maps introduced by Buskes and van Rooij in [16]. We need a brief appearance of orthosymmetric maps in the main result of the theorem above in our last section, and – as we said earlier – order bounded disjointness preserving operators have our interest in Section 5. The study of the geometric mean and square mean in f-algebras in our Section 3, apart from being interesting in its own right, provides exactly a foundation for a convexification procedure in vector lattices that leads to this square of vector lattices (see [5]).

2. Averaging operators

In his celebrated paper [48] written at the end of the 19th century, Reynolds – a pioneer of theoretical fluid dynamics – introduced an operator that maps a function of time and space to its mean over some interval of time. For that operator, Reynolds was led to consider the algebraic identity

$$T\,(aT\,(b) + bT\,(a)) = T\,(a)\,T\,(b) + T\,(T\,(a)\,T\,(b))\,. \tag{$\mathcal{R}$}$$

An operator T with property ($\mathcal{R}$) is called a *Reynolds operator.* In his study, Reynolds also considered *averaging operators*, i.e., operators T that satisfy the identity

$$T(aT\,(b)) = T\,(a)\,T\,(b)\,. \tag{$\mathcal{A}$}$$

There now is an extensive literature on averaging operators, motivated to no small degree from their connection to conditional expectation in probability theory. Kampé de Fériet first recognized the importance of studying averaging and Reynolds operators in general, and substantially advanced the topic in [35]. A more algebraic study of these operators was initiated by Dubreil in [21], while the first study of averaging operators by means of functional analysis is due to

Birkhoff [11]. Interestingly, the averaging identity $(\mathcal{A})$ was being studied at about the same time as Kolmogorov's foundations of probability became known, whereas the connection with conditional expectation was made only many years later by Moy in [43].

Since those early beginnings of the history of averaging operators above, the identities $(\mathcal{R})$ and $(\mathcal{A})$ have been studied by many authors. Some were interested in the logical interdependence of the identities, others examined the relationship between $(\mathcal{R})$, $(\mathcal{A})$, and the differential equations describing the motion of fluids. Further research on the subject was motivated by the fact that both identities abundantly occur in probability theory, and, indeed, conditional expectation operators continue to be a source of inspiration for the general study of averaging and Reynolds operators [4, 10, 19, 22, 42, 49, 51, 53].

In the thirties of the previous century, Kampé de Fériet studied averaging operators on the set of real-valued functions that take only a finite number of values [35], while Birkhoff in [11] investigated them on spaces of real-valued continuous functions on a compact Hausdorff space. We remark that Sopka in [53] independently followed a similar path as Birkhoff, but the latter laced his study with a rather more algebraic point of view, setting the stage for our discussion. Our first proposition below appeared indeed in [11]. Following common terminology, a linear operator $T : A \to A$, where A is a real vector space, is called a *projection* whenever

$$T^2(a) = T(a) \tag{$\mathcal{P}$}$$

holds in A for all $a \in A$.

Proposition 2.1. (Birkhoff [11]) *Let A be an Archimedean f-algebra with unit element e and let $T : A \to A$ be an averaging operator such that $T(e) = e$. Then T is a projection and a Reynolds operator.*

By $C_0(X)$ we denote the (Archimedean and semiprime) f-algebra of all real-valued continuous functions on the locally compact Hausdorff space X that vanish at infinity. In [36], Kelley proved the following result which generalizes the case of compact X, established previously by Birkhoff in [11].

Theorem 2.2. (Kelley [36]) *A norm-one positive projection $T : C_0(X) \to C_0(X)$ is averaging if and only if the range of T is a subalgebra of $C_0(X)$.*

Kelley's proof of Theorem 2.2 is based on an integral representation for T and the fact that X may be decomposed into slices that render $T(a)$ to be the average of the value of a on each slice. Subsequently, Seever in [52] generalized Kelley's theorem as follows.

Theorem 2.3. (Seever [52]) *If $T : C_0(X) \to C_0(X)$ is a norm-one positive projection then*

$$T(aT(b)) = T(T(a)T(b)) \tag{$\mathcal{S}$}$$

holds for all $a, b \in C_0(X)$.

Let A be an (associative) algebra. Following the terminology by Huijsmans and de Pagter in [30], we call a linear operator $T : A \to A$ with property $(\mathcal{S})$ a *Seever operator*. Just like Kelley's proof of Kelley's theorem above, Seever's proof of his Theorem 2.3 uses the machinery of analysis. In [30], Huijsmans and de Pagter gave an f-algebra version of both of these theorems, crafting their proofs from the terrains of positivity and algebra. They restricted their results to semiprime f-algebras with the so-called *Stone condition*, which states that

$$a \wedge I \in A \quad \text{for all} \quad a \in A^+,$$

where I denotes the identity mapping on A and where A is considered as an f-subalgebra of the unital f-algebra Orth (A) of all orthomorphisms of A (see Section 4). Here is there theorem.

Theorem 2.4. (Huijsmans and de Pagter [30]) *Let A be an Archimedean semiprime f-algebra with the Stone condition and $T : A \to A$ be a positive contractive projection. Then T is a Seever operator.*

As a consequence of the previous theorem, Huijsmans and de Pagter also obtained a generalization of Kelley's theorem (see Theorem 2.2).

Theorem 2.5. (Huijsmans and de Pagter [30]) *Let A be an Archimedean semiprime f-algebra with the Stone condition and let $T : A \to A$ be a positive projection. Then the following are equivalent.*

(i) *T is averaging.*
(ii) *The range of T is a subalgebra of A and T is contractive.*

It turns out that the Stone condition in the preceding two theorems can be dropped. This was proved by Triki in [54] via extensions of positive projections. In addition, in his theorem below, A does not even need to be a vector lattice.

Theorem 2.6. (Triki [54]) *Let A be any majorizing subalgebra of the Archimedean semiprime f-algebra B and $T : A \to A$ be a positive contractive projection. Then T is a Seever operator. Moreover, T is averaging if and only if the range of T is a subalgebra of B.*

If in Theorem 2.6 B has a a Riesz norm $\|.\|$ (i.e., $\|a\| \le \|b\|$ whenever $|a| \le |b|$ in B) then the result holds without the extra condition 'A majorizes B'. More precisely, we have the following result.

Theorem 2.7. (Triki [54]) *Let A be a subalgebra of an Archimedean semiprime f-algebra B with a Riesz norm and let $T : A \to A$ be a positive contractive projection. Then T is a Seever operator. Moreover, T is averaging if and only if the range of T is a subalgebra of B.*

More recently, Triki (in [55]) also removed the semiprimeness assumption from the conditions of the theorem by Huijsmans and de Pagter above.

Theorem 2.8. (Triki [55]) *Let A be an Archimedean f-algebra and $T : A \to A$ be a positive contractive projection. Then T is a Seever operator and T is averaging if and only if the range of A is a subalgebra of A.*

It is not true that every positive projection onto a subalgebra is an averaging operator as can be seen from the following example due to Wulbert [59].

Example 2.9. (Wulbert [59]) *Put $X = [0,1] \cup \{2\}$ and let A be the subalgebra of $C(X)$ of all functions that vanish at the point 2. Let h be the function which is identically one on $[0,1]$, and vanishes at 2. Define the linear operator $T : C(X) \to C(X)$ by*

$$T(f)(x) = (f(x) + f(2))\, h(x) \quad \text{for all } f \in C(X), x \in X.$$

Then T is a positive projection on $C(X)$. However, if g is the constant function one on X, then $T(gT(g)) = 2h$ while $T(g)T(g) = 4h$.

Next we bring into focus various relationships between the algebraic identities $(\mathcal{A})$, $(\mathcal{P})$, $(\mathcal{R})$, and $(\mathcal{S})$ for a linear operator T on an f-algebra A. Consider first the properties $(\mathcal{A})$, $(\mathcal{P})$, $(\mathcal{R})$. Every operator T on an Archimedean semiprime f-algebra which satisfies two of these identities, also satisfies the third. This is the content of the following proposition.

Proposition 2.10. *Let A be an Archimedean semiprime f-algebra and let $T : A \to A$ be a linear operator. Then the following hold.*

(i) *If T is averaging and a projection then T is a Reynolds operator.*
(ii) *If T is averaging and a Reynolds operator then T is a projection.*
(iii) *If T is a Reynolds operator and a projection then T is averaging.*

It is easily verified that we can replace 'projection' by 'Seever operator' in the above result. So, if we consider the identities $(\mathcal{A})$, $(\mathcal{R})$, and $(\mathcal{S})$, then every operator T which satisfies two of those identities, also satisfies the third.

Proposition 2.11. *Let A be an Archimedean semiprime f-algebra and $T : A \to A$ be a linear operator. Then the following hold.*

(i) *If T is averaging and a Reynolds operator then T is a Seever operator.*
(ii) *If T is averaging and a Seever operator then T is a Reynolds operator.*
(iii) *If T is a Reynolds and Seever operator then T is averaging.*

Next we will deal with the relationship between the Reynolds identity $(\mathcal{R})$ and the averaging identity $(\mathcal{A})$ in connection with topological properties of certain function algebras. Before doing so, we present an example – due to Scheffold [51] – of a Reynolds operator that is not averaging.

Example 2.12. (Scheffold [51]) *Consider the operator $T : C([0,1]) \to C([0,1])$ defined by*

$$T(f)(x) = \int_0^1 f(tx)\,\mathrm{d}t \quad \text{for all } f \in C(X), x \in X.$$

It is easily verified that T is a Reynolds operator. At the same time, T is of course far from being averaging.

In [50], Rota considered Reynolds operators on the space $L_\infty(S,\Sigma,m)$ with closed range in the L_1-topology and showed that they are automatically averaging.

Theorem 2.13. (Rota [50]) *Let $L_\infty(S,\Sigma,m)$ and $L_1(S,\Sigma,m)$ denote bounded measurable and integrable functions on a σ-finite measure space, respectively. Let $T : L_\infty(S,\Sigma,m) \to L_\infty(S,\Sigma,m)$ be a Reynolds operator which is continuous with respect to the L_1-topology. Then R is averaging if and only if the range of T is closed.*

Rota conjectured that Theorem 2.13 remains valid for Reynolds operators on $C(X)$ with X compact Hausdorff. In his Ph.D. thesis [44], Neeb solved Rota's conjecture.

Theorem 2.14. (Neeb [44]) *Let $T : C_0(X) \to C_0(X)$ be a continuous Reynolds operator. Then the following statements are equivalent.*

(i) *T is averaging.*
(ii) *T is a projection.*
(iii) *The range of T is closed.*

However, the following problem remains open.

Problem 2.15. *Does* Theorem 2.14 *hold for an order bounded Reynolds operator on a semiprime Archimedean f-algebra under the relative uniform topology*?

Returning to Seever's identity $(\mathcal{S})$, we note that since the publication of Seever's paper [52], the identity $(\mathcal{S})$ has been studied by many authors in connection with contractive projections. Besides the results reviewed above, we present several theorems by Hadded that deserve more interest. We begin with the following.

Proposition 2.16. (Hadded [25]) *Let A be a f-algebra with unit element and $T : A \to A$ be a Seever operator. Then T^2 is a projection and a Seever operator.*

A Seever operator T need not be a projection (although T^2 is a projection). Indeed, consider $A = \mathbb{R}^3$ with the pointwise operations and $T : A \to A$ defined by $T(x,y,z) = (0,x,z)$ for all $(x,y,z) \in A$.

At this point, let X be a compact Hausdorff space. We denote the evaluation map at a point $x \in X$ by δ_x, and the restriction of δ_x to a vector subspace B of $C(X)$ is indicated by $\delta_{x,B}$. Recall from [59] that B is said to have a *weakly separating quotient* if for every two distinct points x and y in X and for each scalar $t \neq 1$ such that $\delta_{x,B} = t\delta_{y,B}$, we have that $\delta_{x,B}$ is not an extreme point of $\{\varphi \in B' : ||\varphi|| \leq 1\}$, where B' is the norm dual space of B. In particular, the range of a positive projection has weakly separating quotient. Wulbert improved Seever's theorem (for compact X) by introducing the condition that the range of the norm-one projection T has a weakly separating quotient as follows.

Theorem 2.17. (Wulbert [59]) *Let A denote a subalgebra of $C(X)$ and let $T : A \to A$ be a norm-one projection. If the range of T has a weakly separating quotient then T is A Seever operator.*

Later in [22], Friedman and Russo gave the following example showing that the range of a Seever operator which in addition is a norm-one projection need not have a weakly separating quotient.

Example 2.18. (Friedman and Russo [22]) *Write* $X = [-2,-1] \cup [1,2]$ *and let* $\chi = \chi_{[1,2]}$ *be the characteristic function of the interval* $[1,2]$. *Define a linear operator* $T : A \to A$ *by*

$$T(f)(x) = \frac{1}{2}(\chi(x)\, f(x) - \chi(-x)\, f(-x)) \quad \textit{for all } f \in C(X), x \in X.$$

Then T *is a contractive projection and a Seever operator but the range of* T *does not have a weakly separating quotient.*

In [25], Hadded introduced the notion of an almost positive projection as follows. A projection $T : A \to A$, where A is an f-algebra, is said to be *almost positive* if there exists an order projection $\pi_T : A \to A$ such that

$$T(\pi_T(T(f))) = T(f) \quad \text{for all } f \in A$$

and

$$\pi_T(T(f)) \in A^+ \quad \text{for all } f \in A^+.$$

Of course, a positive projection is almost positive. The following proposition characterizes almost positive projections.

Proposition 2.19. (Hadded [25]) *Let* A *be an* f*-algebra and let* $T : A \to A$ *be a projection. Then* T *is almost positive if and only if there exist linear operators* $T_1, T_2 : A \to A$ *such that* $T = T_1 + T_2$, T_1 *is a positive projection given by* $T_1 = \pi T$ *for some order projection* π, *and* $T_1 T_2 = T_2^2 = 0$.

Hadded additionally linked Seever operators to almost positive projections as follows.

Theorem 2.20. (Hadded [25]) *Let* A *be a* σ*-Dedekind complete* f*-algebra with unit element and let* $T : A \to A$ *be a* σ*-order continuous contractive projection. Then* T *is a Seever operator if and only if* T *is almost positive.*

The assumption that T is σ-order continuous in the above theorem cannot be dropped as the following example shows.

Example 2.21. (Hadded [25]) *Let* A *be the Dedekind completion of* $C([-1,1])$. *Note that the Dedekind completion of* $C([-1,1])$ *equals* $C(X)$ *where* X *is the Gleason projective cover of* $[-1,1]$ (*combine Theorems* 12.9 *and* 14.18 *in* [34] *with* 10.54 *in* [57]). *Then there exists a surjective map from* X *to* $[-1,1]$ *for which no proper subset of* X *maps onto* $[-1,1]$. *Hence (using the Axiom of Choice) there exists a map* $[-1,1] \to X$ *with dense range. Composition of the latter map with elements of* $C(X)$ *yields an algebra and lattice homomorphic embedding of* A *into* $\mathbb{R}^{[-1,1]}$. *Thus we consider* A *as an* f*-subalgebra of* $\mathbb{R}^{[-1,1]}$. *Let* $T : A \to A$ *be the operator defined by*

$$T(f) = f(1)g_1 - f(-1)g_2 \quad \textit{for all } f \in A,$$

where

$$g_1(x) = \begin{cases} 0 & \text{for } -1 \le x \le 1/3 \\ \dfrac{3}{2}x - \dfrac{1}{2} & \text{for } 1/3 \le x \le 1 \end{cases}$$

and

$$g_2(x) = \begin{cases} \dfrac{4}{3}x + \dfrac{1}{3} & \text{for } -1 \le x \le 0 \\ \dfrac{-1}{3}x + \dfrac{1}{3} & \text{for } 0 \le x \le 1 \end{cases}$$

Then T is a contractive projection and it satisfies Seever's identity, but T is not almost positive.

To link Seever's identity to almost positive projections in another way, we have to recall that if A is a semiprime f-algebra then so is its order continuous bidual $(A')'_n$ with respect to the Arens multiplication [7, 28, 29]. The upward directed net $\{a_i : i \in I\}$ in A^+ is said to be an *approximate unit* if $\sup\{a_i b : i \in I\} = b$ for all $b \in A^+$. The approximate unit $[0, I] \cap A$ is said to be $\sigma(A, A')$*-bounded* if $M_f = \sup\{f(a) : a \in [0, I] \cap A\} < \infty$ for all $f \in (A')^+$.

Theorem 2.22. (Hadded [25]) *Let A be a semiprime f-algebra with separating order dual such that A has a $\sigma(A, A')$-bounded approximate unit and let $T : A \to A$ be an order bounded contractive projection. Then T is a Seever operator if and only if $T''_n : (A')'_n \to (A')'_n$ is almost positive, where T''_n is the restriction of the biadjoint T'' of T to $(A')'_n$.*

Let $T : C_0(X) \to C_0(X)$ be a contractive projection. In the proof of [22], Freedman and Russo took an order projection M on the order bidual $C_0(X)''$ which verifies $T''MT'' = T''$ and then proved that T is a Seever operator if and only if MT'' is positive (see [22]). Hence, they actually proved that T is a Seever operator if and only if T'' is almost positive. Since $C_0(X)$ satisfies the hypothesis of Theorem 2.22 and $C_0(X)'' = (C_0(X)')'_n$, the Freedman-Russo result is a consequence of Theorem 2.22.

3. Square-mean closed and geometric-mean closed f-algebras

A vector lattice E is said to be *square-mean closed* if the set

$$\mathfrak{S}(a, b) = \{(\cos x) a + (\sin x) b : x \in [0, 2\pi]\}$$

has a supremum $\mathfrak{s}(a, b)$ in E for every $a, b \in E$ [5]. Notice that if E is square-mean closed then

$$\mathfrak{s}(a, b) = \mathfrak{s}(|a|, |b|) \ge 0 \text{ for all } a, b \in E.$$

In 1968, Lotz [40] proved that any Banach lattice is square-mean closed. Three years later, Luxemburg and Zaanen [39] extended Lotz's theorem to uniformly complete vector lattices. An elementary proof of this result was obtained more than two decades ago by Beukers, Huijsmans, and de Pagter in [8]. However, a square-mean closed Archimedean vector lattice need not be uniformly complete.

For instance, the vector lattice of all step functions on the real interval $[0, 1]$ – equipped with the pointwise operations and ordering – is square-mean closed and not uniformly complete. Obviously, the f-algebra $\mathbb{R}$ of all real numbers is square-mean closed. Moreover,

$$\mathfrak{s}(a, b)^2 = a^2 + b^2 \quad \text{for all } a, b \in \mathbb{R}.$$

The latter identity extends to uniformly complete semiprime f-algebra as was proved by Beukers, Huijsmans, and de Pagter in [8]. Interestingly, their proof actually shows that the identity holds for any square-mean closed Archimedean f-algebra.

Theorem 3.1. *Let A be a square-mean closed Archimedean f-algebra. Then*

$$\mathfrak{s}(a, b)^2 = a^2 + b^2 \textit{ for all } a, b \in A.$$

If A in Theorem 3.1 is semiprime then $\mathfrak{s}(a, b)$ is the unique positive element c in A such that $c^2 = a^2 + b^2$. In fact, we can say more. First, let $N(A)$ denotes the set of all nilpotent elements of the Archimedean f-algebra A. Recall from [60] that

$$N(A) = \{a \in A : a^2 = 0\} = \{a \in A : ab = 0 \text{ for all } b \in A\}.$$

Hence, if a and b are two positive elements in an Archimedean f-algebra A then $a^2 = b^2$ if and only if $a - b \in N(A)$. This observation together with Theorem 3.1 quickly leads to the following.

Corollary 3.2. *Let A be a square-mean closed Archimedean f-algebra and $a, b, c \in A$ with $c \geq 0$. Then $c^2 = a^2 + b^2$ if and only if $c - \mathfrak{s}(a, b) \in N(A)$.*

Now we turn our attention to so-called geometric-mean closed Archimedean f-algebras. A vector lattice E is said to be *geometric-mean closed* if the set

$$\mathfrak{G}(a, b) = \left\{\frac{x}{2}a + \frac{1}{2x}b : x \in (0, \infty)\right\}$$

has an infimum $\mathfrak{g}(a, b)$ in A for every $a, b \in A^+$ [5]. We noticed above that any uniformly complete vector lattice is square-mean closed. However, uniform completeness also implies geometric-mean closedness. Indeed, every $C(X)$ is geometric-mean closed, hence so is every uniformly complete vector lattice.

Theorem 3.3. *Any uniformly complete vector lattice is geometric-mean closed.*

In particular, the f-algbra $\mathbb{R}$ is geometric-mean closed, a fact that goes back to the lever of Archimedes, and

$$\mathfrak{g}(a, b)^2 = ab \text{ for all } a, b \in \mathbb{R}^+ = [0, \infty).$$

Next, we prove that this equality holds in any geometric-mean closed Archimedean f-algebra.

Theorem 3.4. *Let A be a geometric-mean closed Archimedean f-algebra. Then*

$$\mathfrak{g}(a, b)^2 = ab \textit{ for all } a, b \in A^+.$$

Proof. Let $a, b \in A^+$ and notice that, by Lemma 4.1 in [5],

$$\mathfrak{g}(a,b) = \mathfrak{g}(a \vee b, a \wedge b).$$

Moreover,

$$(a \vee b)(a \wedge b) = ab.$$

Hence we may assume that $a \geq b$. Observe now that

$$\mathfrak{g}(a,b)^2 = \frac{1}{4} \inf \left\{ \left(xa + \frac{1}{x} b \right)^2 : x \in (0, \infty) \right\},$$

since the multiplication in A is order continuous. Thus

$$4\left(\mathfrak{g}(a,b)^2 - ab\right) = \inf \left\{ \left(xa - \frac{1}{x} b \right)^2 : x \in (0, \infty) \right\} \geq 0.$$

For convenience, put

$$c := \inf \left\{ \left(xa - \frac{1}{x} b \right)^2 : x \in (0, \infty) \right\}.$$

Take $n \in \{1, 2, \dots\}$ and $k \in \{1, 2, \dots, n\}$. We find that

$$0 \leq c \leq \left(\sqrt{\frac{n}{k}} a - \sqrt{\frac{k}{n}} b \right)^2 = \frac{n}{k} \left(a - \frac{k}{n} b \right)^2 \leq n \left(a - \frac{k}{n} b \right)^2.$$

It follows from Proposition 4.1 in [8] that

$$0 \leq c \leq n \inf \left\{ \left(a - \frac{k}{n} b \right)^2 : k \in \{1, 2, \dots, n\} \right\} \leq \frac{1}{n} b^2.$$

But then $c = 0$ because A is Archimedean and the proof is complete. □

Recall that if a and b are two positive elements in an Archimedean f-algebra A then $a^2 = b^2$ if and only if $a - b \in N(A)$. This leads to the following 'geometric-mean' version of a similar 'square-mean' version above.

Corollary 3.5. *Let A be a geometric-mean closed Archimedean f-algebra and a, b, $c \in A^+$. Then $c^2 = ab$ if and only if $c - \mathfrak{g}(a,b) \in N(A)$.*

We arrive in particular at the fact that if a and b are two positive elements in a geometric-mean closed semiprime Archimedean f-algebra A then $\mathfrak{g}(a,b)$ can be defined as the unique (positive) square-root of ab (compare with Theorem 4.2 in [8]).

In view of Theorems 3.1 and 3.4, and the identity

$$a^2 + b^2 = (a+b)^2 - 2ab$$

which holds for all a, b in the Archimedean f-algebra A, we may also expect that any geometric-mean closed Archimedean f-algebra is square-mean closed. Indeed, this follows from Theorem 4.4 in [5].

Theorem 3.6. *A geometric-mean closed Archimedean f-algebra is square-mean closed.*

We observe here that the identity

$$\mathfrak{s}(a,b)^2 = \left(a+b+\frac{\sqrt{2}}{2}\mathfrak{g}(a,b)\right)\left(a+b-\frac{\sqrt{2}}{2}\mathfrak{g}(a,b)\right)$$

holds for all positive elements a,b in a geometric-mean closed Archimedean f-algebra. Reflecting on that formula, it is natural to ask whether the converse of Theorem 3.6 holds. The answer is no, i.e., there exists a square-mean closed Archimedean f-algebra which is not geometric-mean closed. To that end we give the following example from [5].

Example 3.7. *Let $C(\mathbb{R}^+)$ be the Archimedean f-algebra of all real-valued continuous functions on $\mathbb{R}^+ = [0,\infty)$ and P be the vector subspace of $C(\mathbb{R}^+)$ consisting of all polynomial functions. Define for each $n \in \mathbb{N} = \{1,2,\dots\}$ a vector subspace A_n of $C(\mathbb{R}^+)$ by induction as follows. Let $A_1 = P$ and for each $n \in \mathbb{N}$ let A_{n+1} be the vector subspace of $C(\mathbb{R}^+)$ generated by*

$$A_n \cup \left\{\left(a^2+b^2\right)^{\frac{1}{2}} : a,b \in A_n\right\}$$

We claim that A_n is a subalgebra of $C(\mathbb{R}^+)$ for all $n \in \mathbb{N}$. To this end, we argue by induction. The result being trivial for A_1, let $n \in \mathbb{N}$ and assume that A_n is a subalgebra of $C(\mathbb{R}^+)$. Clearly, to show that A_{n+1} is a subalgebra of $C(\mathbb{R}^+)$, it suffices to prove that

$$a\left(b^2+c^2\right)^{\frac{1}{2}} \in A_{n+1} \text{ and } \left(a^2+b^2\right)^{\frac{1}{2}}\left(c^2+d^2\right)^{\frac{1}{2}} \in A_{n+1} \text{ for all } a,b,c \in A_n$$

Let $a,b,c,d \in A_n$ and put $u = (a+\mathbf{1})^2$ and $v = u-a$. Since A_n is a subalgebra of $C(\mathbb{R}^+)$, we get $0 \le u,v \in A_n$ and

$$a\left(b^2+c^2\right)^{\frac{1}{2}} = (u-v)\left(b^2+c^2\right)^{\frac{1}{2}} = \left((ub)^2+(uc)^2\right)^{\frac{1}{2}} - \left((vb)^2+(vc)^2\right)^{\frac{1}{2}} \in A_{n+1}.$$

On the other hand,

$$\left(a^2+b^2\right)^{\frac{1}{2}}\left(c^2+d^2\right)^{\frac{1}{2}} = \left((ac+bd)^2+(ad-bc)^2\right)^{\frac{1}{2}} \in A_{n+1}.$$

Accordingly, the union

$$A = \bigcup_{n\in\mathbb{N}} A_n$$

is a subalgebra of $C(\mathbb{R}^+)$. Furthermore, if $a \in A$ then there exists $n \in \mathbb{N}$ such that $a \in A_n$. Hence,

$$|a| = \left(a^2+0^2\right)^{\frac{1}{2}} \in A_{n+1} \subset A$$

and A is a vector sublattice of $C(\mathbb{R}^+)$. In summary, A is an Archimedean f-algebra with respect to the pointwise operations and ordering.

To show that A is square-mean closed, let $a,b \in A^+$ and choose $n \in \mathbb{N}$ such that $a,b \in A_n$. Observe that $\left(a^2+b^2\right)^{\frac{1}{2}}$ is the supremum in $C(\mathbb{R}^+)$ of $\mathfrak{S}(a,b)$.

But then the equality

$$\mathfrak{s}(a,b) = \sup \mathfrak{S}(a,b) = \left(a^2+b^2\right)^{\frac{1}{2}}$$

holds in A *because* $\left(a^2+b^2\right)^{\frac{1}{2}} \in A_{n+1} \subset A$. *Thus* A *is square-mean closed.*

Now, we prove by induction that all functions in A *are differentiable at* 0. *Any element of* A_1 *is a polynomial, hence differentiable at* 0. *Let* $n \in \mathbb{N}$ *and assume that all functions in* A_n *are differentiable at* 0. *Pick* $a \in A_{n+1}$ *and write*

$$a = b + \sum_{k=1}^{m} \lambda_k \left(a_k^2 + b_k^2\right)^{\frac{1}{2}}$$

for some $b, a_1, b_1, \ldots, a_m, b_m \in A_n$ *and* $\lambda_1, \ldots, \lambda_m \in \mathbb{R}$. *By the induction hypothesis,* b *and all* a_k, b_k *are differentiable at* 0. *Then so is* $\left(a_k^2+b_k^2\right)^{\frac{1}{2}}$. *It follows that* a *is differentiable at* 0.

Finally, we show that A *is not geometric-mean closed. We argue by contradiction. Let* e *and* u *be the functions in* $C(\mathbb{R}^+)$ *defined respectively by* $e(x) = x$ *and* $u(x) = 1$ *for all* $x \in \mathbb{R}^+$. *Clearly,* $e, u \in A$. *Assume that* $\mathfrak{G}(e,u)$ *has an infimum* $\mathfrak{s}(e,u)$ *in* A. *But* $\mathfrak{G}(e,u)$ *has an infimum in* $C(\mathbb{R}^+)$. *Indeed,*

$$b = \inf \mathfrak{G}(e,u) \ \text{ in } C\left(\mathbb{R}^+\right),$$

where $b(t) = t^{\frac{1}{2}}$ *for all* $t \in \mathbb{R}^+$. *Since* A *is uniformly dense in* $C(\mathbb{R}^+)$, *we get* $b \leq \mathfrak{s}(e,u)$. *Let* $t \in \mathbb{R}^+$ *and observe that*

$$\mathfrak{s}(e,u)(t) \leq xt + x^{-1} \text{ for all } x \in (0,\infty).$$

That is,

$$\mathfrak{s}(e,u)(t) \leq t^{\frac{1}{2}} = b(t)$$

It follows that $\mathfrak{s}(e,u) \leq b$. *Consequently,* $b = \mathfrak{s}(e,u) \in A$. *This contradicts the fact that all functions in* A *are differentiable at* 0.

We derive that A *is an example of an Archimedean* f*-algebra which is square-mean closed but not geometric-mean closed (notice that* A *is even unital).*

Remark 3.8. Interesting as the previous example is, after completing this survey, van Rooij (private communication) pointed out the following much easier and more elegant example.

Example 3.9. *Let* A *be the the Archimedean* f*-algebra of all Lipschitz functions on* $[0,1]$. *We denote the constant function one, the unit in* A, *by* $\mathbf{1}$. *For* $a, b \in A$, *we consider the complex-valued function* $f = a + ib$. *Then* $|f| = (a^2+b^2)^{\frac{1}{2}}$. *Moreover, for* $s, t \in [0,1]$ *it follows that* $||f(s)| - |f(t)|| \leq |f(s) - f(t)| \leq |a(s) - a(t)| + |b(s) - b(t)|$, *hence,* $(a^2+b^2)^{\frac{1}{2}} \in A$. *Thus,* A *is square-mean closed. Of course,* A *is not geometric-mean closed, because* $\sqrt{\mathbf{1}.e} = \sqrt{e}$ *is not in* A, *where* e *is the identity function.*

At the end of this section we remark once more that the geometric mean as studied above gives rise to a concrete construction of what is called the square of a vector lattice. In turn, the square of a vector lattice plays a fundamental role in understanding bilinear maps that are order bounded and separately disjointness preserving. Finally, the construction plays a role in understanding orthosymmetric bilinear maps, i.e., bilinear maps $T : E \times E \to F$ for vector lattices E, F with the property $T(a, b) = 0$ when a and b are disjoint. For more information about squares of vector lattices, we refer the reader to the survey by Bu, Buskes, and Kusarev on page 97 in this volume.

4. Maximal rings of quotients and (extended) orthomorphisms

We will first discuss so-called extended orthomorphisms. Let L be an Archimedean vector lattice. Luxemburg and Schep in [38] defined an order bounded linear operator $\pi : D_\pi \to L$, where D_π is an order dense order ideal in L, to be an *extended orthomorphism* of L if $|a| \wedge |b| = 0$ in D_π implies $|\pi(a)| \wedge |b| = 0$ in L. An extended orthomorphism π of L is called an *orthomorphism* of L if $D_\pi = L$. A natural equivalence relation can be introduced in the set of all extended orthomorphisms of L as follows. Two extended orthomorphisms of L are equivalent whenever they agree on an order dense order ideal in L or, equivalently, they are equal on the intersection of their domains. The intersection of two order dense order ideals in L is of course again an order dense order ideal in L. The set of all equivalence classes of extended orthomorphisms of L is denoted by $\mathrm{Orth}^\infty(L)$. With respect to the pointwise addition, scalar multiplication, and ordering, $\mathrm{Orth}^\infty(L)$ is an Archimedean vector lattice. The lattice operations in the vector lattice $\mathrm{Orth}^\infty(L)$ are given pointwise. It turns out that the vector lattice $\mathrm{Orth}^\infty(L)$ is an f-algebra with respect to composition as multiplication. Moreover, since extended orthomorphisms (and hence orthomorphisms) are order continuous, the set $\mathrm{Orth}(L)$ of all orthomorphisms of L can be embedded naturally in $\mathrm{Orth}^\infty(L)$ as an f-subalgebra. Obviously, the identity operator I_L of L serves as unit element in $\mathrm{Orth}^\infty(L)$ and in $\mathrm{Orth}(L)$. We summarize these facts in the following result, due to Luxemburg and Schep in [38].

Theorem 4.1. (Luxemburg and Schep [38]) *Let L be an Archimedean vector lattice. Then the following hold.*

(i) $\mathrm{Orth}^\infty(L)$ *is an Archimedean f-algebra with I_L as a unit element.*
(ii) $\mathrm{Orth}(L)$ *is an f-subalgebra of* $\mathrm{Orth}^\infty(L)$ *with I_L as a unit element.*

The algebraic properties and order structure of orthomorphisms had also been investigated earlier by Bigard and Keimel in [9], and by Conrad and Diem in [20]. Observe now that the f-algebra $\mathrm{Orth}^\infty(L)$ is commutative since it is Archimedean. Furthermore, due to in de Pagter [46], if L is uniformly complete then $\mathrm{Orth}^\infty(L)$ is von Neumann regular. We remind the reader that a commutative ring R is said to be *von Neumann regular* if for every $r \in R$ there exists $s \in R$ such that $r = r^2 s$.

Theorem 4.2. (de Pagter [46]) *If L is a uniformly complete vector lattice then* $\mathrm{Orth}^\infty(L)$ *is von Neumann regular.*

Next we turn to the maximal ring of quotients of a commutative semiprime ring. Our principal reference on the subject is the classical monograph [37] by Lambek. Let R be a commutative ring and assume that R in addition is semiprime, that is, 0 is the only nilpotent element in R. A ring ideal D of R is said to be *dense* in R if $r = 0$ whenever $r \in R$ and $rd = 0$ for all $d \in D$. Observe that the intersection of two dense ring ideals in R is again a dense ring ideal in R. A mapping $\pi : D_\pi \to R$, where D_π is a dense ring ideal in R, is called *fraction* of R if π is *R-linear*, that is to say, $\pi(c+d) = \pi(c) + \pi(d)$, $\pi(c-d) = \pi(c) - \pi(d)$, and $\pi(rd) = r\pi(d)$ for all $r \in R, c, d \in D_\pi$. Two fractions of R are identified if they coincide on some dense ring ideal of R. An obvious equivalence relation is thus obtained on the set of all fractions of R. The set of all equivalence classes is denoted by $Q(R)$ and called *the maximal ring of quotients of* R. Clearly, $Q(R)$ may be given a ring structure by defining addition and multiplication pointwise on the intersections of domains. Furthermore, $Q(R)$ is commutative and, since R is semiprime, it is von Neumann regular [37]. There is a natural and canonical embedding of R into $Q(R)$, and we accordingly regard R as a subring of $Q(R)$. Moreover, if S is a ring of which the elements are fractions of R then there exists a one-to-one ring homomorphism of S into $Q(R)$ that is induced by the canonical embedding of R into $Q(R)$. Less formally, S can be considered as a subring of $Q(R)$. For this reason, Utumi in [56] has called $Q(R)$ the *maximal* (or *complete*) *ring of quotients* of R (see also [6] by Banaschewski and [41] by Martinez).

Theorem 4.3. (Anderson [1]) *Let A be an Archimedean f-algebra with unit element e. Then the following hold.*

(i) *$Q(A)$ is an Archimedean von Neumann regular f-algebra with e as a unit element.*
(ii) *A is an f-subalgebra of $Q(A)$.*

Now, let A be an Archimedean semiprime f-algebra and consider the linear operator $\iota : A \to \operatorname{Orth}(A)$ defined by

$$\iota(a)(x) = ax \quad \text{for all } a, x \in A.$$

Obviously, ι is a one-to-one lattice and ring homomorphism. Furthermore, it is not hard to see that the range of ι is a ring ideal in $\operatorname{Orth}(A)$. In summary, the elements of $\operatorname{Orth}(A)$ can be considered as fractions of A. It follows that $\operatorname{Orth}(A)$ is (after suitable identifications) an f-subalgebra of $Q(A)$. But then $Q(\operatorname{Orth}(A))$ is contained in $Q(A)$ since elements in $Q(\operatorname{Orth}(A))$ are clearly fractions of A. We derive that $Q(A) = Q(\operatorname{Orth}(A))$. The latter equality together with Theorem 4.3 leads to the following.

Corollary 4.4. *Let A be an Archimedean semiprime f-algebra. Then the following hold.*

(i) *$Q(A)$ is an Archimedean von Neumann f-algebra with unit element.*
(ii) $\operatorname{Orth}(A)$ *(and then A) is an f-subalgebra of $Q(A)$.*

The definition of $\mathrm{Orth}^{\infty}(L)$ for an Archimedean vector lattice L is of course somewhat analogous to the definition of $Q(R)$ for a commutative semiprime ring R. When we add to this the many properties that $\mathrm{Orth}^{\infty}(A)$ and $Q(A)$ share when A is an Archimedean semiprime f-algebra A, one suspects that the two objects are isomorphic. Unfortunately, this is not true in general. An example in this direction is provided by de Pagter in [46].

Example 4.5. (de Pagter [46]) *Let A be the set of all real-valued continuous functions on the real interval $[0,1]$ which are piecewise polynomial. Clearly, A is an Archimedean unital (and then semiprime) f-algebra with respect to the pointwise operation and ordering. Define $a \in A$ by*

$$a(t) = 1 + t \quad \text{for all } t \in [0,1]$$

and $\pi : A \to A$ by

$$\pi(x)(t) = (ax)(t) = a(t)\,x(t) \quad \text{for all } x \in A, t \in [0,1].$$

Clearly, $\pi \in \mathrm{Orth}^{\infty}(A)$ and $\pi \in Q(A)$. The principal ring ideal $aA = \{ax : x \in A\}$ is dense in A. Consider the fraction $\sigma : aA \to A$ defined by

$$\sigma(ax) = x \quad \text{for all } x \in A,$$

that is, σ is the multiplication by the function $1/a$. Obviously, σ is the inverse of π in $Q(A)$. However, one can prove by contradiction that π does not have an inverse in $\mathrm{Orth}^{\infty}(A)$.

In spite of de Pagter's example, $\mathrm{Orth}^{\infty}(A)$ can be embedded in $Q(A)$ as an f-subalgebra. This result was proved by de Pagter [46] in case that A has a unit and Wickstead [58] extended that to Archimedean semiprime f-algebras.

Theorem 4.6. (Wickstead [58]) *Let A be an Archimedean semiprime f-algebra. Then $\mathrm{Orth}^{\infty}(A)$ is an f-subalgebra of $Q(A)$.*

Though we know from de Pagter's example that the converse of Theorem 4.6 fails, Wickstead in [58] proved that the maximal ring of quotients can, in fact, be viewed as consisting of some kind of orthomorphisms. Indeed, an order bounded linear operator $\pi : D_{\pi} \to L$, where D_{π} is an order dense vector sublattice of L, is called a *weak orthomorphism* of L if $|a| \wedge |b| = 0$ in D_{π} implies $|\pi(a)| \wedge |b| = 0$ in L. Hence, a weak orthomorphism of L is an extended orthomorphism of L if and only if D_{π} is an order dense order ideal in L. Unlike extended orthomorphisms, weak orthomorphisms do not, in general, have an additive structure (see [58]). Fortunately, this 'bad' behavior is absent in the case of an Archimedean semiprime f-algebra A. Indeed, amongst those extensions of weak orthomorphisms on A, which are again weak orthomorphisms of A, there is one which has a largest domain. The set of all weak orthomorphisms of A which have maximal domain is denoted by $\mathrm{Orth}^{w}(A)$. It turns out that pointwise operations and ordering make $\mathrm{Orth}^{w}(A)$ into an Archimedean f-algebra with unit element. Actually, we have more.

Theorem 4.7. (Wickstead [58]) *Let A be an Archimedean semiprime f-algebra. Then the following hold.*

(i) $\operatorname{Orth}^{w}(A)$ *is an Archimedean von Neumann regular f-algebra with I_A as a unit element.*

(ii) $\operatorname{Orth}^{\infty}(A)$ *(and hence* $\operatorname{Orth}(A)$*) is an f-subalgebra of* $\operatorname{Orth}^{w}(A)$.

In particular, $\operatorname{Orth}^{w}(A)$ is commutative and has positive squares. The upshot of it all is that $Q(A)$ can indeed be identified with $\operatorname{Orth}^{w}(A)$.

Theorem 4.8. (Wickstead [58]) If *A is an Archimedean semiprime f-algebra then* $Q(A) = \operatorname{Orth}^{w}(A)$.

Under the extra condition of uniform completeness, the extended orthomorphisms and the maximal ring of quotients coincide as well. This result is also due to Wickstead in [58] and, in the unital case, to de Pagter in [46]. In summary, we have the following theorem, the last result of this section.

Theorem 4.9. (Wickstead [58]) *Let A be a uniformly complete semiprime f-algebra A. Then $Q(A) = \operatorname{Orth}^{\infty}(A) = \operatorname{Orth}^{w}(A)$.*

5. Order bounded disjointness preserving operators

Let L and M be vector lattices. A (linear) operator $T : A \to B$ is said to be *disjointness preserving* if $|T(a)| \wedge |T(b)| = 0$ for all $a, b \in A$ with $|a| \wedge |b| = 0$. If A and B are Archimedean semiprime f-algebras, then the operator $T : A \to B$ is disjointness preserving if and only if T is *separating*, meaning that, $T(a)T(b) = 0$ in B whenever $ab = 0$ in A.

In 1983, Arendt proved in [2] that if X and Y are compact Hausdorff spaces and $T : C(X) \to C(Y)$ is an order bounded disjointness preserving operator (a *Lamperti operator* in Arendt's terminology) then T is a weighted composition operator. First, let $\operatorname{coz}(w)$ denote the cozero-set of a real-valued function w on Y, i.e.,

$$\operatorname{coz}(w) = \{y \in Y : w(y) \neq 0\},$$

and denote by $\mathbf{1}$ the function identically equal to one on X.

Theorem 5.1. (Arendt [2]) *Let X and Y be compact Hausdorff spaces. An order bounded operator $T : C(X) \to C(Y)$ is disjointness preserving if and only if there exists a map $h : Y \to X$ such that*

$$T(a)(y) = T(\mathbf{1})(y)\, a(h(y)) \quad \text{for all } a \in C(X), y \in Y.$$

Furthermore, h is continuous and uniquely determined on $\operatorname{coz}(T(\mathbf{1}))$.

We point out that Jarosz in [32] independently obtained Arendt's result. We now look at Theorem 5.1 from a more algebraic point of view. For every $a \in C(X)$, the function $S(a)$ defined by

$$S(a)(y) = 0 \text{ if } y \notin \operatorname{coz}(T(\mathbf{1})) \quad \text{and} \quad S(a)(y) = a(h(y)) \text{ if } y \in \operatorname{coz}(T(\mathbf{1}))$$

need not be a member of $C(Y)$. But $S(a)$ naturally is an element of the maximal ring of quotients $Q(C(Y))$ of $C(Y)$ [26]. Another version of Arendt's result thus arises as follows. If $T : C(X) \to C(Y)$ is an order bounded disjointness preserving operator, then there exists a lattice and ring homomorphism $S : C(X) \to Q(C(Y))$ such that $T(a) = T(\mathbf{1}) S(a)$ for all $a \in C(X)$. Recently Boulabiar proved in [13] that the latter version is true for arbitrary Archimedean unital f-algebras.

Theorem 5.2. (Boulabiar [13]) *Let A and B be Archimedean f-algebras with unit elements. A ordered bounded operator $T : A \to B$ is disjointness preserving if and only if there exists a lattice and ring homomorphism $S : A \to Q(A)$ such that*

$$T(a) = T(e) S(a) \quad \text{for all } a \in A,$$

where e indicates the unit element of A.

The following $C_0(X)$-version of Theorem 5.1 was proved by Jeang and Wong in [33].

Theorem 5.3. (Jeang-Wong [33]) *Let X and Y be locally compact Hausdorff spaces. An order bounded operator $T : C_0(X) \to C_0(Y)$ is a disjointness preserving if and only if there exist a function $w : Y \to \mathbb{R}$, which is continuous on $\operatorname{coz}(w)$, and a function $h : Y \to X$ such that*

$$T(a)(y) = w(y) a(h(y)) \quad \text{for all } a \in C_0(X), y \in Y$$

Moreover, h is continuous and uniquely determined on $\operatorname{coz}(w)$.

One might hope that Theorem 5.3 can be obtained from Theorem 5.1 by extending an order bounded disjointness preserving operator $T : C_0(X) \to C_0(Y)$ to an order bounded disjointness preserving operator $T^\alpha : C(\alpha X) \to C(\alpha Y)$, where αX denotes the one-point compactification of X. However, Jeang and Wong [33] provided the following example of an order bounded disjointness preserving operator T which does not have any such extensions.

Example 5.4. (Jeang-Wong [33]) *Let $X = \mathbb{R}^+$ and $Y = \mathbb{R}$ with the usual topology and define $w, h : \mathbb{R} \to \mathbb{R}$ by*

$$w(y) = \begin{cases} 1 & \text{if } \ y > 2 \\ y - 1 & \text{if } \ 0 \le y \le 2 \\ -1 & \text{if } \ y < 0 \end{cases} \qquad \text{and} \quad h(y) = \begin{cases} y & \text{if } \ y \ge 0 \\ -y & \text{if } \ y < 0. \end{cases}$$

The weighted composition operator $T : C_0(X) \to C_0(Y)$ defined by

$$T(a)(y) = w(y) a(h(y)) \quad \text{for all } a \in C(X), y \in Y$$

is an order bounded disjointness preserving operator. But no order bounded linear extension $T^a : C(\alpha X) \to C(\alpha Y)$ of T can be disjointness preserving.

Now, let X and Y be completely regular spaces. In [3], Araujo, Beckenstein, and Narici proved that if $T : C(X) \to C(Y)$ is a bijective disjointness preserving operator and if the inverse operator T^{-1} of T preserves disjointness as well (such an operator T is said to be *biseparating* in [3]), then T is a weighted composition operator and the realcompactification υX of X is homeomorphic to the realcompactification υY of Y (see the classical book [23] for realcompactification of completely regular spaces).

Theorem 5.5. (Araujo-Beckenstein-Narici [3]) *Let X and Y be completely regular topological spaces and $T : C(X) \to C(Y)$ be a bijective disjointness preserving operator such that T^{-1} also preserves disjointness. Then there exist an homeomorphism $h : \upsilon Y \to \upsilon X$ such that*

$$T(a)(y) = T(\mathbf{1})(y)\, a(h(y)) \quad \textit{for all } a \in C(X), y \in Y.$$

In Theorem 5.5, the composition operator $S : C(X) \to C(Y)$ defined by $S(f) = a \circ h$ for all $a \in C(Y)$ is obviously a lattice and ring isomorphism. Hence, Theorem 5.5 can be stated more algebraically as follows. If $T : C(X) \to C(Y)$ is a bijective disjointness preserving operator with T^{-1} disjointness preserving then there exist a lattice and ring isomorphism $S : C(X) \to C(Y)$ such that

$$T(a) = T(\mathbf{1})\, S(a) \quad \text{for all } a \in C(X).$$

This algebraic version of Theorem 5.5 was obtained by Boulabiar, Buskes, Henriksen in [12] for the more general setting of unital Archimedean f-algebras.

Theorem 5.6. (Boulabiar-Buskes-Henriksen [12]) *Let A and B be Archimedean f-algebras with unit elements. If T is an order bounded disjointness preserving operator $T : A \to B$ with T^{-1} disjointness preserving, then there exists a lattice and ring isomorphism $S : A \to B$ such that*

$$T(a) = T(e)\, S(a) \quad \textit{for all } a \in A,$$

where e denotes the unit element of A.

Bijective disjointness preserving operators on $C_0(X)$-algebras have been studied by Jeang and Wong in [33]. They obtained the following.

Theorem 5.7. (Jeang-Wong [33]) *Let X and Y be locally compact Hausdorff spaces and let $T : C_0(X) \to C_0(Y)$ be a bijective disjointness preserving operator. Then there exist $w \in C_b(Y)$ and an homeomorphism $h : Y \to X$ such that*

$$T(a)(y) = w(y)\, a(h(y)) \quad \textit{for all } a \in C_0(X), y \in Y.$$

Notice that in Theorem 5.7, the operator under consideration is not assumed to be order bounded. Actually, the hypotheses imply automatic order boundedness. This is a particular case of a result by Huijsmans and de Pagter to the effect that any invertible disjointness preserving operator between two Banach lattices is bounded. In [14], Boulabiar and Buskes gave alternative proofs of Theorems 5.5 and 5.7 based on the following theorem by Hart [27]. The vector sublattice of a vector lattice M generated by a subset E of M is denoted by $\mathfrak{R}(E)$.

Theorem 5.8. (Hart [27]) *Let L and M be Archimedean vector lattices and T be an order bounded disjointness preserving operator $T : L \to M$. Then there exists a lattice and ring homomorphism $\widetilde{T} : \mathrm{Orth}\,(L) \to \mathrm{Orth}\,(\mathfrak{R}\,(T\,(L)))$ such that*

$$\widetilde{T}\,(\pi)\,(T\,(a)) = T\,(\pi\,(a)) \quad \textit{for all}\ \pi \in \mathrm{Orth}\,(L)\,, a \in L.$$

Theorem 5.8 leads to the following nice application in [27] to f-algebras. Once more we recall to the reader that if A is an Archimedean semiprime f-algebra then A can be embedded in the unital f-algebra $\mathrm{Orth}\,(A)$ of all orthomorphisms of A as an f-subalgebra and a ring ideal. This identification is taken into consideration below without further ado.

Corollary 5.9. *Let A and B be Archimedean semiprime f-algebras and let T be a bijective order bounded disjointness preserving operator from A onto B. Then there exists a unique algebra and lattice isomorphism $\widetilde{T}$ from $\mathrm{Orth}\,(A)$ onto $\mathrm{Orth}\,(B)$ such that*

$$T\,(fg) = Tf\widetilde{T}g \qquad\qquad (f, g \in A)\,.$$

6. A new representation theorem

Let A be an Archimedean semiprime f-algebra. For $n > 1$ we say that A *is n^{th}-root closed* if for every $f \in A^+$ there exists an element $f^{\frac{1}{n}} \in A^+$ such that $\left(f^{\frac{1}{n}}\right)^n = f$.

The following new theorem implies all of the results about representation of order bounded disjointness preserving operators cited in the previous section. We remind the reader that B^{ru} stands for the uniform completion of B as defined by Quinn in [47].

Theorem 6.1. *Let $n > 1$. Let A be an nth-root closed semiprime f-algebra and let B be a semiprime f-algebra. If $T : A \to B$ is an order bounded disjointness preserving operator then there exist an algebra and lattice homomorphism $S : \mathrm{Orth}\,(A) \to \mathrm{Orth}^{\infty}\,(\mathrm{Orth}\,(B^{\mathrm{ru}}))$ and an element $w \in \mathrm{Orth}^{\infty}\,(\mathrm{Orth}\,(B^{\mathrm{ru}}))$ such that*

$$T(f) = wS(f) \qquad \textit{for all}\ f \in A.$$

The proof consists of three ingredients. It heavily relies on the beautiful theorem by Hart above. Secondly, we need the following result by Buskes and van Rooij about orthosymmetric maps (for the definition of which we refer back to the end of Section 2), introduced in [16].

Theorem 6.2. (Buskes-Van Rooij [16]) *Every orthosymmetric map is symmetric.*

And thirdly, we need the following extension theorem by Buskes and van Rooij in [15].

Theorem 6.3. (Buskes-Van Rooij [15]) *An orthomorphism on a majorizing vector sublattice extends uniquely to an orthomorphism on the whole space.*

We will give an example to show that not for all semiprime f-algebras A a representation like the one above is valid, even when B equals the real numbers and T is a lattice homomorphism. The proof of Theorem 6.1 is now in order.

We start with the following lemma which easily follows from Theorem 6.2 above.

Lemma 6.4. *Let A and B be Archimedean semiprime f-algebras. If $p : A\times A \to B$ is an orthosymmetric map and a, b, and c are elements of A then $p(ab, c) = p(a, bc)$.*

Before we give the proof of our Theorem 6.1, we remark that, by Theorem 4.9, we could alternatively employ (as is also evident from our proof below) the maximal ring of quotients $Q(\mathrm{Orth}(B^{\mathrm{ru}}))$ instead of $\mathrm{Orth}^{\infty}(\mathrm{Orth}(B^{\mathrm{ru}}))$. Now the proof of the main theorem.

Proof of Theorem 6.1. Step 1. We first construct the lattice and algebra homomorphism $S : \mathrm{Orth}(A) \to \mathrm{Orth}^{\infty}(\mathrm{Orth}(B^{\mathrm{ru}}))$. By Hart's theorem 5.8, for every $\pi \in \mathrm{Orth}(A)$ there exists a unique $\widetilde{\pi}$ in $\mathrm{Orth}(\mathcal{R}(T(A)))$ such that

$$\widetilde{\pi}T = T\pi. \tag{1}$$

We denote by $\mathcal{I}(T(A))$ the order ideal generated by $\mathcal{R}(T(A))$ in B^{ru}.

By the Buskes-van Rooij Theorem 6.3, $\widetilde{\pi}$ extends uniquely to an element of $\mathrm{Orth}(\mathcal{I}(T(A)))$. This extension is again called $\widetilde{\pi}$. We extend $\widetilde{\pi}$ once more to an element $S(\pi)$ of $\mathrm{Orth}\left(\mathcal{I}(T(A)) \oplus \mathcal{I}(T(A))^{\mathrm{d}}\right)$ defined by

$$S(\pi)(f) = 0 \qquad \text{for all } f \in \mathcal{I}(T(A))^{\mathrm{d}},$$

where $\mathcal{I}(T(A))^{\mathrm{d}}$ denotes the disjoint complement of $\mathcal{I}(T(A))$ in B^{ru}. We consider $S(\pi)$ as an element of $\mathrm{Orth}^{\infty}(\mathrm{Orth}(B^{\mathrm{ru}}))$. The map S that sends $\pi \in \mathrm{Orth}(A)$ to $S(\pi) \in \mathrm{Orth}^{\infty}(\mathrm{Orth}(B^{\mathrm{ru}}))$ clearly is a lattice and algebra homomorphism.

Step 2. We now show that the equality

$$T(fg) = T(f)\,S(g)$$

holds in $\mathrm{Orth}^{\infty}(\mathrm{Orth}(B^{\mathrm{ru}}))$ for all $f, g \in A$. Take $f, g \in A$ and consider $\pi_f \in \mathrm{Orth}(A)$, the multiplication by f. According to (1), an identity which we henceforth consider in $\mathrm{Orth}^{\infty}(\mathrm{Orth}(B^{\mathrm{ru}}))$, and using the identifications made in the previous section, we obtain

$$S(f)\,T(g) = S(\pi_f)(T(g)) = \widetilde{\pi_f}T(g) = T\pi_f(g) = T(fg).$$

Step 3. In this step we construct the weight w. Let $f \in A^+$ and consider

$$w_f = \left(\frac{\left(T\left(f^{\frac{1}{n}}\right)\right)^n}{T(f)}\right)^{\frac{1}{n-1}}$$

as an element of the formal ring of quotients $q(\mathrm{Orth}(B^{\mathrm{ru}}))$ of $\mathrm{Orth}(B^{\mathrm{ru}})$. This w_f naturally is an element of the maximal ring of quotients $Q(\mathrm{Orth}(B^{\mathrm{ru}}))$ of $\mathrm{Orth}(B^{\mathrm{ru}})$. By Theorem 4.9, w_f is an element of $\mathrm{Orth}^{\infty}(\mathrm{Orth}(B^{\mathrm{ru}}))$.

For $g \in A^+$ we claim that

$$T\left(f^{\frac{1}{n}}\right)^n T(g) = T\left(g^{\frac{1}{n}}\right)^n T(f).$$

To this end, we define for a given $u \in A^+$ the positive bilinear map $\varphi : A \times A \to \operatorname{Orth}^{\infty}(\operatorname{Orth}(B^{\mathrm{ru}}))$ by

$$\varphi(x, y) = T(xu)\, T(y) \qquad \text{for all } x, y \in A.$$

Let $x, y \in A$ such that $x \wedge y = 0$ and observe that $(xu) \wedge y = 0$ so

$$T(xu) \wedge T(y) = 0.$$

Thus

$$\varphi(x, y) = T(xu)\, T(y) = 0$$

and φ is orthosymmetric. It follows by Theorem 6.2 that φ is symmetric and from Lemma 6.4 that

$$T(xu)\, T(y) = T(x)\, T(yu).$$

Therefore,

$$T\left(f^{\frac{1}{n}}\right)^n T(g) = T\left(f^{\frac{1}{n}}\right) \cdots T\left(f^{\frac{1}{n}}\right) T\left(g^{\frac{1}{n}} \cdots g^{\frac{1}{n}}\right) = T\left(g^{\frac{1}{n}}\right)^n T(f),$$

and $w_f = w_g$. Putting $w = w_f$, we have now proved that

$$T(f) = wS(f) \qquad \text{for all } f \in A^+$$

and hence also

$$T(f) = wS(f) \qquad \text{for all } f \in A.$$

Corollary 6.5. *If T is in addition surjective then S maps $\operatorname{Orth}(A)$ to $\operatorname{Orth}(B)$ and $w \in \operatorname{Orth}(B)$. If T is bijective then S maps A to B and w is invertible.* □

Proof. Assume that T is surjective. That S maps $\operatorname{Orth}(A)$ to $\operatorname{Orth}(B)$ is obvious. Remark that

$$wT(f^2) = w^2 S(f^2) = w^2 S(f)^2 = T(f)^2 \text{ for all } f \in A.$$

Therefore, $wg \in B$ for all $g \in B$, i.e., w is in $\operatorname{Orth}(B)$. If T is bijective then so is S (see [27]) and then $w = T \circ S^{-1}$ is invertible as well. Consequently, $S = w^{-1}T$ maps A to B. □

We now observe that all seven Theorems 5.1 through 5.7 immediately follow as consequences from our main result. The condition that A is n^{th}-root closed can not be deleted from the main theorem as the following example shows.

Example 6.6. *Let A be the f-algebra of the piecewise polynomial functions on $[0, 1]$ that are 0 at 0. Then the lattice homomorphism $T : A \to \mathbb{R}$ that assigns to a function its right derivative at 0 is not representable as in the main theorem above. Indeed, denote the idenitity function on $[0, 1]$ by f. Suppose that T has a representation as above with S an algebra and lattice homomorphism $A \to \mathbb{R}$ and α a nonzero real number such that $T = \alpha S$. Then $S(f) \neq 0$, hence $S(f^2) \neq 0$, but $T(f^2) = 0$, a contradiction.*

References

[1] F.W. Anderson, Lattice-ordered rings of quotients. *Canad. J. Math.* **17** (1965), 434–448.

[2] W. Arendt, Spectral properties of Lamperti operators, *Indiana Univ. Math. J.* **32** (1983), 199–215.

[3] J. Araujo, E. Beckenstein and L. Narici, Biseparating maps and homeomorphic real-compactifications, *J. Math. Ana. Appl.* **12** (1995), 258–265.

[4] J. Arbault, Nouvelles propriétés des transformations de Reynolds, *C. R. Acad. Sci. Paris* **239** (1954), 858–860

[5] Y. Azouzi, K. Boulabiar, and G. Buskes, The de Schipper formula and squares of vector lattices, *Indag. Math.* **17**, (2006), 479–496.

[6] B. Banachewski, Maximal rings of quotients of semi-simple commutative rings, *Archiv Math.* **16** (1965), 414–420.

[7] S.J. Bernau and C.B. Huijsmans, The order bidual of almost f-algebras and d-algebras, *Trans. Amer. Math. Soc.* **347** (1995), 4259–4274.

[8] F. Beukers, C.B. Huijsmans and B. de Pagter, Unital embedding and complexification of f-algebras, *Math. Z.* **183** (1983), 131–143.

[9] A. Bigard and K. Keimel, Sur les endomorphismes conservant les polaires d'un groupe réticulé archimédien, *Bull. Soc. Math. France* **97** (1969), 381–153.

[10] M. Billik, Idempotent Reynolds operators, *J. Math. Anal. Appl.* **18** (1967), 486–496.

[11] G. Birkhoff, Moyennes des fonctions bornées, Colloque d'Algèbre et de Théorie des Nombres, pp. 143–153, Centre National de la Recherche Scientifique, Paris, 1949.

[12] K. Boulabiar, G. Buskes, and M. Henriksen, A Generalization of a Theorem on Biseparating Maps, *J. Math. Ana. Appl.* **280** (2003), 334–339.

[13] K. Boulabiar, Order bounded separating linear maps on Φ-algebras, *Houston J. Math.* **30** (2004), 1143–1155.

[14] K. Boulabiar and G. Buskes, A note on bijective disjointness preserving operators, Positivity IV-theory and applications, pp. 29–33, Tech. Univ. Dresden, Dresden, 2006.

[15] G. Buskes and A. van Rooij, Small vector lattices, *Math. Proc. Cambr. Philos. Soc.* **105** (1989), 523–536.

[16] G. Buskes and A. van Rooij, Almost f-algebras: commutativity and Cauchy-Schwarz inequality, *Positivity* **4** (2000), 227–231.

[17] G. Buskes and A. van Rooij, Squares of vector lattices, *Rocky Mountain J. Math.* **31** (2001), 45–56.

[18] K. Boulabiar, G. Buskes, and A. Triki, Recent results in lattice ordered algebras, *Contemporary Math.* **328** (2003), 99–133.

[19] B. Brainerd, On the structure of averaging operators, *J. Math. Anal. Appl.* **5** (1962), 135–144.

[20] P.F. Conrad and J.E. Diem, The ring of polar preserving endomorphisms of an Abelian lattice-ordered group, *Illinois J. Math.* **15** (1971), 222–240.

[21] M.L. Dubreil-Jacotin, Etude algébrique des transformations de Reynolds, Colloque d'algèbre supérieure, pp. 9–27, Bruxelles, 1956.

[22] Y. Friedman and B. Russo, Contractive projections on $C_0(K)$, *Proc. Amer. Math. Soc.* **273** (1982), 57–73.

[23] L. Gillman and M. Jerison, *Rings of Continuous Functions*, Springer Verlag, Berlin-Heidelberg-New York, 1976.

[24] J.J. Grobler, Bivariate and marginal function spaces. Positivity IV-theory and applications, pp. 63–71, Tech. Univ. Dresden, Dresden, 2006.

[25] F. Hadded, Contractive projections and Seever's identity in complex f-algebras, *Comment. Math. Univ. Carolin.* **44** (2003), 203–215.

[26] A.W. Hager, *Isomorphism with a $C(Y)$ of the maximal ring of quotients of $C(X)$*, Fund. Math. **66** (1969), 7–13.

[27] D.R. Hart, Some properties of disjointness preserving operators, *Indag. Math.* **88** (1985), 183–197.

[28] C.B. Huijsmans, The order bidual of lattice ordered algebras II, *J. Operator Theory* **22** (1989), 277–290.

[29] C.B. Huijsmans and B. de Pagter, The order bidual of lattice ordered algebras, *J. Funct. Anal.* **59** (1984), 41–64.

[30] C.B. Huijsmans and B. de Pagter, Averaging operators and positive contractive projections, *J. Math. Anal. Appl.* **113** (1986), 163–184.

[31] C.B. Huijsmans and B. de Pagter, Invertible disjointness preserving operators, *Proc. Edinburgh Math. Soc.* **37** (1993), 125–132.

[32] K. Jarosz, Automatic continuity of separating linear isomorphisms, *Bull. Canadian Math. Soc.* **33** (1990), 139–144

[33] J.S. Jeang and N.C. Wong, Weighted composition of $C_0(X)$'s, *J. Math. Ana. Appl.* **201** (1996), 981–993.

[34] E. de Jonge and A. van Rooij, Introduction to Riesz spaces, Mathematical Centre Tracts, No. 78, Mathematisch Centrum, Amsterdam, 1977.

[35] J. Kampé de Fériet, Sur un problème d'algèbre abstraite posé par la définition de la moyenne dans la théorie de la turbulence, *Ann. Soc. Sci. Bruxelles Sér I.* **63** (1949), 156–172.

[36] J.L. Kelley, Averaging operators on $C_\infty(X)$, *Illinois J. Math.* **2** (1958), 214–223.

[37] J. Lambek, *Lectures on Rings and Modules*, Blaisdell, Toronto, 1966.

[38] W.A.J. Luxemburg and A.R. Schep, A Radon-Nikodym type theorem for positive operators and a dual, *Indag. Math.* **40** (1978), 357–375.

[39] W.A.J. Luxemburg and A.C. Zaanen, The linear modulus of an order bounded linear transformation I, *Indag. Math.* **33** (1971), 422–434.

[40] H.P. Lotz, Über das spektrum positiver operatoren, *Math. Z.* **108** (1968), 15–32.

[41] J. Martinez, The maximal ring of quotients f-ring, *Algebra Univ.* **33** (1995), 355–369.

[42] I. Molinaro, Détermination d'une R-transformation de Reynolds, *C. R. Acad. Sci. Paris* **244** (1957), 2890–2893

[43] S.-T.C. Moy, Characterizations of conditional expectations as a transformation on function spaces, *Pacific J. Math.* **4** (1954), 47–63.

[44] A. Neeb, *Positive Reynolds Operators on $C_0(X)$*, Ph.D. Thesis, Darmstadt 1996.

[45] B. de Pagter, *f-Algebras and Orthomorphisms*, Ph.D. Thesis, Leiden, 1981.

[46] B. de Pagter, The space of extended orthomorphisms on a vector lattice, *Pacific J. Math.* **112** (1984), 193–210.

[47] J. Quinn, Intermidiate Riesz spaces, *Pacific J. Math.* **56** (1975), 225–263.

[48] O. Reynolds, On the dynamical theory of incompressible viscous fluids and the determination of the criterion, *Philos. Trans. Roy. Soc. Ser. A* **186** (1895), 123–164.

[49] G.C. Rota, On the representation of averaging operators, *Rend. Sem. Mat. Univ. Padova* **30** (1960), 52–64.

[50] G.C. Rota, Reynolds operators, *Proc. Sympos. Appl. Math.* **16** (1963), 70–83.

[51] E. Scheffold, Über Reynoldsoperatoren und "Mittelwert bildende" Operatoren auf halbeinfachen F-Banachverbandsalgebren, *Math. Nachr.* **162** (1993), 329–337.

[52] G.L. Seever, Non-negative projections on $C_0(X)$, *Pacific J. Math.* **17** (1966), 159–166.

[53] J. Sopka, On the characterization of Reynolds operators on the algebra of all continuous functions on a compact Hausdorff space, Ph.D. Thesis, Harvard-Cambridge, 1950.

[54] A. Triki, Extensions of positive projections and averaging operators, *J. Math. Anal. Appl.* **153** (1990), 486–496.

[55] A. Triki, A note on averaging operators, *Contemporay Math.* **232** (1999), 345–348.

[56] Y. Utumi, On quotient rings, *Osaka Math. J.* **8** (1956), 1–18.

[57] Walker, Russell C., The Stone-Čech compactification. Ergebnisse der Mathematik und ihrer Grenzgebiete, Band 83, Springer-Verlag, New York – Berlin, 1974.

[58] A.W. Wickstead, The injective hull of an Archimedean f-algebra, *Compositio Math.* **62** (1987), 329–342.

[59] D.E. Wulbert, Averaging projections, *Illinois J. Math.* **13** (1969), 689–693.

[60] A.C. Zaanen, *Riesz Spaces* II, North-Holland, Amsterdam, 1983.

K. Boulabiar
Département du cycle agrégatif
Institut Préparatoire aux Etudes Scientifiques et Techniques
Université du 7 Novembre à Carthage
BP 51
2070 La Marsa, Tunisia

G. Buskes
Department of Mathematics
University of Mississippi
University, MS 38677, U.S.A.

A. Triki
Department of Mathematics
Faculté des Sciences de Tunis
Université des Sciences, des Techniques
et de Médecine de Tunis (Tunis II)
1060 Tunis, Tunisia

Positivity
Trends in Mathematics, 97–126

Bilinear Maps on Products of Vector Lattices: A Survey

Q. Bu, G. Buskes, and A.G. Kusraev

Abstract. This is a survey on bilinear maps on products of vector lattices.

Contents

1. Introduction and terminology 97
2. Fremlin's tensor product 101
2.1. Bilinear maps of order bounded variation 102
2.2. The projective tensor norm 103
2.3. Bilinear maps of bounded semivariation 103
2.4. Bilinear maps of bounded norm variation 104
2.5. Fremlin tensor product and bornologies 105
3. The adjoints of bilinear maps 107
4. Arens regular maps 108
5. Orthosymmetric maps 109
6. Powers of vector lattices 110
6.1. Squares 110
6.2. Higher powers 111
7. Cauchy-Bunyakowski-Schwarz inequalities 113
8. Disjointness preserving bilinear operators 113
9. Injective and projective tensor product 115
10. RNP for Tensor Products 118
References 122

1. Introduction and terminology

The theory of vector lattices and order bounded linear maps between them originated around eighty years ago and the fundamental theory is now contained in a variety of books like [4], [5], [62], [76], [78], [84], [93], and [96]. The study of appropriate bilinear maps on products of vector lattices is of more recent date. It originated in 1953 with the paper [80] by Nakano. Nakano's bilinear maps seem to not have received much attention for at least two decades, though shortly thereafter Birkhoff and Pierce introduced lattice ordered algebras [11], the multiplication map of which provides important examples of the kind of bilinear map that Nakano studied. An extensive literature on lattice ordered algebras is now well established (see, e.g., the survey on certain aspects of f-algebras elsewhere in this book as well as [16]). Nakano's bilinear maps were re-introduced and re-investigated in general (i.e., outside the setting of algebras) by Cristescu in [39], by Fremlin for vector lattices in [48] and then for Banach lattices in [49], by Wittstock for Banach lattices in [94], [95], by Schaefer [86] for Banach lattices, by Kusraev [61] for vector lattices, and by Buskes and van Rooij in [32] for vector lattices and Banach lattices.

In part, motivation for the bilinear maps of this survey derives from lattice ordered algebras. We want to point in particular to the sequence of papers in the setting of lattice ordered Banach algebras by Scheffold [87, 88, 89, 90, 91].

Throughout this paper we assume that all of our vector lattices are over the field of real numbers and that all vector lattices are Archimedean. Though this survey mainly compiles known results, announcements of some new and forthcoming theorems can be found in its Sections 7, 8, and 10.

Our starting point is Fremlin's fundamental construction of the Archimedean tensor product of two Archimidean vector lattices [48]. This tensor product approach lends itself to a systematic transfer of known results on positive linear maps to positive (or, in Fremlin's terminology, bipositive) bilinear maps. Of course, this transfer machinery works also for regular linear maps and differences of positive bilinear operators. The connection between not necessarily regular bilinear maps on products of ordered vector spaces and order bounded maps on their tensor product requires the notion of bilinear maps of order bounded variation. For normed vector lattices we first involve the positive projective norm on Fremlin's tensor product (see [49]), and then study two new norms for bilinear maps, the semivariation norm and the variation norm. Order bounded variation was first introduced in [80], the semivariation norm and the variation norm first appeared in [32]. All this is covered in Section 2.

In Sections 3 and 4 we turn for motivation to the Arens multiplication (introduced in [7]) on the bidual of various lattice ordered algebras (e.g., f-algebras, almost f-algebras and d-algebras) that has been well studied (see, e.g., [10]). More recently in [15] and the thesis by Page [82], the more general questions about triadjoints of bilinear maps on products of vector lattices have been studied (see also the papers by Scheffold [88], [89], [91], and the paper by Boulabiar and Toumi [14]).

In Section 5, we present some results on very special bilinear maps that have recently received attention in the literature, so-called orthosymmetric maps and their connection with squares and powers of vector lattices. The notion of orthosymmetry (and the name) came out of the work on almost f-algebras by Buskes and van Rooij [32] and was then extensively and more abstractly used in vector lattices in [33]. Their importance derives from a similarity with Hilbert space theory, as well as the fundamental fact that they are symmetric. Contributions were made in a sequence of papers by Kusraev [64], Kusraev and Shotaev [65], Kusraev and Tabuev [66], [67], a paper by Boulabiar [12], a paper by Boulabiar and Toumi [14], as well as a paper [17] by Buskes and Boulabiar. One more place where bilinear maps on products of ordered structures appear is in a paper on Hilbert lattices by van Gaans [50].

Sections 6 and 7 are an outflow from Section 5, dealing with the aforementioned squares of vector lattices and some observations about Cauchy-Schwarz inequalities.

In Section 8, we briefly touch on disjointness preserving bilinear maps.

In the final two sections of this paper, we discuss the tensor product of Banach lattices and the Radon-Nikodym property of such tensor products, as studied in a sequence of paper [21], and [22] by Bu and Buskes, and [23] by Bu, Buskes and Lai. In that context we also wish to refer to a forthcoming thesis by Lai [70]. In that final section we look both at the projective and injective tensor product of Banach lattices.

In the rest of this paper (except where clearly labeled otherwise in Sections 2.3 and 2.4) E, F, G, and H are Archimedean vector lattices.

Let $p : E \times F \to G$ be a bilinear map. Then p is called *positive* (respectively a *Riesz bimorphism*) if for each $x \in E^+$, and each $y \in F^+$ the maps

$$e \mapsto p(e, y) \ (e \in E) \qquad \text{and}$$
$$f \mapsto p(x, f) \ (f \in F)$$

are positive maps (respectively Riesz homomorphisms). We note that Fremlin in [48] used the term *bipositive* where we use the term *positive.* Indeed, the bilinear map p is positive if and only if for all $x, y \in E^+$ it follows that $p(x, y) \in F^+$. The bilinear map p is called *regular* if it can be represented as a difference of two positive bilinear operators. The set $BL_r(E, F; G)$ of all regular bilinear operators from $E \times F$ to G is an ordered vector space, where the order is defined by the cone of positive bilinear operators, $BL_+(E, F; G)$. Take $a \in E^+$. A *partition* of a is a finite sequence of elements of E^+ whose sum equals a. The partitions of a form a set $\prod a$. A partition $(x_1, x_2, \ldots, x_n)$ of a will be denoted by just the letter "x". The bilinear map p is said to be of *order bounded variation* if for all $e \in E^+$ and all $f \in F^+$, the set

$$\left\{ \sum_{n,m} \mid p(x_n, y_m) \mid : x \in \prod a, \ y \in \prod b \right\}$$

is order bounded. The set of all bilinear maps $E \times F \to G$ of order bounded variation is denoted by $Bil_{bv}(E, F; G)$. The vector space $Bil_{bv}(E, F; G)$ is a partially ordered vector space under the ordering defined by $p_1 \geq p_2$ if $p_1 - p_2$ is positive. Obviously, $BL_r(E, F; G) \subset Bil_{bv}(E, F; G)$ but the converse inclusion may be false. The map $p \mapsto p^{\otimes}$ defines (see [32]) a bijective bipositive map from $Bil_{bv}(E, F; G)$

to the partially ordered vector space of all order bounded linear maps from the Fremlin tensor product $E\bar{\otimes}F$ to G, denoted by $\mathcal{L}_b(E\bar{\otimes}F,G)$. In particular, when G is Dedekind complete, it follows that $Bil_{bv}(E,F;G)$ is a (Dedekind complete) vector lattice. We use the notation of $x_\alpha \to x$ when the net (x_α) *order converges* to x (for the definition of which we here use the one in [4]). The bilinear map p is called *separately order continuous* if

$$e \mapsto p(e,y)\ (e \in E) \qquad \text{and}$$
$$f \mapsto p(x,f)\ (f \in F)$$

are order continuous for each $x \in E^+$, and each $y \in F^+$. Furthermore, p is *symmetric* if $E=F$ and $p(x,y)=p(y,x)$ for all $x,y \in E$ while p is *anti-symmetric* if $E=F$ and $p(x,y)=-p(y,x)$ for all $x,y\in E$. The *order dual* of a vector lattice E is the vector lattice of all order bounded linear functionals on E and is denoted by $E^\sim$. More generally, the space of all order bounded maps from E to G is denoted by $\mathcal{L}_b(E,G)$, whereas the space of differences of positive maps is denoted by $\mathcal{L}^r(E,G)$. The *order bidual* of a vector lattice E is the order dual of $E^\sim$ and it is denoted by $E^{\sim\sim}$. The order bidual $E^{\sim\sim}$can be decomposed as the orthogonal sum of the projection bands of order continuous linear functionals $(E^\sim)^\sim_n$, and singular linear functionals $(E^\sim)^\sim_s$.

To understand those parts of the paper where lattice ordered algebras appear, we briefly review some of the terminology in lattice ordered algebra. A vector lattice E is called a *lattice ordered algebra* (also called an *ℓ-algebra*) if its positive cone is closed under multiplication. An ℓ-algebra A is said to be to be an *f-algebra* if for every $f,g\in A$, the condition

$$f\wedge g=0 \text{ implies } (fh)\wedge g=(hf)\wedge g=0 \text{ for all } h\in A^+$$

holds. The latter definition is due to Birkhoff and Pierce in [11] and f-algebras continue to play a very important and motivating role for the general theory of bilinear maps on products of ordered vector spaces. We call the ℓ-algebra A an *almost f-algebra* (due to Birkhoff in [29, Section 6]) if

$$f\wedge g=0 \text{ in } A \text{ implies } fg=0.$$

Finally, an ℓ-algebra A in which

$$f\wedge g=0 \text{ in } A \text{ and } h\in A^+ \text{ imply } (fh)\wedge(gh)=(hf)\wedge(hg)=0$$

is called a *d-algebra.* These d-algebras were introduced by Kudláček in [60].

We remark here that as far as tensor products are considered we only treat the projective and injective tensor products. A paper the length of this survey could be written about other tensor products for Banach lattices like the ones developed by Schaefer in [84]. Standard terminology of vector lattices is followed throughout.

2. Fremlin's tensor product

In his fundamental paper [48] Fremlin introduced for every two Archimedean vector lattices E and F a new Archimedean vector lattice $E\bar{\otimes}F$, defined by the following universal property: there exists a Riesz bimorphism $E \times F \to E\bar{\otimes}F$ such that whenever G is a vector lattice and T is a Riesz bimorphism $E \times F \to G$ then there exists a unique Riesz homomorphism $T^{\otimes} : E\bar{\otimes}F \to G$ for which

$$T(x, y) = T^{\otimes}(x \otimes y) \ (x \in E, \ y \in F).$$

He also showed that $E\bar{\otimes}F$, from here on called the Archimedean vector lattice tensor product of E and F, has the following additional universal property.

Theorem 1. *For every positive bilinear map T of $E\times F$ into any uniformly complete (hence Archimedean) vector lattice G there exists a unique positive linear $T^{\otimes}$: $E\bar{\otimes}F \to G$ such that $T(x, y) = T^{\otimes}(x \otimes y)$ $(x \in E, \ y \in F)$.*

It follows immediately that the map $S \mapsto S\circ\otimes$ defines a bijective positive map from $L_r(E\bar{\otimes}F, G)$ onto $BL_r(E, F; G)$, where $L_r(E\bar{\otimes}F, G)$ denotes the partially ordered vector space of all order bounded linear maps from $E\bar{\otimes}F$ to G (see [61], [65]).

Fremlin's construction uses the Axiom of Choice. Buskes and van Rooij gave a construction that is valid in ZF in the more general context of lattice ordered groups in [27].

Now we give an example from [15] and [65] of how Fremlin's tensor product helps to transfer results from linear maps to bilinear maps. The result that we present for bilinear maps was used by Grobler and Labuschagne in [54] to construct the Fremlin tensor product. However, it can in turn be derived from the corresponding known results for positive linear maps: Kantorovich (Theorem 2.8 in [4]), Lipecki (Theorem 2.9 in [4]) and Lipecki-Luxemburg-Schep (Theorem 7.17 in [4]). All that is needed beyond these results is the universal property for the Archimedean vector lattice tensor product as studied by Fremlin in [48] and presented above. As a comment we add that the third author has pointed out in 3.6.3(4) in [62] that the Lipecki-Luxemburg-Schep theorem was first proved by Kutateladze.

Theorem 2. *Let G be a vector lattice, E_1 and F_1 be majorizing vector sublattices of the vector lattices E and F, respectively, and let $p : E_1 \times F_1 \to G$ be a positive bilinear map. If G is Dedekind complete then the set $\mathcal{E}\,(p)$ of positive extensions of p to all of $E \times F$ is a nonempty convex set with extreme points. Moreover, if p is a Riesz bimorphism then $q \in \mathcal{E}\,(p)$ is an extreme point if and only if q is a Riesz bimorphism.*

The fact that $\mathcal{E}\,(p)$ is nonempty was observed in [61]. Moreover, it was proved in the same paper that there exists an order continuous "simultaneous extension" of regular bilinear operators, i.e., there exists an order continuous lattice homomorphism $\varepsilon : BL_r(E_1, F_1; G) \to BL_r(E, F; G)$ such that $\rho \circ \varepsilon$ is an identity mapping on $BL_r(E_1, F_1; G)$, where $\rho : BL_r(E, F; G) \to BL_r(E_1, F_1; G)$ denotes the restriction operator $b \mapsto b|_{E_1 \times F_1}$. This result is an easy corollary of the existence of "simultaneous extension" of regular linear operators, see Theorem 3.4.11 in [62].

One more example taken from [15] and [65] transfers Kutateladze's characterization of lattice homomorphisms (Theorem 3.3.3 in [62]) to bilinear maps via Fremlin's tensor product.

Theorem 3. *Let E, F, and G be vector lattices with G Dedekind complete. A positive bilinear map $p : E \times F \to G$ is a lattice bimorphism if and only if for every bilinear map $T : E \times F \to G$ with $0 \le T \le p$ there exists an orthomorphism $\rho : G \to G$ such that $0 \le \rho \le I_G$ and $T = \rho \circ p$.*

2.1. Bilinear maps of order bounded variation

The start of the theory of vector lattices is marked by F. Riesz' observation that the space of all bounded linear functionals on $C[0,1]$ is itself a Dedekind complete vector lattice. In that direction, by using the idea of bilinear maps of order bounded variation, a complete analogy between operators on vector lattices and bilinear maps on products of vector lattices is obtained as follows (see Theorem 3.1 in [32] and [65]).

Theorem 4. *Let E, F, G be Archimedean vector lattices; let G be Dedekind complete.*

(1) *Let T be a bilinear map $E \times F \to G$ that is of order bounded variation. Then there exists a unique order bounded linear $T^{\otimes} : E \bar{\otimes} F \to G$ for which*

$$T(x, y) = T^{\otimes}(x \otimes y) \ (x \in E,\ y \in F).$$

(2) *$Bil_{bv}(E, F; G)$ is a Dedekind complete vector lattice. The correspondence*

$$T \mapsto T^{\otimes}$$

is a Riesz homomorphism from $Bil_{bv}(E, F; G)$ onto $\mathcal{L}_b(E \bar{\otimes} F, G)$.

(3) *For $T \in Bil_{bv}(E, F; G)$, $|\,T\,|$ is determined by*

$$|\,T\,|\,(a, b) = \sup_{x \in \prod a,\ y \in \prod b} \sum_{n,m} |\,T(x_n, y_m)\,| \ \ (a \in E^+, b \in F^+).$$

(4) *$BL_r(E, F; G) = Bil_{bv}(E, F; G)$.*

As a corollary one immediately obtains (see Corollary 3.1 in [32]).

Corollary 5. *If E, F and G are Archimedean vector lattices and G is Dedekind complete, then the vector lattices $Bil_{bv}(E, F; G)$, $\mathcal{L}_b(E \bar{\otimes} F, G)$ and $\mathcal{L}_b(E, \mathcal{L}_b(F, G))$ are naturally isomorphic.*

The universal property of Fremlin's tensor product stated in Theorem 1 is valid in the context of bilinear maps of order bounded variation as well (see Theorem 3.2 in [32]).

Theorem 6. *Let E, F, G be Archimedean vector lattices; let G be uniformly complete. Then for every bilinear $T : E \times F \to G$ that is of order bounded variation there is a unique order bounded linear map $T^{\otimes} : E\bar{\otimes}F \to G$ for which*

$$T(x,y) = T^{\otimes}(x \otimes y) \ (x \in E, \ y \in F).$$

$T \mapsto T^{\otimes}$ is a linear order isomorphism between $Bil_{bv}(E, F; G)$ and $\mathcal{L}_b(E\bar{\otimes}F, G)$.

2.2. The projective tensor norm

For Banach lattices E and F, Fremlin in [49], defined the *positive projective tensor norm* $\|\|_{|\pi|}$ on the Archimedean vector lattice tensor product $E\bar{\otimes}F$ for all $u \in E\bar{\otimes}F$ by

$$\| u \|_{|\pi|} = \inf\left\{\sum \| a_k \| \| b_k \| : a_1, \dots, a_k \in E^+, b_1, \dots, b_k \in F^+, |u| \leq \sum a_k \otimes b_k\right\}.$$

Fremlin shows in Theorem 1E in [49] that $\| \ \|_{|\pi|}$ indeed is a norm.

Relative to this norm the following properties hold:

(1) *$E \otimes F$ is dense in $E\bar{\otimes}F$.*

(2) *The cone generated by $\{x \otimes y : x \in E^+, \ y \in F^+\}$ is dense in $(E\bar{\otimes}F)^+$; see* [49], 1B(b).

For more information about the above norm in the context of the Radon-Nikodym property we refer the reader to the last two sections of this paper.

For Banach lattices E, F, G and for a bilinear map $T : E \times F \to G$, define the operator norm

$$\| T \| := \sup\{\| T(x,y) \| : x \in E, \ y \in F, \ \| x \| \leq 1, \ \| y \| \leq 1\}.$$

T is continuous if and only if $\| T \|$ is finite. Fremlin proved:

Theorem 7. *Let E, F be Banach lattices. The positive projective norm defined above defines a Riesz norm $\| \ \|_{|\pi|}$ on $E\bar{\otimes}F$. Let T be a positive (and therefore continuous) bilinear map of $E \times F$ into a Banach lattice G. Then $T^{\otimes}$ is continuous relative to $\| \ \|_{|\pi|}$, and $\| T^{\otimes} \| = \| T \|$.*

2.3. Bilinear maps of bounded semivariation

Let E and F be normed vector lattices. Let G be a normed vector space.

We endow $E\bar{\otimes}F$ with Fremlin's projective product norm $\| \ . \ \|_{|\pi|}$ as defined above. For a bilinear map $T : E \times F \to G$ we define its *semivariation* by

$$\begin{aligned} ||| T ||| := \sup\{\| \textstyle\sum_{n,m} \varepsilon_{nm} T(x_n, y_m) \| : x_1, x_2, \dots, x_N \in E^+, \ \| \sum x_n \| \leq 1, \\ y_1, y_2, \dots, y_M \in F^+, \ \| \sum y_n \| \leq 1, \ \varepsilon_{nm} \in \{-1, 1\} \text{ for all } n, m\}. \end{aligned}$$

The bilinear maps $T : E \times F \to G$ for which $||| \ T \ |||$ is finite form a vector space $Bil_{||| \ |||}(E, F; G)$ and $||| \ . \ |||$ is a norm on $Bil_{||| \ |||}(E, F; G)$. All elements of $Bil_{||| \ |||}(E, F; G)$ are norm continuous. Indeed, $\| T \| \leq ||| T |||$ for every T. The elements of $Bil_{||| \ |||}(E, F; G)$ are said to have *finite semivariation* (see [32]).

Theorem 8. *Let E and F be normed vector lattices and let G be a Banach space. The map*

$$S \mapsto S \circ \otimes \ (S \in L(E\bar{\otimes}F, G))$$

is a Banach space isomorphism of $L(E\bar{\otimes}F, G)$ onto $Bil_{|||\ |||}(E, F; G)$. In particular, for every bilinear map $T : E \times F \to G$ that is of finite semivariation there exists a unique continuous linear $T^{\otimes} : E\bar{\otimes}F \to G$ with

$$T(x, y) = T^{\otimes}(x \otimes y) \quad (x \in E,\ y \in F).$$

We have $\| T^{\otimes} \| = ||| T |||$.

2.4. Bilinear maps of bounded norm variation

We next discuss the special case where G is a Banach lattice. Like in Section 2.3, we closely follow the approach of [32]. Fremlin in [49] studied the case where G is an ordered vector space such that

$$(*) \qquad \text{if } x, a \in G \text{ and } -a \le x \le a \text{ then } \| x \| \le \| a \| .$$

The condition $(*)$ above is equivalent to

$$\text{if } a, b \in G^+, \text{ then } \| a - b \| \le \| a + b \|$$

and also to

$$\text{if } x_1, \dots, x_N \in G^+ \text{ and } \varepsilon_1, \dots, \varepsilon_N \in \{-1, 1\}, \text{ then } \left\| \sum \varepsilon_n x_n \right\| \le \left\| \sum x_n \right\| .$$

All closed linear subspaces of Banach lattices satisfy $(*)$, and so does $L(E, F)$ for Banach lattices E and F. The following theorem generalizes one of the main results in [49] (see [32]).

Theorem 9. *Let E and F be normed vector lattices; let G be an ordered Banach space satisfying $(*)$ above. Let $T : E \times F \to G$ be a continuous, positive (i.e., $T(E^+ \times F^+) \subset G^+$) and bilinear map. Then the semivariation of T is finite and, indeed, $\| T \| = ||| T |||$. With $T^{\otimes}$ as in Theorem 8, $T^{\otimes}$ is a positive linear map and $\| T \| = \| T^{\otimes} \|$. Consequently, $L(E\bar{\otimes}F, G)$ and $Bil_{|||\ |||}(E, F; G)$ are isomorphic as ordered Banach spaces.*

The *norm variation* of a bilinear map $T : E \times F \to G$ is

$$Var\ T := \sup \Big\{ \Big\| \textstyle\sum_{n,m} | T(x_n, y_m) | \Big\| : x_1, x_2, \dots, x_N \in E^+,\ \| \sum x_n \| \le 1, \\ y_1, y_2, \dots, y_M \in F^+,\ \| \sum y_n \| \le 1 \Big\};$$

T is of *norm bounded variation* if $Var\ T < \infty$.

For every bilinear map $T : E \times F \to G$ we have $\| T \| \le ||| T ||| \le Var\ T$, whereas $\| T \| = Var\ T$ if T is positive. Also, if $a \in E^+$ and $b \in F^+$, then

$$\left\| \sum | T(x_n, y_m) | \right\| \le \| a \| \| b \| \ Var\ T \quad \left(x \in \prod a,\ y \in \prod b \right).$$

The bilinear maps $E \times F \to G$ of norm bounded variation form a vector space, containing all bilinear maps that are positive and continuous.

To be able to state the Main Universal Theorem for the projective Banach lattice tensor product we need the definition of norm bounded variation for linear maps. Let E and F be normed vector lattices. If S is a linear map $E \to F$ then the *norm variation* of S is defined by

$$Var\ S := \sup\left\{\left\|\sum \mid Sx_n \mid\right\| : x_1, x_2, \ldots, x_N \in E^+,\ \left\|\sum x_n\right\| \leq 1\right\}.$$

S is said to be of norm bounded variation if $Var\ S$ is finite.

If $S : E \to F$ is linear, then $\| S \| \leq Var\ S$, and $\| S \| = Var\ S$ for positive S. Moreover, for all $a \in E^+$,

$$\left\|\sum \mid Sx_n \mid\right\| \leq \| a \|\ Var\ S \quad \left(x \in \prod a\right).$$

The linear maps of order bounded variation form a vector space containing all positive continuous linear maps. Observe that the class of linear maps of order bounded variation between Banach lattices coincides with the class of (p, q)-regular operators with $p = 1$ and $q = 1$ (see 7.2.11 (3) in[62]). The Main Universal Theorem for the projective Banach lattice tensor product follows next (see [32]).

Theorem 10 (Main Universal Theorem for the projective Banach lattice tensor product). *Let E and F be normed vector lattices, let G a Banach lattice, and let T be a bilinear map $E \times F \to G$ that is of norm bounded variation. Then there exists a unique continuous linear $T^{\otimes} : E\bar{\otimes}F \to G$ with*

$$T(x, y) = T^{\otimes}(x \otimes y)\ (x \in E,\ y \in F).$$

We have that $T^{\otimes} \geq 0$ if and only if T is positive. Furthermore, $T^{\otimes}$ is of norm bounded variation and

$$Var\ T^{\otimes} = Var\ T.$$

2.5. Fremlin tensor product and bornologies

One can unify and extend the two Fremlin results as described in the beginning of this section as well the above generalizations of Buskes and van Rooij by using the theory of bornological vector spaces, as we will now do. For more details we refer to [34] and [35] from which we take almost literally (with kind permission of Springer Science and Business Media) what follows in this subsection. For all matters bornological we refer to [58]. This section puts more clearly into focus that it is not the positivity of the bilinear map but the boundedness that matters for the universal property of the tensor product.

A *Riesz disk* in E is a nonempty subset A of E that is convex and solid (i.e., if $a \in A$, $x \in E$, $\mid x \mid \leq \mid a \mid$, then $x \in A$.) If A is such a Riesz disk, then its linear span is $E_A := \bigcup_{\lambda > 0} \lambda A$. This E_A is a order ideal in E. For a set A we define a gauge $p_A : E_A \to [0, \infty)$ by

$$p_A(x) := \inf\{\lambda : \lambda > 0,\ x \in \lambda A\}.$$

Then p_A is a Riesz seminorm on E_A. The disk A is *completant* if p_A is a norm on E_A and E_A with that norm is a Banach lattice.

The Riesz disk *generated* by a nonempty set $S \subset E$ is the convex hull of $\bigcup_{s\in S} [-s, s]$. In a formula, it is the set

$$\Big\{x \in E : |x| \leq \sum_{k=1}^{K} \lambda_k |s_k|,\ \lambda_1, \ldots, \lambda_K \in [0,1],\ \sum_{k=1}^{K} \lambda_k = 1,\ s_1, \ldots, s_K \in S\Big\}.$$

A [*complete*] *Riesz bornology* in E is a bornology (see [58]) having a base that consists of [completant] Riesz disks. We have the following two main examples in mind.

Example 11. *The order bounded subsets of E form a Riesz bornology, the order bornology. It is complete as soon as E is uniformly complete.*

Example 12. *Suppose E is a normed vector lattice. Its von Neumann bornology is the Riesz bornology consisting of all norm bounded sets. It is complete if E is a Banach lattice.*

A [*complete*] *bornological vector lattice* is an Archimedean vector lattice endowed with a [complete] Riesz bornology. For our main result, we need not only completeness, but also the following extra condition on the range space G.

The bornological closure of every bounded subset of G is bounded.

If G is uniformly complete and the bornology is the order bornology or if G is a Banach lattice and the bornology is the von Neumann bornology the extra condition is fulfilled. There exist complete bornological vector lattices that do not satisfy this condition.

Let E, F, G be bornological vector lattices with Riesz bornologies $\mathcal{A}$, $\mathcal{B}$, and $\mathcal{C}$, respectively. Let $T : E \to G$ be linear. T is called *bounded* if $T(A) \in \mathcal{C}$ for all $A \in \mathcal{A}$. We say that T is of *bounded variation* if for every $A \in \mathcal{A}$ the set

$$\left\{\sum_{n=1}^{N} |T(x_n)| : x_1, \ldots, x_N \in E^+,\ \sum_{n=1}^{N} x_n \in A\right\}$$

belongs to $\mathcal{C}$. Clearly, if T is of bounded variation, then it is bounded. The converse is true if T is positive.

Similarly, a bilinear map $T : E \times F \to G$ is *bounded* if $T(A, B) \in \mathcal{C}$ for all $A \in \mathcal{A}$, $B \in \mathcal{B}$, whereas T is *of bounded variation* if for all such A and B the set

$$\Big\{\sum_{n=1}^{N}\sum_{m=1}^{M} |T(x_n, y_m)| : x_1, \ldots, x_N \in E^+,\ \sum_{n=1}^{N} x_n \in A,\ y_1, \ldots, y_M \in F^+,\ \sum_{n=1}^{N} y_m \in B\Big\}$$

belongs to $\mathcal{C}$.

A bilinear map of bounded variation is bounded; for positive bilinear maps the converse is also valid.

The main theorem on bilinear maps of bounded variation now follows after introducing the bornology on the tensor product

Let A and B be the Riesz disks in E and F, respectively. By $[A \otimes B]$ we denote the Riesz disk generated by $\{a \otimes b : a \in A, b \in B\}$. The gauge on $(E\bar{\otimes}F)_{[A\otimes B]}$ determined by $[A \otimes B]$ is given by

$$p_{[A\otimes B]}(u) = \inf\Big\{ \sum_{k=1}^{K} p_A(x_k) p_B(y_k) : \\ x_1, \dots, x_K \in E,\ y_1, \dots, y_K \in F,\ | u | \leq \sum_{k=1}^{K} | x_k | \otimes | y_k | \Big\}$$

This implies that, if E and F are Banach lattices and A, B are their closed unit balls, then $p_{[A\otimes B]}$ equals the norm $\| \ \|_{|\pi|}$ as introduced by Fremlin.

By the *bornological vector lattice tensor product* we mean the vector lattice $E\bar{\otimes}F$ with the Riesz bornology generated by $\{[A \otimes B] : A$ and B are bounded Riesz disks in E and F, respectively$\}$.

Theorem 13. *Let E, F, G be bornological vector lattices. Assume that G is complete and that in G the bornological closure of every bounded set is bounded. Let T be a bilinear map $E \times F \to G$ of bounded variation. Then there exists a unique linear $T^{\otimes} : E\bar{\otimes}F \to G$ that is of bounded variation and satisfies*

$$T(x, y) = T^{\otimes}(x \otimes y)\ (x \in E, y \in F).$$

Moreover, if $A \subset E$, $B \subset F$, and $C \subset G$ are bounded Riesz disks and $T(A, B) \subset C$ then $T^{\otimes}([A \otimes B])$ is contained in the bornological closure of C. If T is positive, then so is $T^{\otimes}$.

Problem 14. *Can the extra condition on G in the theorem above be deleted?*

3. The adjoints of bilinear maps

The adjoint of a bilinear map is defined next. The idea is due to Arens in [7]. If E, F, and G are vector spaces and E', F', and G' are their respective duals (as in dual pairs) and $p : E \times F \to G$ is a bilinear map, then $p^* : G' \times E \to F'$ is defined by

$$(p^*(f, x))(y) = f(p(x, y)) \text{ for all } f \in G',\ x \in E, \text{and } y \in F.$$

We will write p^{**} for $(p^*)^*$.

We will use the notation p^*, p^{**} as above in the situation where E, F, and G are vector lattices and the duals are their order duals $E^\sim$, $F^\sim$, and $G^\sim$. For convenience, we use p^{***} for the restriction of the Arens triadjoint of the bilinear map $p : E \times F \to G$ to $(E^\sim)_n^\sim \times (F^\sim)_n^\sim$.

The following theorem is one of the main results in [15].

Theorem 15. *Let $p : E \times F \to G$ be a bilinear map of order bounded variation. Then the following hold.*

(1) *p^* is of order bounded variation.*
(2) *p^{***} is separately order continuous.*
(3) *$p^{***}((E^\sim)^\sim_n \times (F^\sim)^\sim_n) \subset (G^\sim)^\sim_n$.*
(4) *If p is positive then p^* is positive.*
(5) *If $E = F$ and p is symmetric then p^{***} is symmetric.*
(6) *If $E = F$ and p is anti-symmetric then p^{***} is anti-symmetric. Moreover, the algebraic decomposition of any bilinear map of order bounded variation into its symmetric and antisymmetric parts is preserved when taking the Arens triadjoint map.*
(7) *If p is a Riesz bimorphism then p^{***} is a Riesz bimorphism.*

The following four consequences immediately follow from the above theorem (where the reader needs to keep in mind that p^{***} is assumed to have $(E^\sim)^\sim_n \times (F^\sim)^\sim_n$ as its domain).

(i) (*Bernau and Huijsmans*, 1995, [10]) *If E is a d-algebra then so is $(E^\sim)^\sim_n$ when equipped with the Arens multiplication.*
(ii) (*Scheffold*, 1996, [90]) *If E, F, and G are Banach lattices and $p : E \times F \to G$ is a Riesz bimorphism then so is p^{***}.*
(iii) (*Scheffold*, 1998, [91]) *If E, F, and G are Banach lattices and $p : E \times F \to G$ is symmetric and the difference of two positive bilinear maps then p^{***} is symmetric.*
(iv) (*Grobler*, 1999, [53]) *If E is a commutative lattice ordered algebra then so is $(E^\sim)^\sim_n$ when equipped with the Arens multiplication.*

With a variation on Scheffold's definition in [90], we define the positive bilinear operator $p : E \times F \to G$ to be *left almost interval preserving* if $\overline{p(e,[0,f])} = [0, p(e,f)]$ for every $e \in E^+, f \in F^+$, where the closure is taken in the absolute weak topology $|\sigma|$ (see [5]). Of course, $p : E \times F \to G$ is called *right almost interval preserving* if $\overline{p([0,e],f)} = [0, p(e,f)]$ for every $e \in E^+, f \in F^+$. The notion of almost interval preserving for linear operators was defined in [75]. The following result was proved by R. Page in [82] (also see [15]).

Theorem 16. *Let $E, F,$ and G be vector lattices and assume that H is a Dedekind complete vector lattice and let $p : E \times F \to G$ be a bilinear map.*

(i) *If p is left (or right) almost interval preserving and $H^\sim$ separates the points of H then the operator $S \mapsto S \circ p$ from $\mathfrak{L}_b(G,H)$ to $Bil_{bv}(E,F;H)$ is a Riesz homomorphism.*
(ii) *If p is a Riesz bimorphism then the operator $S \mapsto S \circ p$ from $\mathfrak{L}_b(G,H)$ to $Bil_{bv}(E,F;H)$ is interval preserving.*

4. Arens regular maps

Now let E, F, G be Banach spaces with duals E', F', and G' and let $P : E \times F \to G$ be a bilinear map such that $\sup\{\|P(x,y)\| : \|x\| \le 1, \|y\| \le 1\} < \infty$. The collection of such bilinear maps is called $Bil(E,F;G)$. In case $E = F$, define P^t, the transpose of P, by

$$P^t(x,y) = P(y,x) \text{ for all } x,y \in E.$$

Still under the assumption that $E = F$, the bilinear map P is called Arens regular if

$$P^{t***t} = P^{***}$$

Arens gave an example in [7] of a bilinear map which is not Arens regular. In his example $E = F = \ell_1$ and the bilinear map of his example actually turns out to be positive. The positivity of a bilinear map that is not Arens regular has a lot to do with ℓ_1. The details follow next (see [36]).

Theorem 17. *Let E be a Banach lattice. Then the following are equivalent.*

(1) *Every positive bilinear map $E \times E \to \mathbf{R}$ is Arens regular.*
(2) *ℓ^1 does not embed in E.*
(3) *For every Banach lattice F every bilinear map $E \times E \to F$ of order bounded variation is regular.*
(4) *For every Banach lattice F every positive bilinear map $E \times E \to F$ is regular.*
(5) *Every positive bilinear map $E \times E \to E$ is Arens regular.*
(6) *Every bilinear map of order bounded variation $E \times E \to E$ is Arens regular.*

Though the literature on Arens regularity is extensive, the systematic appearance of positivity in that literature is rather recent.

5. Orthosymmetric maps

The following concept was introduced by Buskes and van Rooij in [31].

Let E and F be vector lattices. A bilinear map $T : E \times E \to F$ is called *orthosymmetric* if whenever $f \wedge g = 0$ for $f, g \in E$ we have $T(f,g) = 0$.

Let X be a vector space. A bilinear operator $b : X \times X \to G$ is said to be *symmetric* if $b(x,y) = b(y,x)$ for all $x, y \in X$, *positively semidefinite* if $b(x,x) \geq 0$ for every $x \in X$, and *positively definite* if it is positively semidefinite and $b(x,x) = 0$ implies $x = 0$. Every orthosymmetric positive bilinear operator is positively semidefinite [50]. More subtle is the fundamental fact that any orthosymmetric positive bilinear operator is symmetric, which was proved in [31].

Theorem 18. *Let E and F be Archimedean vector lattices. Then every orthosymmetric positive bilinear $E \times E \to F$ is symmetric.*

The previous theorem has many consequences. One of them characterizes symmetric Riesz bimorphisms.

Theorem 19. *For any lattice bimorphism $b : E \times E \to F$ the following statements are equivalent.*

(1) *b is symmetric.*
(2) *b is orthosymmetric.*
(3) *b is positively semidefinite.*

Now we state a structural property of orthosymmetric regular bilinear operators due to Kusraev [64]. Denote by $BL_{or}(E,G)$ the ordered space of all orthosymmetric regular bilinear operators from $E \times E$ to G.

Theorem 20. *Let E, F, and G be vector lattices with G uniformly complete. Let $\langle \cdot, \cdot \rangle : E \times E \to F$ be a positively definite lattice bimorphism and F_0 be the smallest vector sublattice in F containing the set $\{\langle x, y \rangle : x, y \in E\}$. Then for every orthosymmetric regular bilinear operator $b : E \times E \to G$ there exists a unique regular linear operator $\Phi_b : F_0 \to G$ such that*

$$b(x,y) = \Phi_b(\langle x, y \rangle) \quad (x, y \in E).$$

The correspondence $b \mapsto \Phi_b$ is a linear and order isomorphism of $BL_{or}(E,G)$ and $L_r(F_0, G)$.

6. Powers of vector lattices

6.1. Squares

The following construction has roots in convexification of Banach lattices. The results are mostly taken from [33].

Definition 21. *Let E be a vector lattice. $(E^{\odot}, \odot)$ is called a square of E, if $E^{\odot}$ is a vector lattice and if*

(1) *$\odot : E \times E \to E^{\odot}$ is an orthosymmetric Riesz bimorphism.*
(2) *For every vector lattice F, whenever $T : E \times E \to F$ is an orthosymmetric bimorphism there exists a unique Riesz homomorphism $T^{\odot} : E^{\odot} \to F$ such that $T^{\odot} \circ \odot = T$.*

Remarkably, the square of any Archimedean vector lattice exists.

Theorem 22. *Let E be an Archimedean vector lattice. Then*

(1) *E has a square $(E^{\odot}, \odot)$.*
(2) *$(E^{\odot}, \odot)$ is (essentially) unique.*

There are various concrete ways to look at the square of a vector lattice. We offer two of these representations.

Theorem 23. *Let E be a uniformly complete vector sublattice of an Archimedean semiprime f-algebra G whose multiplication is indicated by a period $\bullet$. Put $E^2 := \{x \bullet y : x, y \in E\}$. Then E^2 is a vector sublattice of G and $(E^2, \bullet)$ is a square of E.*

The function $\vartheta : t \mapsto t \mid t \mid$ is an order isomorphism of $\mathbf{R}$. We define $H, J : \mathbf{R}^2 \to \mathbf{R}$ by

$$H(s,t) = \vartheta^{-1}(\vartheta(s) + \vartheta(t)) \text{ and}$$
$$J(s,t) = \vartheta^{-1}(st) \text{ for all } s, t \in \mathbf{R}.$$

Then $H, J \in \mathcal{H}(\mathbf{R}^2)$. Thus, $H(x,y)$ and $J(x,y)$ exist for all $x, y \in E$. We use the map H to define an addition $\tilde{+}$ and a scalar multiplication $\cdot$ on E as follows

$$x \tilde{+} y := H(x,y) \text{ and } \lambda \cdot x := \vartheta^{-1}(\lambda)x \text{ for all } x, y \in E \text{ and all } \lambda \in \mathbf{R}.$$

Then we show that the resulting vector space is a vector lattice that satisfies the universal property of the square.

Theorem 24. *Let E be a uniformly complete vector lattice. Define an addition $\tilde{+}$ and a scalar multiplication $\cdot$ on E as above. Then the following statements hold.*

(1) *Under these operations and with the given ordering, E is a vector lattice $E^\bullet$ and J, considered as a map from $E \times E$ into $E^\bullet$, is a surjective orthosymmetric Riesz bimorphism.*
(2) *If F is any vector lattice and if $T : E \times E \to F$ is bilinear, orthosymmetric and order bounded, there exists a unique $T^\bullet : E^\bullet \to F$ with $T = T^\bullet J$; this $T^\bullet$ is linear and order bounded. $T^\bullet$ is positive if and only if T is positive. $T^\bullet$ is a Riesz homomorphism if and only if T is a Riesz bimorphism.*
(3) *In particular $(E^\bullet, J)$ is a square of E.*

Corollary 25. *If E is Dedekind complete [or σ-Dedekind complete, or laterally complete] then so is its square.*

One can prove that the Fremlin tensor product

$$c_0 \bar{\otimes} c_0$$

is not uniformly complete. But for squares of course the following holds.

Corollary 26. *If E is uniformly complete then so is its square.*

One of the constructions of the square of E is as an appropriate quotient of the Fremlin tensor product of E with itself ([33]). From that construction, other nice properties of the square of a vector lattice can be derived as well. We offer some of them in the following result, see [63].

Theorem 27. *Let an Archimedean vector lattice E and the lattice bimorphism $\odot : (x,y) \mapsto x \odot y$ from $E \times E$ to $E^\odot$ be like before. Then we have the following.*

(1) *If b is an orthosymmetric regular bilinear map from $E \times E$ to some uniformly complete vector lattice F then there is a unique regular linear map $\Phi_b : E^\odot \to F$ with $b = \Phi_b \odot$.*
(2) *Given an arbitrary $u \in E^\odot$, there is $e_0 \in E_+$ such that, for every $\varepsilon > 0$, one can choose $x_1, \ldots, x_n, y_1, \ldots, y_n \in E$ with*

$$\left| u - \sum_{i=1}^{n} x_i \odot y_i \right| \leq \varepsilon e_0 \odot e_0.$$

(3) *For any $x, y \in E$ we have $x \odot y = 0$ if and only if $|x| \wedge |y| = 0$.*
(4) *Given an element $0 < u \in E^\odot$, there exits an $e \in E_+$ with $0 < e \odot e \leq u$.*

6.2. Higher powers

Now we turn our attention to higher powers of E, following the exposition in [17].

An s-linear map $T : \times_s E \to F$ is called *positive* if $T(u_1, \dots, u_s) \in F^+$ for all $u_1, \dots, u_s \in E^+$, where $\times_s E$ denotes the Cartesian product $E \times \cdots \times E$ (s-times). A *Riesz s-morphism* from $\times_s E$ into F is an s-linear map $T : \times_s E \to F$ such that for each $i \in \{1, \dots, s\}$ and $u_j \in E^+$ ($j \in \{1, \dots, s\}, j \neq i$), the equality

$$\begin{aligned} &T(u_1, \dots, u_{i-1}, u, u_{i+1}, \dots, u_s) \wedge T(u_1, \dots, u_{i-1}, v, u_{i+1}, \dots, u_s) \\ &= T(u_1, \dots, u_{i-1}, u \wedge v, u_{i+1}, \dots, u_s). \end{aligned}$$

holds in F. Riesz 1-morphisms are referred to as *Riesz homomorphisms.* It is obvious that every Riesz s-morphism is positive. Besides, it is proven in [12] that the s-linear map $T : \times_s E \to F$ is a Riesz s-morphism if and only if $|T(u_1, \dots, u_s)| = T(|u_1|, \dots, |u_s|)$ for all $u_1, \dots, u_s \in E$.

We now focus on orthosymmetric s-linear maps on vector lattices. Assume that $s \geq 2$. The s-linear map $T : \times_s E \to F$ is said to be *orthosymmetric* if T is positive and $T(u_1, \dots, u_s) = 0$ whenever $u_1, \dots, u_s \in E^+$ and $u_i \wedge u_j = 0$ for some i, j in $\{1, \dots, s\}$. In Proposition 2.1 of [13] it is shown that if in addition E and F are Archimedean then any orthosymmetric s-linear map $T : \times_s E \to F$ is *symmetric*, that is, $T(u_1, \dots, u_s) = T\left(u_{\sigma(1)}, \dots, u_{\sigma(s)}\right)$ for every permutation σ of the set $\{1, \dots, s\}$ and all $u_1, \dots, u_s \in E$. Next, from [17] we provide a useful necessary and sufficient condition for a lattice s-morphism to be orthosymmetric.

Lemma 28. *Let $s \in \{2, 3, \dots\}$, E and F be Archimedean vector lattices, and $T : \times_s E \to F$ be a lattice s-morphism. Then T is orthosymmetric if and only if T is symmetric.*

The formal definition of the s-power for an Archimedean vector lattice is as follows.

Definition 29. *Let E be a vector lattice and $s \in \{2, 3, \dots\}$. The pair $\left(E^{\circledS}, \circledS\right)$ is called an s-power of E if*

(1) *$E^{\circledS}$ is a vector lattice,*
(2) *$\circledS : \times_s E \to E^{\circledS}$ is a symmetric Riesz s-morphism, and*
(3) *for every vector lattice F and every symmetric Riesz s-morphism $T : \times_s E \to F$, there exists a unique Riesz homomorphism $T^{\circledS} : E^{\circledS} \to F$ such that $T = T^{\circledS} \circ \circledS$.*

Every Archimedean vector lattice has an s-power. The construction uses the Fremlin tensor product with more than two components like presented in [92]

Theorem 30. *Let $s \in \{2, 3, \dots\}$ and E be an Archimedean vector lattice. Then E has a unique (up to a lattice isomorphism) s-power $\left(E^{\circledS}, \circledS\right)$.*

An alternative formulation, just like for Fremlin's tensor product, can be offered when considering the range space to be uniformly complete.

Theorem 31. *Let $s \in \{2, 3, \dots\}$ and E be an Archimedean vector lattice. If E^{Φ} is an Archimedean vector lattice and $\Phi : \times_s E \to E^{\Phi}$ is a symmetric lattice s-morphism then the following statements are equivalents.*

(1) *The pair* (E^Φ, Φ) *is the s-power of* E.
(2) *For every uniformly complete vector lattice* F *and every orthosymmetric s-linear map* $T : \times_s E \to F$, *there exists a unique positive operator* $T^\Phi : E^\Phi \to F$ *such that* $T = T^\Phi \circ \Phi$.

7. Cauchy-Bunyakowski-Schwarz inequalities

In this section and the next we state a couple of results that are not yet avaible in the literature. In [31] the following general form of the classical Cauchy–Bunyakowski–Schwarz inequality was proved: if X is a real vector space and $b : X \times X \to F$ is a positively semidefinite symmetric bilinear operator with values in an almost f-algebra F then

$$b(x,y)b(x,y) \leq b(x,x)b(y,y) \quad (x,y \in X).$$

In the case of a semiprime f-algebra F this fact was established earlier in [59] and it was shown in [9] that the semiprimeness assumption can be omitted. In this survey we point to another improvement by Kusraev [63], replacing the almost f-algebra multiplication by an arbitrary positive orthosymmetric bilinear operator, the proof of which we will provide in a future paper. One can find a review of different generalizations and refinements of the classical Cauchy–Bunyakowski–Schwarz inequality in [45].

Theorem 32. *Let* X *be a real vector space,* E *be a vector lattice, and* $\langle \cdot, \cdot \rangle$ *be a positively semidefinite symmetric bilinear operator from* $X \times X$ *to* E. *Let* F *be another vector space and* $\circ : E \times E \to F$ *be a positive orthosymmetric bilinear operator. Then*

$$\langle x,y \rangle \circ \langle x,y \rangle \leq \langle x,x \rangle \circ \langle y,y \rangle \quad (x,y \in X).$$

8. Disjointness preserving bilinear operators

Let E, F, and G be vector lattices. A bilinear map $b : E \times F \to G$ is called *disjointness preserving* if

$$x_1 \perp x_2 \text{ implies } b(x_1,y) \perp b(x_2,y),$$
$$y_1 \perp y_2 \text{ implies } b(x,y_1) \perp b(x,y_2)$$

for arbitrary $x \in E$ and $y \in F$.

In many ways disjointness preserving bilinear maps behave like disjointness preserving linear maps.

Theorem 33.

(1) *A disjointness preserving bilinear map is order bounded if and only if it is regular (if and only if it is of order bounded variation).*
(2) *A Meyer type theorem is true for bilinear maps (cf. [46], first proved in [79]): For an order bounded disjointness preserving bilinear map* $p : E \times F \to G$

there exist two lattice bimorphisms p^+ and p^- from $E \times F$ to G such that $p = p^+ - p^-$ and $p^+(x,y) = p(x,y)^+$, $p^-(x,y) = p(x,y)^-$ for all $x \in E^+$ and $y \in F^+$. Moreover,

$$|p|(|x|,|y|) = |p(x,y)| \text{ for all } x \in E,\ y \in F.$$

(3) *If $p : E \times F \to G$ is an order bounded disjointness preserving bilinear map, G is uniformly complete, and $\phi = |p|$ then the principal ideal and the principal band generated by ϕ in $\mathit{Bil}_{bv}(E,F;G)$ can be described as $A \circ \phi = \{\alpha \circ \phi : \alpha \in A\}$, where A is an appropriate set of orthomorphisms.*

(4) *The correspondence $S \mapsto S\otimes$ is a bijection between the set of order bounded disjointness preserving linear maps from $E\bar{\otimes}F$ to G and the set of order bounded disjointness preserving bilinear maps from $E \times F$ into G.*

The results (1) and (2) were obtained independently in [15], [66], and [82], whereas (3) and (4) can be found in [66].

Order bounded disjointness preserving bilinear maps admit a nice analytic representation. Let E, F, and G be order-dense ideals of some universally complete vector lattices $\mathcal{E}$, $\mathcal{F}$, and $\mathcal{G}$. In $\mathcal{E}$, $\mathcal{F}$, and $\mathcal{G}$ we fix order units $1_{\mathcal{E}}$, $1_{\mathcal{F}}$, and $1_{\mathcal{G}}$ and consider multiplications that make these spaces f-algebras with units $1_{\mathcal{E}}$, $1_{\mathcal{F}}$, and $1_{\mathcal{G}}$ respectively. We recall that orthomorphisms in $\mathcal{E}$, $\mathcal{F}$, and $\mathcal{G}$ are multiplication operators and we identify them with the corresponding multipliers. For every $f \in \mathcal{E}$, there exists a unique element $g \in \mathcal{E}$ such that $fg{=}[f]\,1_{\mathcal{E}}$ and $[f] = [g]$, where $[f]$ stands for the band projection onto $f^{\perp\perp}$. We denote such an element g by $1_{\mathcal{E}}/f$ and the orthomorphism $g \mapsto g/f$ is also denoted by $1_{\mathcal{E}}/f$.

Consider order dense ideals $E' \subset \mathcal{E}$, $F' \subset \mathcal{F}$, $G' \subset \mathcal{G}$, and $G'' \subset \mathcal{G}$. Denote by $G' \cdot G''$ vector sublattice in $\mathcal{G}$ generated by the set $\{g'g'' : g' \in G', g'' \in G''\}$. If $w : E \to E'$, $v : F \to F'$, $S : E' \to G'$, and $T : F' \to G''$ are linear maps then $w \times v : E \times F \to E' \times F'$ and $S \bullet T : E' \times F' \to G' \cdot G''$ denote the linear and bilinear maps defined by $(x,y) \mapsto (wx, vy)$ and $(x,y) \mapsto S(x)T(y)$, respectively.

A *shift operator* from E' to G' is a restriction to $E'$of a positive linear map $\hat{S} : \hat{E} \to \mathcal{G}$ satisfying the properties: 1) $\hat{E}$ is an order dense ideal in $\mathcal{E}$ containing E' and $1_{\mathcal{E}}$; 2) $\hat{S}$ sends any component of $1_{\mathcal{E}}$ into a component of $1_{\mathcal{G}}$; 3) $\hat{S}$ is disjointness preserving; 4) $\hat{S}(1_{\mathcal{E}})^{\perp\perp} = \hat{S}(\hat{E})^{\perp\perp}$.

Now we phrase a representation result for order bounded disjointness preserving bilinear operators by Kusraev and Tabuev [67] corresponding to a weight-shift-weight representation result for linear operators due to A.E. Gutman, see [57] and [62]. The symbol ρS denotes an extension of $\rho \circ S$ as defined in [57].

Theorem 34. *Let $b : E \times F \to G$ be a regular disjointness preserving bilinear map. Then there exist order dense ideals $E' \subset \mathcal{E}$, $F' \subset \mathcal{F}$, $G' \subset \mathcal{G}$, and $G'' \subset \mathcal{G}$, shift operators $S : E' \to G'$ and $T : F' \to G''$, a family of pair-wise disjoint band projections $(\rho_\xi)_{\xi\in\Xi}$ in G, and families $(e_\xi)_{\xi\in\Xi}$ in E and $(f_\xi)_{\xi\in\Xi}$ in F such that*

$$\rho_\xi \circ b = \rho_\xi W \circ (\rho_\xi S \bullet \rho_\xi T) \circ (1_{\mathcal{E}}/e_\xi \times 1_{\mathcal{F}}/f_\xi) \quad (\xi \in \Xi)$$

where $W : \mathcal{G} \to \mathcal{G}$ is the orthomorphism of multiplication by $\sum_{\xi\in\Xi} \rho_\xi b(e_\xi, f_\xi)$.

9. Injective and projective tensor product

Let X and Y be Banach spaces and $X \otimes Y$ denote their algebraic tensor product. The *injective tensor norm* $\|\cdot\|_\varepsilon$ and the *projective tensor norm* $\|\cdot\|_\pi$ are, respectively, defined by

$$\|u\|_\varepsilon = \sup\left\{\left|\sum_{k=1}^{n} x^*(x_k)y^*(y_k)\right| : u = \sum_{k=1}^{n} x_k \otimes y_k \in X \otimes Y,\ x^* \in B_{X^*},\ y^* \in B_{Y^*}\right\}$$

and

$$\|u\|_\pi = \sup\left\{\left|\sum_{k=1}^{n} \varphi(x_k, y_k)\right| : u = \sum_{k=1}^{n} x_k \otimes y_k \in X \otimes Y,\ \varphi \in A\right\},$$

where A is the set of all bilinear functionals on $X \times Y$ with their norms ≤ 1. The completion of $X \otimes Y$ with respect to $\|\cdot\|_\varepsilon$ and $\|\cdot\|_\pi$ are, respectively, denoted by $X\check{\otimes}_\varepsilon Y$ and $X\hat{\otimes}_\pi Y$, called the *injective tensor product* and the *projective tensor product* of X and Y, respectively (see [40][44, Chapter 10][55, 56, 83]). The projective tensor norm $\|\cdot\|_\pi$ has another equivalent form:

$$\|u\|_\pi = \inf\left\{\sum_{k=1}^{n} \|x_k\| \cdot \|y_k\| : u = \sum_{k=1}^{n} x_k \otimes y_k \in X \otimes Y\right\}.$$

From the positivity perspective, it is interesting to know whether or not $X\check{\otimes}_\varepsilon Y$ and $X\hat{\otimes}_\pi Y$ are Banach lattices if both X and Y are Banach lattices. It is known that for a finite measure space (Ω, Σ, μ), $L_1(\mu)\hat{\otimes}_\pi X = L_1(\mu, X)$, which is a Banach lattice, and for a Hausdorff compact topological space K, $C(K)\check{\otimes}_\varepsilon X = C(K, X)$, which is also a Banach lattice. However, in general, $X\check{\otimes}_\varepsilon Y$ and $X\hat{\otimes}_\pi Y$ need not be Banach lattices. To explain why, let us first introduce the concepts of local unconditional structure and Gordon-Lewis spaces.

A Banach space X is said to have *a local unconditional structure* if there is a constant $c \geq 1$ such that for every finite-dimensional subspace E of X, the canonical embedding from E to X has a factorization $E \xrightarrow{v} Y \xrightarrow{u} X$, where Y is a Banach space with an unconditional basis, and u and v are continuous linear operators satisfying $\|u\| \cdot \|v\| \cdot ub(Y) \leq c$, where $ub(Y)$ is the unconditional basis constant of Y (see [42, p. 345]). Maurey in [77] showed that every Banach lattice has a local unconditional structure. A continuous linear operator T from a Banach space X to a Banach space Y is called 1-*summing* if there is a constant $c > 0$ such that for any finite sequence $x_1, \ldots, x_n$ in X,

$$\sum_{k=1}^{n} \|T(x_k)\| \leq c \cdot \sup\left\{\sum_{k=1}^{n} |x^*(x_k)| : x^* \in B_{X^*}\right\}.$$

T is called 1-*factorable* if there exists a measure space (Ω, Σ, μ) and operators $a : L_1(\mu) \longrightarrow Y^{**}$, $b : X \longrightarrow L_1(\mu)$ such that $k_Y T$ has a factorization $k_Y T : X \xrightarrow{b} L_1(\mu) \xrightarrow{a} Y^{**}$, where $k_Y : Y \longrightarrow Y^{**}$ is the canonical embedding. A Banach space X is said to have the *Gordon-Lewis property*, or said to be a *GL-space*, if every

1-summing operator from X to ℓ_2 is 1-factorable. Gordon and Lewis in [52] showed that every Banach space with a local unconditional structure is a GL-space. Thus every Banach lattice is a GL-space (also see [42, chapter 17]).

Let X be a Banach space with a Schauder basis $(x_m)_{m\in\mathbb{N}}$ and Y be a Banach space with a Schauder basis $(y_n)_{n\in\mathbb{N}}$. Gelbaum and Lamadrid in [51] showed that $(x_m \otimes y_n)_{(m,n)\in\mathbb{N}\times\mathbb{N}}$, taken in the rectangular order, forms a Schauder basis for both $X\check{\otimes}_\varepsilon Y$ and $X\hat{\otimes}_\pi Y$. In case that both $(x_m)_{m\in\mathbb{N}}$ and $(y_n)_{n\in\mathbb{N}}$ are unconditional basis, Diestel, Jarchow, and Tonge in [42, p. 365] showed that $(x_m\otimes y_n)_{(m,n)\in\mathbb{N}\times\mathbb{N}}$ is an unconditional basis for $X\check{\otimes}_\varepsilon Y$ and $X\hat{\otimes}_\pi Y$ if and only if $X\check{\otimes}_\varepsilon Y$ and $X\hat{\otimes}_\pi Y$ are GL-spaces, respectively. It is known that $\ell_2\check{\otimes}_\varepsilon\ell_2$ and $\ell_2\hat{\otimes}_\pi\ell_2$ (due to Gelbaum and Gil de Lamadrid [51]), and $\ell_p\check{\otimes}_\varepsilon\ell_q$ and $\ell_{p'}\hat{\otimes}_\pi\ell_{q'}$ (due to Kwapien and Pelczynski [68]), where $1 \le p,q < \infty$, $1/p + 1/p' = 1$, $1/q + 1/q' = 1$, and $1 \le 1/p + 1/q$, do not have an unconditional basis. Consequently, $\ell_2\check{\otimes}_\varepsilon\ell_2$, $\ell_2\hat{\otimes}_\pi\ell_2$, $\ell_p\check{\otimes}_\varepsilon\ell_q$, and $\ell_{p'}\hat{\otimes}_\pi\ell_{q'}$ are not Banach lattices.

Remark 35. *Cartwright and Lotz* [37] *showed that if X is a Banach space with a normalized unconditional basis $(x_n)_{n\in\mathbb{N}}$ and $1 \le p, p' < \infty$ such that $1/p+1/p' = 1$, then $(e_m \otimes x_n)_{(m,n)\in\mathbb{N}\times\mathbb{N}}$ is an unconditional basis for $\ell_p\check{\otimes}_\varepsilon X$ (respectively, for $\ell_{p'}\hat{\otimes}_\pi X$, $\ell_{p'} = c_0$ if $p = 1$) if and only if $(x_n)_{n\in\mathbb{N}}$ is equivalent to the usual basis in c_0 (respectively, in ℓ_1). They also in* [37] *showed that if X and Y are Banach lattices and $X\check{\otimes}_\varepsilon Y$ (respectively, $X\hat{\otimes}_\pi Y$) is a Banach lattice, and X (respectively, X^*) contains a closed sublattice which is lattice isomorphic to ℓ_p for $1 \le p < \infty$, then Y is lattice isomorphic to an AM-space (respectively, to an AL-space).*

Abramovich, Chen, and Wickstead [2] showed that a Banach lattice Y is isometrically lattice isomorphic to $C(K)$ for some Stonean space K, if and only if, for every Banach lattice X, $\mathcal{L}(X, Y)$ is a Banach lattice under the operator norm. Note that $(X\hat{\otimes}_\pi Y)^* = L(X, Y^*)$ (see [44, p. 230, Corollary 2]). The following proposition follows.

Proposition 36. *If one of the Banach lattices X or Y is an AL-space then $X\hat{\otimes}_\pi Y$ is a Banach lattice. On the other hand, if $X\hat{\otimes}_\pi Y$ is a Banach lattice for every Banach lattice X then Banach lattice Y is isometrically lattice isomorphic to an AL-space.*

Remark 37. *From Proposition* 39 *below, if one of X and Y is a Dedekind complete AM-space with an order unit then $X\check{\otimes}_\varepsilon Y$ is a Banach lattice.*

For Banach lattices X and Y, let $\mathcal{L}^r(X, Y)$ denote the space of continuous linear regular operators from X and Y and let $\mathcal{K}^r(X, Y)$ denote the space of compact regular operators from X and Y. For each $T \in \mathcal{L}^r(X, Y)$, the r-norm of T is defined as follows:

$$\|T\|_r = \inf\{\|S\| : S \in \mathcal{L}(X, Y)^+, |T(x)| \le S(x)\ \forall\ x \subset X^+\}.$$

Then $(\mathcal{L}^r(X, Y), \|\cdot\|_r)$ is a Banach space. If, moreover, Y is Dedekind complete then $(\mathcal{L}^r(X, Y), \|\cdot\|_r)$ is a Banach lattice (see [78, §1.3]).

For Banach lattices X and Y, let $X \otimes Y$ denote the algebraic tensor product of X and Y. For each $u = \sum_{k=1}^{n} x_k \otimes y_k \in X \otimes Y$, define $T_u : X^* \to Y$ by $T_u(x^*) = \sum_{k=1}^{n} x^*(x_k) y_k$ for each $x^* \in X^*$. The injective cone on $X \otimes Y$ is defined by

$$C_i = \{u \in X \otimes Y : T_u(x^*) \in Y^+ \text{ for all } x^* \in X^{*+}\}.$$

Wittstock in [94, 95] introduced the positive injective tensor norm on $X \otimes Y$ as follows:

$$\|u\|_{|\varepsilon|} = \inf\{\{\sup \|T_v(x^*)\| : x^* \in B_{X^{*+}}\} : v \in C_i,\ v \pm u \in C_i\}.$$

Let $X \check{\otimes}_{|\varepsilon|} Y$ denote the completion of $X \otimes Y$ with respect to $\|.\|_{|\varepsilon|}$. Then $X \check{\otimes}_{|\varepsilon|} Y$ with C_i as its positive cone is a Banach lattice (also see Section 3.8 in [78]), called the *Wittstock injective tensor product*. Moreover, by Proposition 3.8.7 in [78], $\|u\|_{|\varepsilon|} = \|T_u\|_r$. Thus $X \check{\otimes}_{|\varepsilon|} Y$ is a closed subspace of $\mathcal{L}^r(X^*, Y)$ with the regular operator norm $\|.\|_r$. We also know that $X \check{\otimes}_{|\varepsilon|} Y$ is a closed subspace of $\mathcal{L}(X^*, Y)$ with the operator norm $\|.\|$, that as vector spaces $\mathcal{L}^r(X^*, Y) \subseteq \mathcal{L}(X^*, Y)$ and that for each $T \in \mathcal{L}^r(X^*, Y)$ we have $\|T\| \leq \|T\|_r$. Thus (as vector spaces) $X \check{\otimes}_{|\varepsilon|} Y \subseteq X \check{\otimes}_{\varepsilon} Y$ and for each $u \in X \check{\otimes}_{|\varepsilon|} Y$, $\|u\|_{\varepsilon} \leq \|u\|_{|\varepsilon|}$. To get the equality (isometrically and isomorphically) $X \check{\otimes}_{|\varepsilon|} Y = X \check{\otimes}_{\varepsilon} Y$, we need the following proposition due to Cartwright and Lotz [37].

Proposition 38. *Let Y be a Dedekind complete Banach lattice such that there is a positive projection from Y^{**} to Y. Then $\mathcal{L}^r(X, Y)$ is (isometrically) isomorphic to $\mathcal{L}(X, Y)$ if and only if X is (isometrically) isomorphic to an AL-space or Y is (isometrically) isomorphic to an AM-space.*

Note that for every Dedekind complete AM-space Y with an order unit, there is a positive projection from Y^{**} to Y. Therefore, if Y is (isometrically) isomorphic to a Dedekind complete AM-space with an order unit, then $\mathcal{L}^r(X, Y)$ is (isometrically) isomorphic to $\mathcal{L}(X, Y)$. Note that $X \otimes Y$ is dense in $X \check{\otimes}_{|\varepsilon|} Y$ and $X \check{\otimes}_{\varepsilon} Y$ respectively. Then we have the following proposition.

Proposition 39. *Let X and Y be Banach lattices. If one of X and Y is (isometrically) isomorphic to a Dedekind complete AM-space with an order unit, then $X \check{\otimes}_{|\varepsilon|} Y$ is (isometrically) isomorphic to $X \check{\otimes}_{\varepsilon} Y$.*

The partial converse to Proposition 36 is contained in [37] and [78, p. 196, Corollary 3.2.2] as follows.

Proposition 40. *Let $1 < p < \infty$. If*

$$\ell_p \check{\otimes}_{|\varepsilon|} Y = \ell_p \check{\otimes}_{\varepsilon} Y \quad \text{or} \quad L_p[0,1] \check{\otimes}_{|\varepsilon|} Y = L_p[0,1] \check{\otimes}_{\varepsilon} Y,$$

then Y is isomorphic to an AM-space.

Remark 41. *Abramovich and Wickstead* [3] *showed that if Y is a Dedekind complete Banach lattice then $\mathcal{L}^r(X,Y) = \mathcal{L}(X,Y)$ for all Banach lattices X if and only if Y has an order unit. In this case, the regular operator norm and the operator norm on $\mathcal{L}^r(X,Y) = \mathcal{L}(X,Y)$ are equivalent. Thus, $X\check{\otimes}_{|\varepsilon|}Y$ is isomorphic to $X\check{\otimes}_{\varepsilon}Y$.*

For Banach lattices X and Y, the *projective cone* on $X \otimes Y$ is defined to be (see [78, p. 229])

$$C_p = \left\{ \sum_{k=1}^{n} x_k \otimes y_k : n \in \mathbb{N}, x_k \in X^+, y_k \in Y^+ \right\}.$$

Fremlin [48, 49] introduced the *positive projective tensor norm* on $X \otimes Y$ as follows:

$$\|u\|_{|\pi|} = \sup \left\{ \left| \sum_{k=1}^{n} \varphi(x_k, y_k) \right| : u = \sum_{k=1}^{n} x_k \otimes y_k \in X \otimes Y, \varphi \in B \right\},$$

where B is the set of all bipositive bilinear functionals on $X \times Y$ with their norms ≤ 1. An equivalent form of $\|\cdot\|_{|\pi|}$, also introduced by Fremlin, was already used in the beginning of Section 2.

Let $X\hat{\otimes}_{|\pi|}Y$ denote the completion of $X \otimes Y$ with respect to the positive projective tensor norm $\|\cdot\|_{|\pi|}$. Then $X\hat{\otimes}_{|\pi|}Y$ with C_p as its positive cone is a Banach lattice, called the *Fremlin projective tensor product* of X and Y.

If X and Y are Banach lattices, then $X\hat{\otimes}_{|\pi|}Y$ and $X\hat{\otimes}_{\pi}Y$ are both well defined, where $X\hat{\otimes}_{|\pi|}Y$ is a Banach lattice and $X\hat{\otimes}_{\pi}Y$ is a Banach space (but it may not be a lattice). It follows from definitions that $\|u\|_{|\pi|} \leq \|u\|_{\pi}$ for each $u \in X \otimes Y$. Thus as vector spaces, $X\hat{\otimes}_{\pi}Y \subseteq X\hat{\otimes}_{|\pi|}Y$ and for each $u \in X\hat{\otimes}_{\pi}Y$, $\|u\|_{|\pi|} \leq \|u\|_{\pi}$. Fremlin in [49] showed that $(X\hat{\otimes}_{|\pi|}Y)^* = \mathcal{L}^r(X, Y^*)$. Note that $(X\hat{\otimes}_{\pi}Y)^* = \mathcal{L}(X, Y^*)$, every dual Banach lattice Y^* is Dedekind complete, and there is a positive projection from Y^{***} to Y^*. Thus by Proposition 36, $\mathcal{L}^r(X, Y^*)$ is (isometrically) isomorphic to $\mathcal{L}(X, Y^*)$ if and only if X is (isometrically) isomorphic to an AL-space or Y^* is (isometrically) isomorphic to an AM-space. Therefore, we have

Proposition 42. *Let X and Y be Banach lattices. Then $X\hat{\otimes}_{|\pi|}Y$ is (isometrically) isomorphic to $X\hat{\otimes}_{\pi}Y$ if and only if one of X and Y is (isometrically) isomorphic to an AL-space.*

10. RNP for Tensor Products

In 1976, Diestel and Uhl in [43] showed that the Radon-Nikodym property (hereafter called RNP) is inherited from the dual Banach spaces X^*, Y^* to their projective tensor product $X^*\hat{\otimes}_{\pi}Y^*$ if X^* has the approximation property (hereafter called AP). They then asked in their classic monograph [44] whether RNP can be inherited from any two Banach spaces to their projective tensor product.

In 1983, Bourgain and Pisier in [18] constructed a Banach space X with RNP for which the projective tensor product $X\hat{\otimes}_\pi X$ fails to have RNP. Thus in general RNP is *not* inherited from any two Banach spaces to their projective tensor product. However, Diestel and Uhl's result in [43] tells us that RNP is inherited under special circumstances. In 1983, Andrews in [6] improved Diestel and Uhl's result and showed that RNP is inherited from any Banach dual space X^* and any Banach space Y to their projective tensor product $X^*\hat{\otimes}_\pi Y$ if X^* has AP. Moreover, Bu and Lin in 2004 in [25] showed that RNP is inherited from any two Banach lattices E and F to their projective tensor product $E\hat{\otimes}_\pi F$. Recently, Diestel, Fourie, and Swart in [41] improved Bu and Lin's result and showed that RNP is inherited from any Banach lattice E and any Banach space X to their projective tensor product $E\hat{\otimes}_\pi X$.

Moreover, not only RNP, but other types of RNP as well, such as the analytic RNP, the near RNP, the non-containment of a copy of c_0 (due to [24]), and the weakly sequential completeness (due to [73]) are inherited from Banach spaces X and Y to their projective tensor product $X\hat{\otimes}_\pi Y$ if one of them has an unconditional basis.

If X and Y are Banach lattices, is RNP inherited from X and Y to their Fremlin projective tensor product? Fremlin in [49] showed that $L_2[0,1]\hat{\otimes}_{|\pi|}L_2[0,1]$ is not Dedekind complete. Thus $L_2[0,1]\hat{\otimes}_{|\pi|}L_2[0,1]$ does not have RNP. This counterexample tells us that RNP is not always inherited from Banach lattices X and Y to their Fremlin projective tensor product $X\hat{\otimes}_{|\pi|}Y$, even if both X and Y have very nice properties as $L_2[0,1]$ does. However, in case that one of X, Y is ℓ_p for $1 \le p < \infty$, Bu and Buskes in [21] showed that RNP is inherited from Banach lattices X, Y to their Fremlin projective tensor product $X\hat{\otimes}_{|\pi|}Y$, which we now more closely examine.

Let X be a Banach lattice. Bu and Buskes in [21] introduced the Banach sequence lattices $\ell_p^\varepsilon(X)$ $(1 \le p < \infty)$ and $\ell_p^\pi(X)$ $(1 < p < \infty)$ as follows:

$$\ell_p^\varepsilon(X) = \left\{ \bar{x} = (x_n)_n \in X^{\mathbb{N}} : \sum_{n=1}^{\infty} [|x^*|(|x_n|)]^p < \infty,\ \forall\, x^* \in X^* \right\}$$

with a lattice norm

$$\|\bar{x}\|_{\ell_p^\varepsilon(X)} = \sup\left\{ \left(\sum_{n=1}^{\infty} [|x^*|(|x_n|)]^p \right)^{1/p} : x^* \in B_{X^*} \right\},$$

and

$$\ell_p^\pi(X) = \left\{ \bar{x} = (x_n)_n \in X^{\mathbb{N}} : \sum_{n=1}^{\infty} |x_n^*|(|x_n|) < \infty,\ \forall\, (x_n^*)_n \in \ell_{p'}^\varepsilon(X^*) \right\}$$

with a lattice norm

$$\|\bar{x}\|_{\ell_p^\pi(X)} = \sup\left\{ \sum_{n=1}^{\infty} |x_n^*|(|x_n|) : (x_n^*)_n \in B_{\ell_{p'}^\varepsilon(X^*)} \right\},$$

where $1 < p' < \infty$ such that $1/p+1/p' = 1$. Let $\ell_p^{\varepsilon,0}(X)$ denote the closed subspace of $\ell_p^{\varepsilon}(X)$ consisting of all elements whose tails converge to 0, i.e.,

$$\ell_p^{\varepsilon,0}(X) = \left\{\bar{x} \in \ell_p^{\varepsilon}(X) : \lim_n \|\bar{x}(> n)\|_{\ell_p^{\varepsilon}(X)} = 0\right\},$$

where $\bar{x}(> n) = (0, \dots, 0, x_{n+1}, x_{n+2}, \dots)$ for $\bar{x} = (x_1, x_2, \dots)$. Then $\ell_p^{\varepsilon,0}(X)$ is a sublattice and ideal of $\ell_p^{\varepsilon}(X)$. For each $\bar{x} = (x_n)_n \in \ell_p^{\varepsilon}(X)$, define $T_{\bar{x}} : \ell_{p'} \longrightarrow X$ by $T_{\bar{x}}(t) = \sum_{n=1}^{\infty} t_n x_n$ for each $t = (t_n)_n \in \ell_{p'}$. Following these notations, Bu and Buskes in [21] showed the following propositions.

Proposition 43. *Let X be a Banach lattice, $1 \le p, p' < \infty$ such that $1/p+1/p' = 1$, and $\ell_{p'} = c_0$ if $p = 1$. Then $\mathcal{L}^r(\ell_{p'}, X)$ is a Banach lattice which is isometrically lattice isomorphic to $\ell_p^{\varepsilon}(X)$ under the mapping: $\bar{x} \longleftrightarrow T_{\bar{x}}$. Moreover, $T_{\bar{x}} \in \mathcal{K}^r(\ell_{p'}, X)$ if and only if $\bar{x} \in \ell_p^{\varepsilon,0}(X)$.*

Proposition 44.

(1) *Let X be a Banach lattice and $1 \le p < \infty$. Then $\ell_p \check{\otimes}_{|\varepsilon|} X$ is isometrically lattice isomorphic to $\ell_p^{\varepsilon,0}(X)$.*
(2) *Let X be a Banach lattice and $1 < p < \infty$. Then $\ell_p \hat{\otimes}_{|\pi|} X$ is isometrically lattice isomorphic to $\ell_p^{\pi}(X)$.*

Recall that a continuous linear operator between Banach spaces is called a *semi-embedding* if it is one to one and takes the closed unit ball in the domain to a closed subset (see [74]). For a Banach lattice X and $1 \le p < \infty$, define

$$\ell_p^{strong}(X) = \left\{\bar{x} = (x_n)_n \in X^{\mathbb{N}} : \sum_{n=1}^{\infty} \|x_n\|^p < \infty\right\}$$

and

$$\|\bar{x}\|_{\ell_p^{strong}(X)} = \left(\sum_{n=1}^{\infty} \|x_n\|^p\right)^{1/p}.$$

Then $\ell_p^{strong}(X)$ with this norm is a Banach lattice. Bu and Buskes in [21] showed that if $1 < p < \infty$ then the inclusion map $\ell_p^{\pi}(X) \hookrightarrow \ell_p^{strong}(X)$ is a semi-embedding and also showed that the tail of each element in $\ell_p^{\pi}(X)$ converges to 0 in $\ell_p^{\pi}(X)$. It is known that $\ell_p^{strong}(X)$ $(1 \le p < \infty)$ has RNP if X has RNP (see [71]) and that if there is a semi-embedding from a separable Banach space X to a Banach space with RNP then X has RNP (see [19]). Thus Bu and Buskes showed that $\ell_p^{\pi}(X)$ has RNP if X has. They also showed that $\ell_p^{\varepsilon}(X)$ has RNP if and only if $\ell_p^{\varepsilon,0}(X)$ has RNP if and only if X has RNP and $\ell_p^{\varepsilon}(X) = \ell_p^{\varepsilon,0}(X)$. Combining these results with Propositions 43 and 44, we have the following.

Theorem 45.

(1) *Let X be a Banach lattice and $1 < p < \infty$. Then $\ell_p \hat{\otimes}_{|\pi|} X$ has RNP if and only if X has RNP.*
(2) *Let X be a Banach lattice, $1 \leq p, p' < \infty$ such that $1/p + 1/p' = 1$, and $\ell_{p'} = c_0$ if $p = 1$. Then $\ell_p \check{\otimes}_{|\varepsilon|} X$ has RNP if and only if $\mathcal{L}^r(\ell_{p'}, X)$ has RNP if and only if X has RNP and each positive operator from $\ell_{p'}$ to X is compact.*

Remark 46.

(1) *In case $p = 1$, $\ell_1 \hat{\otimes}_{|\pi|} X = \ell_1 \hat{\otimes}_\pi X = \ell_1^{\mathrm{strong}}(X)$. It follows from [75] that $\ell_1 \hat{\otimes}_{|\pi|} X$ has RNP if and only if X has RNP.*
(2) *Similar to the proof of Theorem 12 in [20], one can show that $\ell_p \check{\otimes}_\varepsilon X$ has RNP if and only if $\mathcal{L}(\ell_{p'}, X)$ has RNP if and only if X has RNP and each continuous linear operator from $\ell_{p'}$ to X is compact, where X is a Banach space, $1 \leq p, p' < \infty$ such that $1/p + 1/p' = 1$, and $\ell_{p'} = c_0$ if $p = 1$. Thus in case that X is a Banach lattice, if $\ell_p \check{\otimes}_\varepsilon X$ has RNP then $\ell_p \check{\otimes}_{|\varepsilon|} X$ also has RNP. However, the converse is not true for all Banach lattices X. Indeed, let p and q be numbers such that $1 \leq q < p' \leq 2$ where $1/p + 1/p' = 1$. Then combining Theorem 45 with Theorem 4.9 in [39], we infer that $\ell_p \check{\otimes}_{|\varepsilon|} L_q[0,1]$ has RNP. But from Theorem 4.7 in [39], it follows that there is a continous linear operator from $\ell_{p'}$ to $L_q[0,1]$ which is not compact. Therefore $\ell_p \check{\otimes}_\varepsilon L_q[0,1]$ does not have have RNP.*

Using the same ideas, Bu, Buskes, and Lai in [23] (also see [70]) extended results in [21] to Orlicz sequence spaces ℓ_φ. They introduced Banach sequence lattices $\ell_\varphi^{strong}(X), \ell_\varphi^{\epsilon}(X), \ell_\varphi^{\epsilon,0}(X)$, and $\ell_\varphi^{\pi}(X)$ where ℓ_φ is an Orlicz sequence space and X is a Banach lattice. They used the lower and upper Matuszewska-Orlicz indices of an Orlicz function φ to show that the inclusion map $\ell_\varphi^{\pi}(X) \hookrightarrow \ell_\varphi^{strong}(X)$ is a semi-embedding and that the tail of each element of $\ell_\varphi^{\pi}(X)$ converges to 0 in $\ell_\varphi^{\pi}(X)$. They also showed that the Fremlin projective tensor product $\ell_\varphi \hat{\otimes}_{|\pi|} X$ is isometrically isomorphic and lattice homomorphic to $\ell_\varphi^{\pi}(X)$ and that the Wittstock injective tensor product $\ell_\varphi \check{\otimes}_{|\varepsilon|} X$ is isometrically isomorphic and lattice homomorphic to $\ell_\varphi^{\epsilon,0}(X)$. Finally, they proved the following theorem.

Theorem 47. *Let X be a Banach lattice, and φ and φ^* be Orlicz functions that are complementary to each other. Then*

(1) *$\ell_\varphi \hat{\otimes}_{|\pi|} X$ has the RNP if and only if both ℓ_φ and X have RNP.*
(2) *$\ell_\varphi \check{\otimes}_{|\varepsilon|} X$ has RNP if and only if $\mathcal{L}^r(h_{\varphi^*}, X)$ has RNP if and only if both ℓ_φ and X have RNP and each positive continuous linear operator from h_{φ^*} to X is compact, where h_{φ^*} is the order continuous part of ℓ_{φ^*}.*

Remark 48. *Recently, Bu and Buskes in [22] have used a different technique to show that if one of the Banach lattices X, Y is atomic then the Fremlin projective tensor product $X \hat{\otimes}_{|\pi|} Y$ has RNP if and only if both X and Y have RNP.*

References

[1] Abramovich, Y.A, When each continuous operator is regular, in *Functional analysis, optimization, and mathematical economics*, 133–140, Oxford Univ. Press, New York, 1990.

[2] Abramovich, Y.A., Chen, Z.L., and Wickstead, A.W., Regular-norm balls can be closed in the strong operator topology, *Positivity* **1** (1997), 75–96.

[3] Abramovich, Y.A. and Wickstead, A.W., When each continuous linear operator is regular II, *Indag. Math. (N.S.)* **8** (1997), 281–294.

[4] Aliprantis, C.D. and Burkinshaw, O., *Positive Operators,* Academic Press, Orlando, 1985.

[5] Aliprantis, Charalambos D. and Burkinshaw, Owen, *Locally Solid Riesz spaces with Applications to Economics*, Second edition, Mathematical Surveys and Monographs, **105**, American Mathematical Society, Providence, RI, 2003.

[6] Andrews, K.T., The Radon-Nikodym property for spaces of operators, *J. London Math. Soc.* **28** (1983), 113–122.

[7] Arens, R., The adjoint of a bilinear operation, *Proc. Amer. Math. Soc.*, **2** (1951), 839–848.

[8] Arikan, Nilgün, A simple condition ensuring the Arens regularity of bilinear mappings, *Proc. Amer. Math. Soc.* **84** (1982), 525–531.

[9] Bernau, S.J. and Huijsmans, C.B., The Schwarz inequality in Archimedean f-algebras. *Indag. Math. (N.S.)* **7** (1996), no. 2, 137–148.

[10] Bernau, S.J. and Huijsmans, C.B., The order bidual of almost f-algebras and d-algebras, *Trans. Amer. Math. Soc.* **347** (1995), 4259–4274.

[11] Birkhoff, Garrett and Pierce, R.S., Lattice-ordered rings, *An. Acad. Brasil. Ci.* 28 (1956), 41–69.

[12] Boulabiar, K., Some aspect of Riesz multimorphisms, *Indag. Math.* **13** (2002), 419–432.

[13] Boulabiar, K., On products in lattice-ordered algebras, *J. Aust. Math. Soc.*, **75** (2003), 23–40.

[14] Boulabiar, K. and Toumi, M.A., Lattice bimorphisms on f-algebras, *Algebra Univ.* **48** (2002), 103–116.

[15] Boulabiar, Karim, Buskes, Gerard, and Page, Robert On some properties of bilinear maps of order bounded variation. *Positivity* **9** (2005), no. 3, 401–414.

[16] Boulabiar, K., Buskes G., and Triki, A., Some recent trends and advances in certain lattice ordered algebras. (English. English summary) *Function spaces* (Edwardsville, IL, 2002), 99–133, Contemp. Math., **328**, Amer. Math. Soc., Providence, RI, 2003.

[17] Boulabiar, Karim and Buskes, Gerard, Vector lattice powers: f-algebras and functional calculus. *Comm. Algebra* **34** (2006), no. 4, 1435–1442.

[18] Bourgain, J. and Pisier, G., A construction of $\mathcal{L}_\infty$-spaces and related Banach spaces, *Bol. Soc. Brasil. Mat.* **14** (1983), 109–123.

[19] Bourgain, J. and Rosenthal, H.P., Applications of the theory of semi-embeddings to Banach space theory, *J. Funct. Anal.* **52** (1983), 149–188.

[20] Bu, Q., Some properties of the injective tensor product of $L^p[0,1]$ and a Banach space, *J. Funct. Anal.* **204** (2003), 101–121.

[21] Bu, Q. and Buskes, G., The Radon-Nikodym property for tensor products of Banach lattices, *Positivity* **10** (2006), 365–390.

[22] Bu, Q. and Buskes, G., Schauder decompositions and the Fremlin projective tensor product of Banach lattices, preprint.

[23] Bu, Q., Buskes, G., and Lai, W.K., The Radon-Nikodym property for tensor products of Banach lattices II, to appear in *Positivity.*

[24] Bu, Q., Diestel, J., Dowling, P.N., and Oja, E., Types of Radon-Nikodym properties for the projective tensor product of Banach spaces, *Illinois J. Math.* **47** (2003), 1303–1326.

[25] Bu, Q. and Lin, P.K., Radon-Nikodym property for the projective tensor product of Köthe function spaces, *J. Math. Anal. Appl.* **294** (2004), 149–159.

[26] Buskes, G.J.H.M. and van Rooij, A.C.M.: Hahn–Banach for Riesz homomorphisms, *Indag. Math.* **51** (1993), 25–34.

[27] Buskes, G.J.H.M.; van Rooij, A.C.M. The Archimedean l-group tensor product, *Order* **10** (1993), no. 1, 93–102.

[28] Buskes, G. and van Rooij, A., Bounded variation and tensor products of Banach lattices, *Positivity* **7** (2003), 47–59.

[29] Buskes, Gerard, Commutativity of certain algebras using a Mazur-Orlicz idea, *Proc. Orlicz Mem. Conf., Oxford/MS (USA)* 1991, Exp. No.1, 1-3 (1991).

[30] Buskes, G. and Kusraev, A., Extension and representation of orthoregular maps, *Vladikavkaz Math. J.* **9** (2007), no. 2, 16–29.

[31] Buskes, G. and van Rooij, A., Almost f-algebras: commutativity and Cauchy-Schwarz inequality, *Positivity* **4** (2000), 227–231.

[32] Buskes, G. and van Rooij, A., Bounded variation and tensor products of Banach lattices. Positivity and its applications (Nijmegen, 2001).*Positivity* **7** (2003), no. 1–2, 47–59.

[33] Buskes, G.; van Rooij, A., Squares of Riesz spaces. *Rocky Mountain J. Math.* **31** (2001), no. 1, 45–56.

[34] Buskes, G.; van Rooij, A., The bornological tensor product of two Riesz spaces. *Ordered algebraic structures*, 3–9, Dev. Math., **7**, Kluwer Acad. Publ., Dordrecht, 2002.

[35] Buskes, G.; van Rooij, A., The bornological tensor product of two Riesz spaces: proof and background material. *Ordered algebraic structures*, 189–203, Dev. Math., 7, Kluwer Acad. Publ., Dordrecht, 2002.

[36] Buskes, Gerard; Page, Robert, A positive note on a counterexample by Arens. *Quaest. Math.* **28** (2005), no. 1, 117–121.

[37] Cartwright, D.I. and Lotz, H.P., Some characterizations of AM- and AL-spaces, *Math. Z.* **142** (1975), 97–103.

[38] Chen, Z.L. and Wickstead, A.W., Some applications of Rademacher sequences in Banach lattices, *Positivity* **2** (1998), 171–191.

[39] Cristescu, R., *Ordered vector spaces and linear operators*, Translated from the Romanian, Abacus Press, Tunbridge, 1976.

[40] Defant, A. and Floret, K., *Tensor Norms and Operator Ideals*, Noth-Holland, Amsterdam, 1993.

[41] Diestel, J., Fourie, J., and Swart, J., The projective tensor product II: The Radon-Nikodym property, *Rev. R. Acad. Cien. Serie A. Mat.* **100** (2006), 1–26.

[42] Diestel, J., Jarchow, H., and Tonge, A., *Absolutely Summing Operators*, Cambridge University Press, 1995.

[43] Diestel, J. and Uhl, J.J., The Radon-Nikodym theorem for Banach space valued measures, *Rocky Mountain J. Math.* **6** (1976), 1–46.

[44] Diestel, J. and Uhl, J.J., *Vector Measures*, Mathematical Surveys, vol. **15**, American Mathematical Society, Providence, R.I., 1977.

[45] Dragomir, S.S., *A Survey on Cauchy–Buniakowsky–Schwartz Type Discrete Inequalities.* Victoria University, RGMIA Monographs (2000).

[46] Duhoux, M. and Meyer, M., A new proof of the lattice structure of orthomorphisms, *J. London Math. Soc.*, **25**, No. 2 (1982), 375–378.

[47] Fremlin, D.H., Abstract Köthe spaces I, *Proc. Cambridge Philos. Soc.* **63** (1967), 653–660.

[48] Fremlin, D.H., Tensor products of Archimedean vector lattices, *Amer. J. Math.* **94** (1972), 778–798.

[49] Fremlin, D.H., Tensor products of Banach lattices, *Math. Ann.* **211** (1974), 87–106.

[50] Gaans, O.W. van.: The Riesz part of a positive bilinear form, In: *Circumspice*, Katholieke Universiteit Nijmegen, Nijmegen (2001), 19–30.

[51] Gelbaum, B.R. and Gil de Lamadrid, J., Bases of tensor products of Banach spaces, *Pacific J. Math.* **11** (1961), 1281–1286.

[52] Gordon, Y. and Lewis, D.R., Absolutely summing operators and local unconditional structure, *Acta. Math.* **133** (1974), 27–48.

[53] Grobler, J.J., Commutativity of the Arens product in lattice ordered algebras, *Positivity,* **3** (1999), 357–364.

[54] Grobler, J.J. and Labuschagne, C., The tensor product of Archimedean ordered vector spaces, *Math. Proc. Cambridge Philos. Soc.* **104** (1988), no. 2, 331–345.

[55] Grothendieck, A., 'Résumé de la théorie métrique des produits tensoriels topologiques, *Bol. Soc. Mat. São Paulo* **8** (1953/1956), 1–79.

[56] Grothendieck, A., Produits tensoriels topologiques et espaces nucléaires, *Mem. Amer. Math. Soc.* **16** (1955).

[57] Gutman, A.G., Disjointness preserving operators, In: *Vector Lattices and Integral Operators* (Ed. S. S. Kutateladze), Kluwer, Dordrecht etc., 361–454 (1996).

[58] Hogbe-Nlend, Henri, *Bornologies and Functional Analysis*, Mathematics Studies 26, North Holland Publishing Company, Amsterdam-New York-Oxford, 1977.

[59] Huijsmans, C.B. and de Pagter, B.: Averaging operators and positive contractive projections, *J. Math. Anal. and Appl.* **107** (1986), 163–184.

[60] Kudláček, V., On some types of ℓ-rings. *Sb. Vysoké. Učení Tech. Brno* **1–2** (1962), 179–181.

[61] Kusraev, A.G., *On a Property of the Base of the K-space of Regular Operators and Some of Its Applications* [in Russian], Sobolev Inst. of Math., Novosibirsk, (1977).

[62] Kusraev, A.G., *Dominated Operators.* Kluwer, Dordrecht (2000).

[63] Kusraev, A.G., On the structure of orthosymmetric bilinear operators in vector lattices [in Russian], Dokl. RAS, **408**, No 1 (2006), 25–27.

[64] Kusraev, A.G., On the representation of orthosymmetric bilinear operators in vector lattices [in Russian], *Vladikavkaz Math. J.* **7**, No 4 (2005), 30–34.

[65] Kusraev, A.G. and Shotaev, G.N., Bilinear majorizable operators [in Russian], *Studies on Complex Analysis, Operator Theory, and Mathematical Modeling* (Eds. Yu. F. Korobeĭnik and A. G. Kusraev). Vladikavkaz Scientific Center, Vladikavkaz (2004), 241–262.

[66] Kusraev, A.G. and Tabuev, S.N., On disjointness preserving bilinear operators [in Russian], Vladikavkaz Math. J. **6**, No. 1 (2004), 58–70.

[67] Kusraev, A.G. and Tabuev, S.N., On multiplicative representation of disjointness preserving bilinear operators [in Russian], *Sib. Math. J.*, to appear.

[68] Kwapień, S. and Pelczyński, A., The main triangle projection in matrix spaces and its applications, *Studia Math.* **34** (1970), 43–68.

[69] Labuschagne, C.C.A., Riesz reasonable cross norms on tensor products of Banach lattices. *Quaest. Math.* **27** (2004), no. 3, 243–266.

[70] Lai, W.K., *The Radon-Nikodym property for the Fremlin and Wittstock tensor products of Orlicz sequence spaces with Banach lattices*, Ph.D. dissertation, the University of Mississippi, 2007.

[71] Leonard, I.E., Banach sequence spaces, *J. Math. Anal. Appl.* **54** (1976), 245–265.

[72] Levin, V.L., Tensor products and functors in Banach space categories defined by KB-lineals. (Russian) *Dokl. Akad. Nauk SSSR* **163** (1965), 1058–1060.

[73] Lewis, D.R., Duals of tensor products, in Banach Spaces of Analytic Functions (Proc. Pelczynski Conf., Kent State Univ., Kent, Ohio, 1976), pp. 57–66. *Lecture Notes in Math.*, Vol. **604**, Springer, Berlin, 1977.

[74] Lotz, H.P., Peck, N.T., and Porta, H., Semi-embedding of Banach spaces, *Proc. Edinburgh Math. Soc.* **22** (1979), 233–240.

[75] Lotz, Heinrich P., Extensions and liftings of positive linear mappings on Banach lattices, *Trans. Amer. Math. Soc.* **211** (1975), 85–100.

[76] Luxemburg, W.A.J.; Zaanen, A.C., *Riesz spaces. Vol. I.*, North-Holland Mathematical Library. North-Holland Publishing Co., Amsterdam-London; American Elsevier Publishing Co., New York, 1971.

[77] Maurey, B., Type et cotype dans les espaces munis de structures locales inconditionelles, École Polyt. Palaiseau, *Sém. Maurey-Schwartz* 1973/74, Exp. XXIV-XXV.

[78] Meyer-Nieberg, P., *Banach Lattices*, Springer, 1991.

[79] Meyer, Y., Quelques propriétés des homomorphismes d'espaces vectoriels réticulés. *Equipe d'Analyse, E.R.A.* 294, Université Paris VI, 1978.

[80] Nakano, H., Product spaces of semi-ordered linear spaces, *J. Fac. Sci. Hokkaido Univ. Ser. I.*, **12**, (1953), 163–210.

[81] de Pagter, B., *f-algebras and orthomorphisms* (Thesis, Leiden, 1981).

[82] Page, Robert, *On bilinear maps of order bounded variation* (Thesis, University of Mississippi, 2005).

[83] Ryan, R.A., *Introduction to Tensor Products of Banach Spaces*, Springer, 2002.

[84] Schaefer, Helmut H., *Banach lattices and positive operators,* Die Grundlehren der mathematischen Wissenschaften, Band 215. Springer-Verlag, New York-Heidelberg, 1974.

[85] Schaefer, Helmut H., Normed tensor products of Banach lattices. *Proceedings of the International Symposium on Partial Differential Equations and the Geometry of Normed Linear Spaces* (Jerusalem, 1972). Israel J. Math. **13** (1972), 400–415.

[86] Schaefer, H.H., Positive bilinear forms and the Radon-Nikodym theorem. *Funct. Anal.: Survey and Recent Results.* **3**: Proc 3rd Conf. Paderborn, 24–29 May, 1983. Amsterdam e.a., 1984, 135–143.

[87] Scheffold, E., FF-Banachverbandsalgebren, *Math. Z.* **177** (1981), 183–205.

[88] Scheffold, Egon, Über Bimorphismen und das Arens-Produkt bei kommutativen *D*-Banachverbandsalgebren, *Rev. Roumaine Math. Pures Appl.* **39** (1994), no. 3, 259–270.

[89] Scheffold, Egon, Über die Arens-Triadjungierte Abbildung von Bimorphismen, *Rev. Roumaine Math. Pures Appl.* **41** (1996), no. 9–10, 697–701.

[90] Scheffold, E., Maßalgebren mit intervallerhaltender linksseitiger Multiplikation, *Acta Math. Hungar.* **76** (1997), no. 1–2, 59–67.

[91] Scheffold, Egon, Über symmetrische Operatoren auf Banachverbänden und Arens-Regularität, *Czechoslovak Math. J.* **48**(123) (1998), no. 4, 747–753.

[92] Schep, A.R., Factorization of positive multilinear maps, *Illinois J. Math.*, **28** (1984), 579–591.

[93] Vulikh, B.Z., *Introduction to the theory of partially ordered spaces*, Wolters-Noordhoff Scientific Publications, Ltd., Groningen, 1967

[94] Wittstock, Gerd, Ordered normed tensor products, Foundations of quantum mechanics and ordered linear spaces (Advanced Study Inst., Marburg, 1973), *Lecture Notes in Phys., Vol.* 29*, Springer*, *Berlin*, pp. 67–84, 1974.

[95] Wittstock, Gerd, Eine Bemerkung über Tensorprodukte von Banachverbänden, *Arch. Math. (Basel)*, **25** (1974), 627–634.

[96] Zaanen, A.C., *Riesz Spaces* II, North Holland, Amsterdam-New York-Oxford, 1983.

G. Buskes and Q. Bu
Department of Mathematics
University of Mississippi
University, MS 38677, U.S.A.
e-mail: `mmbuskes@olemiss.edu`
e-mail: `qbu@olemiss.edu`

Anatoly G. Kusraev
Institute of Applied
Mathematics and Informatics
Vladikavkaz Science Center of the RAS
Vladikavkaz, 362040, RUSSIA
e-mail: `kusraev@alanianet.ru`

Positivity
Trends in Mathematics, 127–160

Vector Measures, Integration and Applications

G.P. Curbera and W.J. Ricker

Introduction

We will deal exclusively with the integration of scalar (i.e., $\mathbb{R}$ or $\mathbb{C}$)-valued functions with respect to *vector measures*. The general theory can be found in [36, 37, 32], [44, Ch. III] and [67, 124], for example. For applications beyond these texts we refer to [38, 66, 80, 102, 117] and the references therein, and the survey articles [33, 68]. Each of these references emphasizes its own preferences, as will be the case with this article. Our aim is to present some theoretical developments over the past 15 years or so (see §1) and to highlight some recent applications. Due to space limitation we restrict the applications to two topics. Namely, the extension of certain operators to their *optimal domain* (see §2) and aspects of *spectral integration* (see §3). The interaction between order and positivity with properties of the integration map of a vector measure (which is defined on a *function space*) will become apparent and plays a central role.

Let Σ be a σ-algebra on a set $\Omega \neq \emptyset$ and E be a locally convex Hausdorff space (briefly, lcHs), over $\mathbb{R}$ or $\mathbb{C}$, with continuous dual space E^*. A σ-additive set function $\nu : \Sigma \to E$ is called a *vector measure*. By the Orlicz-Pettis theorem this is equivalent to the scalar-valued function $x^*\nu : A \mapsto \langle \nu(A), x^* \rangle$ being σ-additive on Σ for each $x^* \in E^*$; its variation measure is denoted by $|x^*\nu|$. A set $A \in \Sigma$ is called *ν-null* if $\nu(B) = 0$ for all $B \in \Sigma$ with $B \subseteq A$. A scalar-valued, Σ-measurable function f on Ω is called *ν-integrable* if

$$f \in L^1(x^*\nu), \qquad x^* \in E^*, \tag{0.1}$$

and, for each $A \in \Sigma$, there exists $x_A \in E$ such that

$$\langle x_A, x^* \rangle = \int_A f \, dx^*\nu, \qquad x^* \in E^*. \tag{0.2}$$

The first author acknowledges gratefully the support of D.G.I. # BFM2003-06335-C03-01 (Spain).

We denote x_A by $\int_A f\, d\nu$. Two ν-integrable functions are identified if they differ on a ν-null set. Then $L^1(\nu)$ denotes the linear space of all (equivalence classes of) ν-integrable functions (modulo ν-a.e.).

Let Q denote a family of continuous seminorms determining the topology of E. Each $q \in Q$ induces a seminorm in $L^1(\nu)$ via

$$\|f\|_q := \sup_{x^* \in U_q^\circ} \int_\Omega |f|\, d|x^*\nu|, \qquad f \in L^1(\nu), \tag{0.3}$$

where $U_q^\circ \subseteq E^*$ is the polar of $q^{-1}([0,1])$. The Σ-simple functions are always dense in the lcHs $L^1(\nu)$. For E a Banach space, we have the single norm

$$\|f\|_{L^1(\nu)} = \sup_{\|x^*\| \le 1} \int_\Omega |f|\, d|x^*\nu|, \qquad f \in L^1(\nu). \tag{0.4}$$

Whenever E is a Fréchet space, $L^1(\nu)$ is metrizable and complete. We point out that (0.3) is a *lattice seminorm*, i.e., $\|f\|_q \le \|g\|_q$ whenever $f, g \in L^1(\nu)$ satisfy $|f| \le |g|$. Actually, $L^1(\nu)$ is also *solid*, i.e., $f \in L^1(\nu)$ whenever $|f| \le |g|$ with f measurable and $g \in L^1(\nu)$. If we wish to stress that we are working over $\mathbb{R}$ or $\mathbb{C}$, then we write $L^1_{\mathbb{R}}(\nu)$ or $L^1_{\mathbb{C}}(\nu)$, resp. Of course, $L^1_{\mathbb{C}}(\nu) = L^1_{\mathbb{R}}(\nu) + iL^1_{\mathbb{R}}(\nu)$ is the complexification of $L^1_{\mathbb{R}}(\nu)$, with the order in the positive cone $L^1(\nu)^+$ of $L^1(\nu)$ being that defined pointwise ν-a.e. on Ω. The dominated convergence theorem ensures that the topology in the (lc-lattice =) lc-Riesz space $L^1(\nu)$ has the *σ-Lebesgue property* (also called *σ-order continuity* of the norm if E is Banach), i.e., if $\{f_n\} \subseteq L^1(\nu)$ satisfies $f_n \downarrow 0$ with respect to the order, then $\lim_{n\to\infty} f_n = 0$ in the topology of $L^1(\nu)$. Moreover, χ_Ω is a *weak order unit* in $L^1(\nu)$; see [3] for the definition.

If a Σ-measurable function f satisfies only (0.1), then (0.3) is still finite for each $q \in Q$, [128]; we then say that f is *weakly ν-integrable* and denote the space of all (classes of) such functions by $L^1_w(\nu)$. Equipped with the seminorms $\{\|\cdot\|_q : q \in Q\}$, this is a *lc-lattice* (Fréchet whenever E is Fréchet) containing $L^1(\nu)$ as a closed subspace. If E does not contain an isomorphic copy of c_0, then necessarily $L^1_w(\nu) = L^1(\nu)$; see [67, II 5 Theorem 1] and [73, Theorem 5.1]. For all of the above basic facts (and others) concerning $L^1_{\mathbb{R}}(\nu)$ we refer to [67] and for $L^1_{\mathbb{C}}(\nu)$ to [50], for example. If E is not complete, then various subtleties may arise in passing from the case of $\mathbb{R}$ to $\mathbb{C}$, [91, 59, 119].

Given $1 \le p < \infty$, a Σ-measurable function f belongs to $L^p(\nu)$ if $|f|^p \in L^1(\nu)$. When E is a Banach space, then $L^p(\nu)$ is a *p-convex Banach lattice* relative to the norm

$$\|f\|_{L^p(\nu)} := \sup_{\|x^*\| \le 1} \left(\int_\Omega |f|^p\, d|x^*\nu| \right)^{1/p}, \qquad f \in L^p(\nu), \tag{0.5}$$

and satisfies $L^p(\nu) \subseteq L^1(\nu)$ continuously. Such spaces and operators defined in them have recently been studied in some detail; see [26, 122, 56, 101, 55], for example, and the references therein. Of course, the spaces $L^p_w(\nu)$ can also be defined in the obvious way (i.e., by requiring $|f|^p \in L^1_w(\nu)$), [26, 56, 101]. More generally, a study of Orlicz spaces with respect to a vector measure, which stems from a detailed study of the Banach function subspaces of $L^1(\nu)$, has been made in [27].

Of central importance to this article will be the *integration operator* I_ν : $L^1(\nu) \to E$ defined by $f \mapsto \int_\Omega f\, d\nu$. For brevity, we also write $\int f\, d\nu$ for $\int_\Omega f\, d\nu$. According to (0.2) and (0.3) we have

$$q\left(\int_\Omega f\, d\nu\right) = \sup_{x^* \in U_q^\circ} |\langle \int_\Omega f\, d\nu, x^* \rangle| \leq \|f\|_q, \qquad f \in L^1(\nu),$$

for each $q \in Q$. This shows that I_ν is continuous, with $\|I_\nu\| \leq 1$ if E is a Banach space. Given a lc-lattice E, a vector measure $\nu : \Sigma \to E$ is called *positive* if it takes its values in the positive cone E^+ of E. In this case, it is an easy approximation argument using the denseness of the Σ-simple functions in $L^1(\nu)$ to see that I_ν is a *positive operator*, i.e., $I_\nu(L^1(\nu)^+) \subseteq E^+$. For recent aspects of the theory of vector measures and integration in a Banach space E see [16, 17, 18, 19, 101] and for E a Fréchet space or lattice, we refer to [54, 53, 52, 49, 31, 51, 48, 8, 125, 126, 127], and the references therein.

1. Representation theorems

When E is a Banach space and ν is an E-valued measure, the correct framework for interpreting both $L^1(\nu)$ and $L^1_w(\nu)$ is that of *Banach function spaces* (briefly, B.f.s.). Let $(\Omega, \Sigma, \lambda)$ be a σ-finite measure space, $\mathcal{M}$ be the space of all Σ-measurable functions on Ω (functions equal λ-a.e. are identified), and $\mathcal{M}^+$ be the cone of those elements of $\mathcal{M}$ which are non-negative λ-a.e. A *function norm* is a map $\rho\colon \mathcal{M}^+ \to [0, \infty]$ satisfying

(a) $\rho(f) = 0$ iff $f = 0$ λ-a.e.,
$\rho(af) = a\rho(f)$ for every $a \geq 0$ and $f \in \mathcal{M}^+$,
$\rho(f + g) \leq \rho(f) + \rho(g)$ for all $f, g \in \mathcal{M}^+$,

(b) If $f, g \in \mathcal{M}^+$ and $f \leq g$ λ-a.e., then $\rho(f) \leq \rho(g)$.

The function space L_ρ is defined as the set of all $f \in \mathcal{M}$ satisfying $\rho(|f|) < \infty$; it is a linear space and ρ is a norm. Moreover, whenever ρ has the Riesz-Rischer property (cf. [136, Ch. 15]) the space L_ρ is a Banach lattice for the λ-a.e. order and it is always an ideal of measurable functions, that is, if $f \in L_\rho$ and $g \in \mathcal{M}$ satisfies $|g| \leq |f|$ λ-a.e., then $g \in L_\rho$. The *associate space* $L_{\rho'}$ of L_ρ is generated by the function norm $\rho'(g) := \sup\{\int |fg|\, d\lambda : \rho(f) \leq 1, f \in \mathcal{M}^+\}$. We also denote $L_{\rho'}$ by L'_ρ. If $g \in L'_\rho$ and $G(f) := \int fg\, d\lambda$ for every $f \in L_\rho$, then $G \in L^*_\rho$ and $\|G\| = \rho'(g)$. In this sense, L'_ρ is identified with a closed subspace of L^*_ρ. Applying the same procedure to L'_ρ, we obtain the second associate space L''_ρ. The B.f.s. L_ρ satisfies the *Fatou property* if $0 \leq f_n \uparrow f$ in $\mathcal{M}^+$ implies that $\rho(f_n) \uparrow \rho(f)$. The *Lorentz function norm* ρ_L associated to any given function norm ρ is defined by

$$\rho_L(f) := \inf\{\lim \rho(f_n) : 0 \leq f_n \uparrow f \text{ with } f_n \in \mathcal{M}^+\}.$$

Then ρ_L is the largest norm majorized by ρ and having the Fatou property, [136, Ch. 15, §66]. It follows that L_{ρ_L} is the *minimal* B.f.s. (over λ) with the Fatou property and continuously containing (with norm ≤ 1) L_ρ. Since $\rho_L = \rho''$ [136, Ch. 15, §71, Theorem 2], we see that L''_ρ is the minimal B.f.s. (over λ) with the Fatou property and continuously containing (with norm ≤ 1) L_ρ. The *order continuous part* $(L_\rho)_a$ of a B.f.s. L_ρ consists of all $f \in L_\rho$ such that $\rho(f_n) \downarrow 0$ whenever $\{f_n\} \subseteq L^+_\rho$ satisfies $|f| \geq f_n \downarrow 0$ (equivalently, increasing sequences order bounded by $|f|$ are norm convergent, [136, Ch. 15, §72, Theorem 2]). B.f.s.' were studied by Luxemburg and Zaanen; see [79, 75, 76, 77, 78] and [136, Ch. 15]. Caution should be taken since some authors consider different definitions of B.f.s.' which are more restrictive; see [6, Definition I.1.1] and [74, Definition 1.b.17].

It was observed in [17, Theorem 1] that $L^1(\nu)$ is a σ-order continuous (briefly, σ-o.c.) B.f.s. with weak order unit with respect to the measure space $(\Omega, \Sigma, \lambda)$, where λ is a Rybakov control measure for the vector measure $\nu : \Sigma \to E$, that is, a finite measure of the form $\lambda = |x_0^*\nu|$ for some suitable $x_0^* \in B_{E^*}$ (the closed unit ball of E^*) such that λ and ν have the same null sets; see [32, Ch. IX, §2, Theorem 2]. The crucial property is σ-order continuity, which leads to the converse result (cf. Theorem 1.1 below); this is the main representation theorem for the class of spaces $L^1(\nu)$, [17, Theorem 8]. Note that in Banach lattices, σ-order continuity and σ-Dedekind completeness (which is satisfied by $L^1(\nu)$) is equivalent to order continuity (i.e., $\|x_\tau\| \downarrow 0$ whenever $\{x_\tau\}$ decreases to zero in order).

Theorem 1.1. *Let E be any Banach lattice with o.c. norm and possessing a weak order unit. Then there exists a (E^+-valued) vector measure ν such that E is order and isometrically isomorphic to $L^1(\nu)$.*

This theorem characterizes the spaces $L^1(\nu)$, for Banach space-valued measures, and explains the diversity of spaces arising as $L^1(\nu)$. For example, for a finite measure space $(\Omega, \Sigma, \lambda)$, $1 \leq p < \infty$, and the vector measure $A \mapsto \nu_p(A) := \chi_A \in L^p(\lambda)$ on Σ, we obtain $L^1(\nu_p) = L^p(\lambda)$.

The theory of integrating scalar-valued functions with respect to a vector measure defined on a σ-algebra can be extended to vector measures defined on δ-rings; [73, 81, 82]. A study of the corresponding space of integrable functions has been undertaken in [28]. In this context, Theorem 1.1 can be generalized as follows, [16, pp. 22–23].

Theorem 1.2. *Let E be any Banach lattice with o.c. norm. Then there exists an E-valued vector measure ν defined on a δ-ring such that E is order and isometrically isomorphic to $L^1(\nu)$.*

A version of these representation theorems in a more general setting is also known, [41, Proposition 2.4(vi)]. For the definition of a spectral measure, see Section 3.

Theorem 1.3. *Let E be a Dedekind complete, complex Riesz space with locally solid, Lebesgue topology. Assume E is quasicomplete and has a weak order unit $e \geq 0$*

and that the space $\mathcal{L}(E)$ of continuous linear operators in E is sequentially complete for the strong operator topology. Then there exists a closed, equicontinuous $\mathcal{L}(E)$-valued spectral measure P for which e is a cyclic vector and such that the integration map $f \mapsto \int f \, dPe$ is a topological and Riesz homomorphism of $L^1(Pe)$ onto E.

What is the connection between the spaces $L^1(\nu)$ and $L^1_w(\nu)$? For the function norm (relative to $\lambda = |x_0^*\nu|$ as above) given by

$$\rho_w(f) := \sup_{x^* \in B_{E^*}} \int_\Omega f \, d|x^*\nu|, \qquad f \in \mathcal{M}^+,$$

we have $L_{\rho_w} = L^1_w(\nu)$. So: what is the connection between the B.f.s.' $L^1(\nu)$ and $L^1_w(\nu)$? In this regard the role of the Fatou property is relevant, [23, Propositions 2.1, 2.3 and 2.4].

Theorem 1.4. *Let ν be any vector measure.*

(a) *The B.f.s. $L^1_w(\nu)$ has the Fatou property.*
(b) $L^1(\nu)'' = L^1_w(\nu)$.
(c) *$L^1(\nu)$ has the Fatou property iff $L^1_w(\nu)$ has o.c.-norm.*

The answer to the question above can now be given. Since $L^1(\nu)$ has o.c.-norm and the Σ-simple functions are dense, it can be verified that $(L^1_w(\nu))_a = L^1(\nu)$. So, $L^1(\nu)$ is the *maximal* B.f.s. inside $L^1_w(\nu)$ (with the same norm) which has o.c.-norm. Since the Lorentz function norm ρ_L is the largest norm majorized by ρ and having the Fatou property, [136, Ch. 15, §66], it follows that $L^1_w(\nu)$ is the *minimal* B.f.s. (over λ) with the Fatou property and continuously containing (with norm ≤ 1) $L^1(\nu)$. Accordingly, $L^1_w(\nu)$ can be interpreted as the "Fatou completion" of $L^1(\nu)$. Statement (c) in Theorem 1.4 now follows: if $L^1(\nu)$ has the Fatou property, then the minimal property of $L^1_w(\nu)$ forces $L^1(\nu) = L^1_w(\nu)$; on the other hand, if $L^1_w(\nu)$ has o.c.-norm, then the maximal property of $L^1(\nu)$ forces $L^1_w(\nu) = L^1(\nu)$.

Theorem 1.5 below characterizes all Banach lattices which arise as $L^1_w(\nu)$ for some vector measure ν, [23, Theorem 2.5]. Note what we have called the Fatou property, technically speaking, should be called the σ-Fatou property but, since B.f.s.' are super Dedekind complete and $\mathcal{M}$ is order separable, there is no distinction between using increasing sequences or increasing nets, [137, Theorem 112.3].

Theorem 1.5. *Let E be any Banach lattice with the σ-Fatou property and possessing a weak order unit which belongs to E_a. Then there exists a (E_a^+-valued) vector measure ν such that E is order and isometrically isomorphic to $L^1_w(\nu)$.*

For E a B.f.s., the previous result gives the following

Theorem 1.6. *Let L_ρ be a B.f.s. over a finite measure space $(\Omega, \Sigma, \lambda)$ such that L_ρ has the Fatou property and $\chi_\Omega \in (L_\rho)_a$. Then L_ρ is order and isometrically isomorphic to $L^1_w(\nu)$ for some ν.*

For p-convex Banach lattices with $1 \leq p < \infty$ (see [74, Definition 1.d.3]) the following extension of Theorem 1.1 holds, [56, Theorem 2.4].

Theorem 1.7. *Let E be a p-convex Banach lattice with o.c. norm and a weak order unit. Then there exists a vector measure ν such that E is order isomorphic to $L^p(\nu)$.*

For $1 \leq p < \infty$, the space $L^p_w(\nu)$ is generated by the function norm

$$\rho^p_w(f) := \sup_{x^* \in B_{E^*}} \left(\int_\Omega f^p \, d|x^*\nu| \right)^{1/p}, \qquad f \in \mathcal{M}^+.$$

The result corresponding to Theorem 1.4, now relating the spaces $L^p(\nu)$ and $L^p_w(\nu)$ is also known, [26, Propositions 1, 2 and 4]. For $E = L^p_w(\nu)$, we know that E is p-convex, has the σ-Fatou property, $E_a = L^p(\nu)$ and χ_Ω is a weak order unit for E which belongs to E_a. These properties of $L^p_w(\nu)$ characterize a large class of abstract Banach lattices, [26, Theorem 4].

Theorem 1.8. *Let $1 \leq p < \infty$ and E be any p-convex Banach lattice with the σ-Fatou property and possessing a weak order unit which belongs to E_a. Then there exists an E_a-valued vector measure ν such that E is Banach lattice isomorphic to $L^p_w(\nu)$.*

We conclude with a different kind of representation result, not of a space, but of an operator. The *variation measure* of a Banach space-valued vector measure ν, denoted by $|\nu|$, can be defined via the "partition process" as for scalar measures; see [32, pp. 2–3]. It turns out that always $L^1(|\nu|) \subseteq L^1(\nu)$ with a continuous inclusion, [73]. For the notion of *Bochner integrals* we refer to [32, Ch. II].

Theorem 1.9. *Let E be a Banach space and $\nu : \Sigma \to E$ be a vector measure with finite variation (i.e., $|\nu|(\Omega) < \infty$). The integration map $I_\nu : L^1(\nu) \to E$ is compact iff ν possesses an E-valued, Bochner $|\nu|$-integrable Radon-Nikodým derivative $G = d\nu/d|\nu|$ which has $|\nu|$-essentially relatively compact range in E.*

In this case, $L^1(\nu) = L^1(|\nu|)$ and (with Bochner integrals) we have

$$I_\nu f = \int_\Omega fG \, d|\nu|, \qquad f \in L^1(\nu).$$

Remark 1.10.

(i) This result occurs in [96, Theorem 1].

(ii) It is also true that if I_ν is compact, then ν has finite variation, [96, Theorem 4]. Examples of vector measures which do not have finite variation arise via a *Pettis integrable* density $G : \Omega \to E$ (see [32, Ch. II, §3] for the definition), i.e., $\nu(A) := \int_A G \, d\lambda$, where $\lambda : \Sigma \to [0, \infty]$ is an infinite measure; see [46, Proposition 5.6(iv)] where it is shown that $|\nu|$ is σ-finite. More precisely, if G is strongly measurable, [32, p. 41], and Pettis λ-integrable (but *not* Bochner λ-integrable), then ν has σ-finite but, not finite, variation. For the existence of such functions G on $\Omega = [0, \infty)$ for Lebesgue measure λ, see the proof of [120, Theorem 3.3]. A characterization of vector measures with σ-finite variation occurs in [121, Theorem 2.4]. In every infinite-dimensional space E there also exist vector measures with infinite but, not σ-finite, variation;

they can even be chosen to have relatively compact range, [132, p. 90]. In particular, relative compactness of the range of ν does not suffice for I_ν to be compact. For concrete examples of ν (arising in classical analysis) which fail to have σ-finite variation we refer to [87, Lemma 2.1] and [99, Proposition 4.1], for example.

(iii) Let $\dim(E) = \infty$. Then there exists an E-valued measure ν with $|\nu|(\Omega) < \infty$, the range of ν is not contained in any finite-dimensional subspace of E, and I_ν is compact, [96, Theorem 2]. There also exists an E-valued measure μ with $|\mu|(\Omega) < \infty$ satisfying $L^1(|\mu|) = L^1(\mu)$ and having an E-valued Bochner $|\mu|$-integrable Radon-Nikodým derivative $d\mu/d|\mu|$ such that I_μ is not compact [96, Theorem 3].

(iv) For $1 < p < \infty$, the compactness properties of I_ν, restricted to $L^p_{\mathbb{R}}(\nu) \subseteq L^1_{\mathbb{R}}(\nu)$ and $L^p_w(\nu)_{\mathbb{R}} \subseteq L^1_{\mathbb{R}}(\nu)$, are studied in [56, 122].

(v) An extension of Theorem 1.9 to Fréchet spaces E occurs in [97]. Given extra properties of E, more can be said. For instance, if E is Fréchet-Montel, then $I_\nu : L^1(\nu) \to E$ is compact iff the Fréchet lattice $L^1(\nu)$ is order and topologically isomorphic to a Banach AL-lattice, [98, Theorem 2]. Nuclearity of E also has some consequences, [98, Theorem 1].

2. Optimal domains

Let X be a B.f.s. over a finite measure space $(\Omega, \Sigma, \lambda)$, E be a Banach space and $T\colon X \to E$ be a linear operator. There arise situations when T has a natural extension (still with values in E) to a larger space Y into which X is continuously embedded. This is the case for the Riesz representation theorem: a positive linear operator $\Lambda\colon C(K) \to \mathbb{C}$ can be extended to the space $L^1(\lambda)$, where λ is a scalar measure associated to Λ. Under certain conditions, we associate to the operator T a vector measure ν_T with values in E. We will say that the operator T is λ-*determined* if the additive set function (with values in E)

$$\nu_T : A \mapsto T(\chi_A), \qquad A \in \Sigma$$

has the same null sets as λ. For the next result see [20, Theorem 3.1].

Theorem 2.1. *Let X be a B.f.s. over a finite measure space $(\Omega, \Sigma, \lambda)$, E be a Banach space and $T\colon X \to E$ be a λ-determined linear operator with the property that $Tf_n \to Tf$ weakly in E whenever $\{f_n\} \subset X$ is a positive sequence increasing λ-a.e to $f \in X$. Then the measure ν_T is countably additive, X is continuously embedded in $L^1(\nu_T)$ and the integration operator from $L^1(\nu_T)$ into E extends T.*

Remark 2.2.

(i) The assumptions of Theorem 2.1 hold if X has order continuous norm and T is continuous and linear. They also hold for $X = L^\infty(\lambda)$ and T weak*-to-weak continuous. Continuity of T alone does not suffice in general (e.g., the identity operator on $L^\infty([0,1])$). The result can be extended to σ-finite measure spaces, or even general measure spaces provided X contains a weak order unit, that is, a function $\varphi > 0$, λ-a.e. In this case the measure ν is defined by $\nu_T(A) := T(\varphi\chi_A)$, $A \in \Sigma$, and the embedding from X into $L^1(\nu_T)$ is $f \mapsto f/\varphi$.

(ii) Theorem 2.1 provides an integral representation for certain operators via integration with respect to a vector measure, even in cases where the Bochner or Pettis integrals do not exist. For instance, the fractional integral of order α, $0 < \alpha < 1$, of a function f at a point $x \in [0,1]$ is given by

$$I_\alpha f(x) = \frac{1}{\Gamma(\alpha)} \int_0^1 \frac{f(t)}{|x-t|^{1-\alpha}}\, dt$$

whenever it is defined.
We can consider I_α as an operator $I_\alpha \colon L^\infty([0,1]) \to L^p([0,1])$. By Theorem 2.1, $I_\alpha(f) = \int f \, d\nu_p$, where ν_p denotes the measure given by $\nu_p(A)(x) = \int_A |x-y|^{\alpha-1}\, dy \in L^p([0,1])$. This can be done for any $1 \le p \le \infty$. However, unless $(1-\alpha)p < 1$, there is no Bochner or Pettis integrable function $G\colon [0,1] \to L^p([0,1])$ such that $I_\alpha(f) = \int_{[0,1]} f(t)G(t)\, dt$, [20, Remark 3.5].

A basic problem is to identify the optimal space to which the operator T can be extended, within a particular class of spaces, but keeping the codomain space of T fixed. This is sometimes considered within the theory of integral operators; see [4, 71, 88, 129, 130], for example, and the references therein. We will denote by $[T,E]$ the maximal B.f.s. (containing X) to which T can be extended as a continuous linear operator, still with values in E. This maximality is to be understood in the following sense. There is a continuous linear extension of T (which we still denote by T) $T\colon [T,E] \to E$, and if T has a continuous, linear extension $\tilde{T}\colon Y \to E$, where Y is a B.f.s. containing X, then Y is continuously embedded in $[T,E]$, and the extended operator T coincides with $\tilde{T}$ on Y. The space $[T,E]$ is then the *optimal lattice domain* for T. If we consider the class of B.f.s.' with order continuous norm (briefly, o.c.), then we have the space $[T,E]_o$, which is the *o.c. optimal lattice domain* for T. Theorem 2.1 shows that the space $L^1(\nu_T)$ is the o.c. optimal lattice domain for T.

We identify situations in which $[T,X] = L^1(\nu_T)$; this has the advantage that the properties of ν_T and E, needed in determining the space $L^1(\nu_T)$ hence, also $[T,E]$, are well understood. As we will see, this procedure for identifying optimal domains is extremely fruitful; see also [101].

Let $K\colon [0,1]\times[0,1] \to [0,\infty)$ be a measurable function. We associate to K an operator T via the formula

$$Tf(x) := \int_0^1 f(y)K(x,y)\, dy, \qquad x \in [0,1], \tag{2.1}$$

for any function f for which it is meaningful to do so for m-a.e. $x \in [0,1]$, where m is Lebesgue measure in [0,1]. We will say that the kernel K is *admissible* if it satisfies the following three conditions: (i) for every $x \in [0,1]$, the function $K_x\colon y \mapsto K(x,y), y \in [0,1]$, is Lebesgue integrable in [0,1]; (ii) $\int_0^1 K_y(x)\, dx > 0$

for m-a.e. $y \in [0,1]$ where, for every $y \in [0,1]$, the function K_y is defined by $x \mapsto K(x,y)$, for $x \in [0,1]$; and (iii) $K_{x_n} \to K_{x_0}$ weakly in $L^1([0,1])$ whenever $x_0 \in [0,1]$ and $x_n \to x_0$. These conditions guarantee that

$$\nu(A)(\cdot) := \int_A K(\cdot\,,y)\,dy, \qquad A \in \mathcal{B},$$

($\mathcal{B}$ is the Borel σ-algebra of [0,1]) is a $C([0,1])$-valued, σ-additive measure, which has the same null sets as m, [20, Proposition 4.1].

Let E be a B.f.s. over $([0,1],\mathcal{B},m)$ for which $L^\infty([0,1]) \subseteq E \subseteq L^1([0,1])$. Under the above conditions on K, we have $T\colon L^\infty([0,1]) \to E$ continuously. It turns out that $[T,E] = \{f : T|f| \in E\}$, [20, Proposition 5.2], and $\|f\|_{[T,E]} := \|\,T|f|\,\|_E$ is a complete function norm in $[T,E]$. Since $C([0,1])$ is continuously embedded in E, the measure ν is also E-valued and σ-additive with $\nu(A) = T(\chi_{_A})$ for $A \in \mathcal{B}$. We denote it by ν_E in this case. Moreover, because $K \geq 0$, it is clear that $T\colon [T,E] \to E$ is a *positive operator* and that ν_E takes its values in E^+.

The relationships between the three B.f.s.' associated to K and E, namely, $L^1(\nu_E)$, $L^1_w(\nu_E)$ and $[T,E]$, are precise, [23, p. 199].

Theorem 2.3. *Let K be a non-negative admissible kernel and E be a B.f.s. satisfying the above conditions. The following inclusions hold:*

$$L^1(\nu_E) \subseteq [T,E] \subseteq [T,E]'' = L^1_w(\nu_E) \subseteq [T,E'']. \tag{2.2}$$

The first inclusion is an isometric imbedding, and the norms of the spaces $[T,E]''$ and $L^1_w(\nu_E)$ coincide.

Remark 2.4. (i) The inclusions in (2.2) can be strict; see [20, Remark 5.3] and [23, Example 3.4].

(ii) The optimal domain $[T,E]$ has the Fatou property iff $[T,E] = L^1_w(\nu_E)$, [23, p. 199].

(iii) If E' is a norming subspace of E^*, then all inclusions in (2.2) are isometric imbeddings and $L^1_w(\nu_E) = [T,E'']$; see Corollary 3.5 and Theorem 3.6 of [23].

(iv) If E has o.c. norm, then $L^1(\nu_E) = [T,E]$ and

$$L^1(\nu_E) = [T,E] \subseteq [T,E]'' = L^1_w(\nu_E) = [T,E''],$$

with the imbedding $[T,E] \subseteq L^1_w(\nu_E)$ isometric, [23, Corollary 3.7].

(v) If E has the Fatou property, then $L^1_w(\nu_E) = [T,E]$ and

$$L^1(\nu_E) \subseteq [T,E] = [T,E]'' = L^1_w(\nu_E) = [T,E''];$$

see [23, Corollary 3.7].

(vi) If E is weakly sequentially complete, then all spaces in (2.2) are equal, [23, Corollary 3.7].

An interesting result concerning the relationships between the spaces $L^1(\nu_E)$, $L^1_w(\nu_E)$ and $[T,E]$ is the following one, [23, Proposition 3.12]. Recall, for a B.f.s. E, that E_a (cf. Section 1) is the *o.c. part* of E and E_b is the closure of the $\mathcal{B}$-simple functions in E.

Theorem 2.5. *Let Y be any one of $L^1(\nu_E)$, $L^1_w(\nu_E)$ or $[T,E]$. Then*

$$Y_b = Y_a = L^1(\nu_E).$$

Additional information on the optimal domain $[T,E]$ is available in the case when the B.f.s. E is rearrangement invariant (briefly, r.i.). Recall that a B.f.s. E is r.i. if it satisfies the Fatou property and $f \in E$ implies that $g \in E$ with $\|g\| = \|f\|$ whenever g and f are equimeasurable, [6, II.4.1]. Every r.i. space E on [0,1] is an interpolation space between the spaces $L^1([0,1])$ and $L^\infty([0,1])$, arising via the the K-functional of Peetre (as $E = (L^1, L^\infty)_\rho$ for a suitable r.i. norm ρ); see [6, V.1]. It turns out, under certain conditions, that the optimal domain $[T,E]$ is an interpolation space between the optimal domains $[T, L^1([0,1])]$ and $[T, L^\infty([0,1])]$ (both of these spaces being weighted L^1-spaces) in the same way that E is an interpolation space between $L^1([0,1])$ and $L^\infty([0,1])$ (by a technical result of Gagliardo, we can substitute here $L^\infty([0,1])$ with $C([0,1])$). Let us be more precise.

Theorem 2.6. *Let K be a non-negative admissible kernel.*

(a) *Let ν be the associated $C([0,1])$-valued measure. Then we have $[T, C([0,1])] = L^1(\nu)$.*
(b) *If, in addition, K is non-decreasing (i.e., $K_{x_1} \le K_{x_2}$ a.e. on $[0,1]$ whenever $0 \le x_1 \le x_2 \le 1$), then $[T, C([0,1])] = L^1_\xi$ where the weight $\xi(y) := K(1,y)$.*
(c) *Let ν_{L^1} denote the measure ν considered as being $L^1([0,1])$-valued. Then $[T, L^1([0,1])] = L^1(\nu_{L^1}) = L^1_\omega$, where the weight $\omega(y) := \int_0^1 K(x,y)\,dx$.*

This identification (see Propositions 5.1 and 5.4 of [20]) makes it possible, with the aid of interpolations techniques, to obtain the following result, [20, Proposition 5.5 & Theorem 5.11].

Theorem 2.7. *Let K be a non-negative admissible kernel and $E = (L^1, L^\infty)_\rho$ be a r.i. B.f.s. on $[0,1]$.*

(a) *The space $(L^1_\omega, L^1(\nu))_\rho$ is continuously embedded in $[T,E]$.*
(b) *If K is non-decreasing with the property that there exists a constant $\beta > 0$ such that, for every $t > 0$ and every $y \in [0,1]$,*

$$\int_{\max\{0,1-t\}}^1 K(x,y)\,dx \ge \beta \cdot \min\left\{\int_0^1 K(x,y)dx;\ t \cdot K(1,y)\right\}, \qquad (*)$$

then, with equivalence of norms,

$$[T,E] = (L^1_\omega, L^1_\xi)_\rho.$$

(c) *If, in addition, E has o.c. norm, then $[T,E] = (L^1_\omega, L^1_\xi)_\rho = L^1(\nu_E)$.*

Many interesting kernels satisfying the above properties occur. For example, the classical Volterra operator given by the kernel $K(x,y) = \chi_\Delta(x,y)$, where $\Delta = \{(x,y) \in [0,1]^2 : 0 \le y \le x\}$, is non-decreasing and satisfies condition $(*)$ with $\beta = 1$. There is a corresponding result for non-increasing kernels, [20, Theorem 5.12], an example of which is given by the kernel arising from nilpotent left translation semigroups; see [20, Example 4.4 and Remark 5.14].

The extension of the previous results to kernel operators defined for functions on $[0,\infty)$ requires a consideration of optimal domains for operators defined on B.f.s.' over $[0,\infty)$. For this, the required tool is the theory of L^1-spaces for vector measures on δ-rings. Such a study is made in [29]; for applications to the Hardy operator, see [30].

The above results on optimal domains for kernel operators can be applied to the study of refinements of the classical Sobolev inequality. This inequality, valid for differentiable functions f on a bounded domain Ω in $\mathbb{R}^n$ with $n \geq 2$, states that

$$\|f\|_{L^q(\Omega)} \leq C\| \, |\nabla f| \, \|_{L^p(\Omega)}, \qquad f \in C_0^1(\Omega), \tag{2.3}$$

where $1 < p < n$, $q := np/(n-p)$ and $C > 0$ depends only on p, n. Edmunds, Kerman and Pick studied the optimal domain problem, for the inequality (2.3), within the class of r.i. spaces, [47]. They consider r.i. spaces E and F on $[0,1]$ and a generalized Sobolev inequality

$$\|f^*\|_E \leq C\||\nabla f|^*\|_F, \qquad f \in C_0^1(\Omega), \tag{2.4}$$

where f^* and $|\nabla f|^*$ are, respectively, the decreasing rearrangements of f and the norm of its gradient $|\nabla f|$. They show (for $|\Omega| = 1$) that (2.4) is equivalent to boundedness of the kernel operator T associated with Sobolev's inequality, namely

$$Tf(t) = \int_t^1 f(s)s^{(1/n)-1}\, ds, \qquad t \in [0,1], \tag{2.5}$$

acting between the r.i. spaces E and F, that is, $\|Tf\|_E \leq K\|f\|_F$, [47, Theorem 6.1]. Since the kernel in (2.5) satisfies the non-increasing versions of Theorems 2.6 and 2.7 (see [20, Theorem 5.12]) it follows, for a r.i. space $E = (L^1, L^\infty)_\rho$, that the optimal domain $[T, E]$ can be identified as the interpolation space

$$[T, E] = \left(L^1(s^{1/n}ds), L^1(s^{(1/n)-1}ds)\right)_\rho,$$

since $[T, L^\infty([0,1])] = L^1(s^{(1/n)-1}ds)$ and $[T, L^1([0,1])] = L^1(s^{1/n}ds)$, [21, Proposition 2.1(d) and Corollary 4.3].

For the kernel operator associated to Sobolev's inequality a thorough study of the optimal domains has been made; see the following result, [21, Proposition 3.1(a), Theorem 4.2 and Corollary 4.3].

Theorem 2.8. *The optimal domain $[T, E]$ is order isomorphic to an AL-space if and only if E is a Lorentz Λ-space, in which case*

$$[T, E] = L^1(\nu_E) = L^1(|\nu_E|).$$

Moreover, the variation measure $|\nu_E|$ is given by

$$|\nu_E|(A) = \int_A s^{(1/n)-1}\varphi_E(s)\, ds, \qquad A \in \mathcal{B},$$

with $\varphi_E(s) := \|\chi_{[0,s]}\|_E$ being the fundamental function of the r.i. space E.

The question of whether or not the spaces $[T, E]$ are r.i. is an important one. The answer is given in [24, Theorem 3.4].

Theorem 2.9. *Let E be a r.i. space on $[0,1]$. The optimal domain $[T, E]$ is itself r.i. if and only if E is the Lorentz space $L^{n',1}([0,1])$, where n' is the conjugate index of n.*

This result focuses the investigation on the *largest* r.i. space continuously contained in the optimal domain $[T, E]$, denoted by $[T, E]^{ri}$. An example illustrates the importance of this issue. For $E = L^p([0,1])$ and $n' < p < \infty$, the optimal r.i. domain $[T, L^p]^{ri}$ is the Lorentz $L^{p,q}$-space $L^{p_0,p}([0,1])$, where $p_0 := np/(n+p)$, [108, Theorem 3.20]. Note that $p = np_0/(n-p_0)$ is precisely the exponent corresponding to p_0 in the classical Sobolev inequality (2.3):

$$\|f\|_p \leq C \| \, |\nabla f| \, \|_{p_0}.$$

Hence, Sobolev's inequality is actually sharpened, since $[T, L^p]^{ri} = L^{p_0,p}([0,1])$ implies that

$$\|f\|_p \leq C \| \, |\nabla f| \, \|_{p_0,p},$$

with $\| \, |\nabla f| \, \|_{p_0,p} \leq \| \, |\nabla f| \, \|_{p_0,p_0} = \| \, |\nabla f| \, \|_{p_0}$ (as $p > p_0$). Moreover, this sharpening is optimal within the class of r.i. norms.

The following result identifies $[T, E]^{ri}$ for certain classes of r.i. spaces E, [21, Proposition 4.7, Theorem 5.7 and Theorem 5.11]. For technical details on r.i. spaces E and their lower (resp. upper) dilation exponent γ_{φ_E} (resp. δ_{φ_E}) we refer to [6] and [70].

Theorem 2.10. *Let E a r.i. B.f.s. on $[0,1]$.*

(a) *Suppose that φ_E is $(1/n')$-quasiconcave. Then, for $\Theta(t) := \int_0^t s^{(1/n)-1} \varphi_E(s)\, ds$, the Lorentz space Λ_Θ is the largest r.i. space inside $L^1(|\nu_E|)$.*
(b) *Let E be a Marcinkiewicz space M_φ with φ satisfying $(1/n) < \gamma_\varphi \leq \delta_\varphi < 1$. Then the largest r.i. space inside $[T, M_\varphi]$ is the Marcinkiewicz space M_Ψ, where $\Psi(t) := t^{-1/n}\varphi(t)$.*
(c) *Suppose that φ_E is $(1/n')$-quasiconcave and $0 < \gamma_{\varphi_E} \leq \delta_{\varphi_E} < 1/n'$. Then the largest r.i. space inside $[T, E]$ has fundamental function equivalent to $\Gamma(t) = t^{1/n}\varphi_E(t)$.*

These results allow the formulation of an extended version of the classical Rellich-Kondrachov theorem on compactness of the Sobolev imbedding (for suitable $\Omega \subset \mathbb{R}^n$), which asserts that the imbedding

$$W_0^{1,p}(\Omega) \hookrightarrow L^q(\Omega) \tag{2.6}$$

is compact for $1 \leq q < np/(n-p)$ whenever $1 \leq p < n$. In the case $q = np/(n-p)$, although Sobolev's theorem ensures boundedness, the imbedding is not compact. This can be interpreted in the following way: if we *fix* the range space to be some $L^q(\Omega)$ smaller than $L^{n'}(\Omega)$, then the imbedding remains compact as long as the domain space $W_0^{1,p}(\Omega)$ does not reach the space $W_0^{1,nq/(n+q)}(\Omega)$ which is "too large" (i.e., the endpoint $nq/(n+q)$ is avoided). But, if we fix the range space to be some $L^q(\Omega)$ larger than $L^{n'}(\Omega)$, then the imbedding is compact for all domain spaces $W_0^{1,p}(\Omega)$ (since it is so for $W_0^{1,1}(\Omega)$), i.e., no endpoint occurs.

Setting $E(\Omega) := \{u\colon \Omega \to \mathbb{R} : u^* \in E\}$, and $\|u\|_{E(\Omega)} := \|u^*\|_E$ (which is a norm because E is r.i.), the Sobolev space $W_0^1E(\Omega)$ is defined as the closure of $C_0^1(\Omega)$ with respect to the norm $\|u\|_{W_0^1E(\Omega)} := \|u\|_{E(\Omega)} + \|\,|\nabla u|\,\|_{E(\Omega)}$ (see, e.g., [15]). It follows that the inequality (2.4) is equivalent (by a generalized Poincaré inequality, [15, Lemma 4.2]) to boundedness of the inclusion

$$j\colon W_0^1F(\Omega) \hookrightarrow E(\Omega). \tag{2.7}$$

Hence, (2.7) is equivalent to boundedness of the kernel operator T in (2.5) from F to E. In view of the above results on optimal domains for the kernel operator T in (2.5), the *optimal r.i. Sobolev imbedding* is

$$j : W_0^1[T,E]^{ri}(\Omega) \hookrightarrow E(\Omega). \tag{2.8}$$

It turns out that compactness/noncompactness of the optimal r.i. Sobolev imbedding (2.8) is intimately connected to that of the associated kernel operator $T\colon [T,E]^{ri} \to E$. This is rather interesting, given that the extended operator $T\colon [T,E] \to E$ is *never* compact; see [20, Proposition 5.2(a)] and [21, Propositions 2.2(c) and 3.6(d)]. In this regard, we have the following result, [24, Theorems 3.7 and 3.9].

Theorem 2.11. *Let E be an r.i. space.*

(a) *If $t^{-1/n'}\varphi_E(t)$ is decreasing, then $T\colon [T,E]^{ri} \to E$ is not compact.*
(b) *$[T,E]^{ri} = L^1([0,1])$ and $T\colon [T,E]^{ri} \to E$ is compact if and only if $\lim_{t\to 0^+} t^{-1/n'}\varphi_E(t) = 0$.*

For $E = L^p([0,1])$, say, the condition (a) of Theorem 2.11 is satisfied whenever $p \geq n'$, so that $T\colon [T,E]^{ri} \to E$ is noncompact. Condition (b) is satisfied for $p < n'$, so that $T\colon [T,E]^{ri} \to E$ is compact in this case.

Theorem 2.11(a) can be "lifted" to obtain the following result, [24, Theorem 4.1].

Theorem 2.12. *Let E and F be r.i. spaces such that $F \subset [T,E]^{ri}$. If $T\colon F \to E$ is noncompact, then the Sobolev imbedding $j\colon W_0^1F(\Omega) \hookrightarrow E(\Omega)$ is bounded, but not compact.*

With the aid of this result, we have the extended version of the Rellich-Kondrachov theorem for the optimal r.i. Sobolev imbedding (2.8), [24, Theorems 4.3 and 4.4].

Theorem 2.13. *Let E be an r.i. space.*

(a) *If $t^{-1/n'}\varphi_E(t)$ is decreasing, then the optimal r.i. Sobolev imbedding* (2.8) *fails to be compact.*
(b) *If $\lim_{t\to 0^+} t^{-1/n'}\varphi_E(t) = 0$, then $[T,E]^{ri} = L^1([0,1])$ and we have compactness of the optimal r.i. Sobolev imbedding* (2.8).

We end the current part concerning optimal Sobolev imbeddings by discussing the possibility of extending the previous results to the non-r.i. setting. For the operator T in (2.5) associated to Sobolev's inequality we know, by Theorem 2.9, that $T\colon [T,E]^{ri} \to E$ has a further genuine extension to $T\colon [T,E] \to E$ only in the case $E \neq L^{n',1}([0,1])$. Hence, if this is the case, then we may consider an *optimal Sobolev imbedding* more general than (2.8), namely

$$j : W_0^1[T,E](\Omega) \hookrightarrow E(\Omega). \tag{2.9}$$

However, difficulties arise in this attempt. Firstly, because $[T,E]$ will not be r.i., it is unclear how the spaces $[T,E](\Omega)$ and hence, also $W_0^1[T,E](\Omega)$, should even be defined. It turns out, due to specific properties of the kernel operator T and of the particular B.f.s. $[T,E]$, that the space $[T,E]$ is always a r.i. *quasi-Banach function space* and hence, that $W_0^1[T,E](\Omega)$ is always a quasi-Banach space (containing $W_0^1[T,X]^{ri}(\Omega)$), [25, Proposition 2.1].

Secondly, it is unclear whether the Sobolev imbedding (2.9) exists or not. The following result shows, at least by following this approach, that there is no possibility of extending the optimal result for r.i. Sobolev imbeddings to the non-r.i. case; [25, Theorems 1.1, 1.2 and 1.3].

Theorem 2.14. *Let E be a r.i. space.*

(a) *If $\lim_{t\to 0}\varphi_E(t)/t^{1/n'} = 0$, then the optimal Sobolev imbedding* (2.9) *fails to exist for the space E.*

(b) *Let $E = \Lambda_\varphi$ be a Lorentz Λ-space such that $\varphi(t)/t^{1/n'}$ is equivalent to a decreasing function. Then the optimal Sobolev imbedding* (2.9) *exists for $E = \Lambda_\varphi$ but, it is not a further extension of the optimal r.i. Sobolev imbedding* (2.8).

(c) *Let E be a r.i. space whose Boyd indices satisfy*

$$0 < \underline{\alpha}_E \le \overline{\alpha}_E < \frac{1}{n'}.$$

The optimal Sobolev imbedding (2.9) *exists for E but, it is not a further extension of the optimal r.i. Sobolev imbedding* (2.8).

Two of the most important operators acting in harmonic analysis are the Fourier transform and convolutions. Both are integral operators corresponding to $\mathbb{C}$-valued kernels and hence, the question of their optimal extension is again relevant. Let G be a compact abelian group with dual group Γ. Recall that $T \in \mathcal{L}(L^p(G))$, for $1 \le p < \infty$, is a *Fourier p-multiplier operator* if it commutes with all translations, where $L^p(G)$ denotes the complex B.f.s. $L^p_{\mathbb{C}}(\lambda)$ with λ being Haar measure on G. Equivalently, there exists $\psi \in \ell^\infty(\Gamma)$ such that

$$\widehat{(Tf)} = \psi\widehat{f}, \qquad f \in L^2 \cap L^p(G), \tag{2.10}$$

where $\widehat{\cdot}$ denotes the Fourier transform, [72]. Since ψ is unique, T is typically denoted by T_ψ. Translations correspond to $\psi(\gamma) = \langle x, \gamma\rangle$ on Γ, for some $x \in G$, and convolutions to $\psi(\gamma) = \widehat{\mu}(\gamma)$ on Γ for some $\mu \in M(G)$, the space of all regular,

$\mathbb{C}$-valued Borel measures on G. Here $\widehat{\mu}(\gamma) := \int_G \overline{\langle x, \gamma \rangle}\, d\mu(x)$, for $\gamma \in \Gamma$, is the *Fourier-Stieltjes transform* of μ. The (continuous) *convolution operator* $T_{\widehat{\mu}}$ acting in $L^p(G)$ is denoted by $C_\mu^{(p)}$ and is defined by $f \mapsto f * \mu$, for $f \in L^p(G)$, where $(f * \mu)(x) := \int_G f(x-y)\, d\mu(y)$ for λ-a.e. $x \in G$, belongs to $L^p(G)$ and satisfies $\|f * \mu\|_p \le |\mu|(G)\|f\|_p$. The vector measure $\nu_{T_{\widehat{\mu}}}$ corresponding to $T_{\widehat{\mu}} = C_\mu^{(p)}$ will be denoted more suggestively by $\nu_\mu^{(p)}$, that is, $\nu_\mu^{(p)}(A) = \mu * \chi_{_A}$ for $A \in \mathcal{B}(G)$. The subclass corresponding to absolutely continuous measures $\mu \ll \lambda$ (i.e., $\mu = \lambda_h$, for some $h \in L^1(G)$, where $\lambda_h(A) := \int_A h\, d\lambda$ on $\mathcal{B}(G)$) is quite different to that for general $\mu \in M(G)$ and so we consider this first. We abbreviate $\nu_{\lambda_h}^{(p)}$ simply to $\nu_h^{(p)}$. For $\varphi \in L^{p'}(G) = L^p(G)^*$, where $p' := p/(p-1)$ is the conjugate index to p, it turns out that

$$(\varphi\nu_h^{(p)})(A) := \langle \nu_h^{(p)}(A), \varphi \rangle = \int_A \varphi * \widetilde{h}\, d\lambda, \qquad A \in \mathcal{B}(G),$$

where $\widetilde{h}(x) := h(-x)$, for $x \in G$, is the *reflection* of h. The next result, [99, Lemma 2.2], collects together some basic properties of the vector measure $\nu_h^{(p)}$.

Theorem 2.15. *Let $1 \le p < \infty$ and fix $h \in L^1(G)$.*

(a) *The range of $\nu_h^{(p)}$ is a relatively compact subset of $L^p(G)$.*
(b) *Given any $A \in \mathcal{B}(G)$ its semivariation (cf. [32, p. 2]) equals*

$$\|\nu_h^{(p)}\|(A) = \sup\left\{ \int_A |\varphi * \widetilde{h}|\, d\lambda : \varphi \in L^{p'}(G),\ \|\varphi\|_{p'} \le 1 \right\},$$

and satisfies

$$\|\widehat{h}\|_\infty\, \lambda(A) \le \|\nu_h^{(p)}\|(A) \le \|h\|_1\, (\lambda(A))^{1/p}.$$

(c) *If $h \ne 0$, then $\lambda \ll \nu_h^{(p)}$. Conversely, always $\nu_h^{(p)} \ll \lambda$.*

It follows from Theorem 2.15(c) that $C_h^{(p)}$ is λ-*determined* whenever $h \in L^1(G) \setminus \{0\}$. The following result, which is a combination of Theorem 1.1, Lemma 3.1 and Proposition 3.4 of [99], summarizes the essential properties of the o.c. optimal lattice domain space $L^1(\nu_h^{(p)})$ of $C_h^{(p)}$; see also [101, Proposition 7.46].

Theorem 2.16. *Let $1 \le p < \infty$ and fix $h \in L^1(G) \setminus \{0\}$.*

(a) *The inclusions*

$$L^p(G) \subseteq L^1(\nu_h^{(p)}) = L^1_w(\nu_h^{(p)}) \subseteq L^1(G)$$

hold and are continuous. Indeed,

$$\|f\|_{L^1(\nu_h^{(p)})} \le \|h\|_1\, \|f\|_p, \qquad f \in L^p(G),$$

and also

$$\|f\|_1 \le \|\widehat{h}\|_\infty^{-1}\, \|f\|_{L^1(\nu_h^{(p)})}, \qquad f \in L^1(\nu_h^{(p)}).$$

(b) $L^1(\nu_h^{(p)}) = \{f \in L^1(G) : \int_G |f| \cdot |\varphi * \widetilde{h}| \, d\lambda < \infty, \ \forall \varphi \in L^{p'}(G)\}$ *and also* $L^1(\nu_h^{(p)}) = \Big\{f \in L^1(G) : \Big((\chi_{_A} f) * h\Big) \in L^p(G), \ \forall A \in \mathcal{B}(G)\Big\}$.

Moreover, the norm of $f \in L^1(\nu_h^{(p)})$ *is given by*

$$\|f\|_{L^1(\nu_h^{(p)})} = \sup\left\{\int_G |f| \cdot |\varphi * \widetilde{h}| \, d\lambda : \varphi \in L^{p'}(G), \ \|\varphi\|_{p'} \leq 1\right\}.$$

(c) $L^1(\nu_h^{(p)})$ *is a translation invariant subspace of* $L^1(G)$ *which is stable under formation of reflections and complex conjugates. Moreover, the extension* $I_{\nu_h^{(p)}} : L^1(\nu_h^{(p)}) \to L^p(G)$ *of* $C_h^{(p)}$ *to its o.c. optimal lattice domain* $L^1(\nu_h^{(p)})$ *is given by*

$$I_{\nu_h^{(p)}}(f) = h * f, \qquad f \in L^1(\nu_h^{(p)}). \tag{2.11}$$

It is known that $C_h^{(p)}$ is a compact operator in $L^p(G)$ for all $1 \leq p < \infty$ and $h \in L^1(G)$. For $h \neq 0$, it turns out that the extended operator $I_{\nu_h^{(p)}}$ (see (2.11)) of $C_h^{(p)}$ to its o.c. optimal lattice domain $L^1(\nu_h^{(p)})$ is a compact operator iff $h \in L^p(G)$ iff the vector measure $\nu_h^{(p)} : \mathcal{B}(G) \to L^p(G)$ has finite variation iff $L^1(\nu_h^{(p)}) = L^1(G)$ is as large as possible; see [99, Theorem 1.2], where Theorem 1.9 above plays a crucial role in the proof. The following result, essentially Proposition 4.1 of [99], shows that more information about $I_{\nu_h^{(p)}}$ is available.

Theorem 2.17. *Let* $1 \leq p < \infty$ *and* $h \in L^1(G) \setminus \{0\}$.

(a) *If* $h \notin L^p(G)$, *then* $I_{\nu_h^{(p)}} : L^1(\nu_h^{(p)}) \to L^p(G)$ *is not compact and both the following inclusions are proper:*

$$L^p(G) \subseteq L^1(\nu_h^{(p)}) \subseteq L^1(G). \tag{2.12}$$

The first inclusion in (2.12) *is proper for every* $h \in L^1(G) \setminus \{0\}$. *In particular, the o.c. optimal lattice domain* $L^1(\nu_h^{(p)})$ *of* $C_h^{(p)}$ *is always genuinely larger than* $L^p(G)$.

(b) *There exists* $h \in L^1(G)$ *with* $\bigcup_{1<p<\infty} L^1(\nu_h^{(p)}) \subsetneqq L^1(G)$.

For further results we refer to [99] and [101, Ch. 7, §7.3].

We now turn our attention to $C_\mu^{(p)}$ with $\mu \in M(G) \setminus L^1(G)$. Of relevance are the measures in $M_0(G) := \{\mu \in M(G) : \widehat{\mu} \in c_0(\Gamma)\}$. Indeed, it turns out for $1 < p < \infty$ and $\mu \in M(G)$ that $C_\mu^{(p)}$ is compact in $L^p(G)$ iff $\mu \in M_0(G)$ iff the range of the vector measure $\nu_\mu^{(p)}$ is a relatively compact subset of $L^p(G)$, [100, Proposition 2.3]. For arbitrary $\mu \in M(G)$ it is the case that $\nu_\mu^{(p)} \ll \lambda$ and, if $\mu \neq 0$, then also, $\lambda \ll \nu_\mu^{(p)}$, [100, Proposition 2.4]. In particular, $C_\mu^{(p)}$ is λ-*determined* whenever $\mu \in M(G) \setminus \{0\}$. Moreover, the statement of Theorem 2.16 remains valid (throughout) if we replace $h \simeq \lambda_h$ (resp. $\nu_h^{(p)}$) with $\mu \in M(G)$ (resp. $\nu_\mu^{(p)}$); see [100, Theorem 1.1 & Corollary 3.2] and [101, Proposition 7.60]. Compactness of the operators $C_\mu^{(p)}$ was characterized above (e.g., in terms of $M_0(G)$, say). For their optimal extension $I_{\nu_\mu^{(p)}}$ we have the following result, which is a combination of Theorem 1.2, Remark 4.2(b) and Proposition 4.3 of [100].

Theorem 2.18. *Let $1 \leq p < \infty$ and fix $\mu \in M(G) \setminus \{0\}$. Then the following assertions are equivalent.*

(a) *The extension $I_{\nu_\mu^{(p)}} : L^1(\nu_\mu^{(p)}) \to L^p(G)$ of $C_\mu^{(p)}$ to its o.c. optimal lattice domain $L^1(\nu_\mu^{(p)})$ is a compact operator.*

(b) *$\mu = \lambda_h$ for some $h \in L^p(G)$.*

(c) *The vector measure $\nu_\mu^{(p)} : \mathcal{B}(G) \to L^p(G)$ has finite variation.*

(d) *$L^1(\nu_\mu^{(p)}) = L^1(G)$.*

We now take a closer look at the spaces $L^1(\nu_\mu^{(p)})$ for arbitrary $\mu \in M(G)$; phenomena quite different to the case of $\mu \ll \lambda$ can occur (which is covered by Theorem 2.17 above). For $a \in G$, we denote the Dirac point mass at a by δ_a.

Theorem 2.19. *Let $1 < p < \infty$ and $\mu \in M(G)$.*

(a) *If $\operatorname{supp}(\mu) \neq G$ and $a \notin \operatorname{supp}(\mu)$, then $L^1(\nu_{\mu+\delta_a}^{(p)}) = L^p(G)$.*

(b) *If $a \in G$ and $\mu \in M_0(G)$, then $L^1(\nu_{\mu+\delta_a}^{(p)}) = L^p(G)$.*

(c) *If there exists $\eta \in M(G)$ satisfying $\mu * \eta = \delta_0$, then $L^1(\nu_\mu^{(p)}) = L^p(G)$.*

(d) *If $\mu \geq 0$ and $L^1(\nu_\mu^{(p)}) \neq L^p(G)$ for some $1 < p < \infty$, then μ is a continuous measure (i.e., $\mu(\{a\}) = 0$ for all $a \in G$).*

(e) *The inclusion $L^p(G) \subseteq L^1(\nu_\mu^{(p)})$ is proper if $\mu \in M_0(G) \setminus \{0\}$.*

For part (a) we refer to [101, Remark 7.75], for (b) to [101, Proposition 7.77], for (c) to [101, Corollary 7.79] and for (d) to [101, Corollary 7.76]. Part (e) is [100, Proposition 4.5]. Cases (a)-(c) in Theorem 2.19 show that $C_\mu^{(p)}$ may already be defined on its o.c. optimal lattice domain and no further extension is possible. Case (e) in Theorem 2.19 exhibits a large class of measures μ where the optimal extension is always *genuine*. Theorem 2.18 characterizes those μ for which the optimal extension is to the largest possible domain space, namely $L^1(G)$.

Consider now $G := \mathbb{T}$, the circle group, in which case $\Gamma = \mathbb{Z}$. For $1 \leq p \leq 2$, it is known that the Fourier transform $F_p : L^p(\mathbb{T}) \to \ell^{p'}(\mathbb{Z})$ is injective and that it is continuous, because of the *Hausdorff-Young inequality*

$$\|\widehat{f}\|_{p'} \leq \|f\|_p, \qquad f \in L^p(\mathbb{T}).$$

Of course, $F_p f := \widehat{f}$ for each $f \in L^p(\mathbb{T}) \subseteq L^1(\mathbb{T})$. As for Sobolev's inequality, discussed in Section 2, one may ask whether the Hausdorff-Young inequality is optimal. Now, for $X := L^p(\mathbb{T})$ and $E := \ell^{p'}(\mathbb{Z})$ and $T := F_p$, we have $\nu_{F_p}(A) = F_p(\chi_A) = \widehat{\chi_A}$ for $A \in \mathcal{B}(\mathbb{T})$. Note that the B.f.s. X has o.c. norm and that the codomain space E is reflexive for $1 < p \leq 2$. Moreover, ν_{F_p} is σ-additive and $L^1(\nu_{F_p}) = L^1_w(\nu_{F_p})$ for all $1 \leq p \leq 2$. It is known that the vector measure $\nu_{F_p} : \mathcal{B}(\mathbb{T}) \to \ell^{p'}(\mathbb{Z})$ is mutually absolutely continuous with respect to Haar measure

λ on $\mathbb{T}$ (hence, F_p is a λ-*determined* operator), that it has infinite variation for $1 < p \le 2$ (with ν_{F_1} having finite variation), and that the range of ν_{F_p}, for $1 \le p < \infty$, is not a relatively compact subset of $\ell^{p'}(\mathbb{Z})$; see Lemma 2.1, Remark 2.2 and Corollary 2.5 of [87]. Moreover, $L^1(\nu_{F_p}) \subseteq L^1(\mathbb{T})$ with $\|f\|_1 \le \|f\|_{L^1(\nu_{F_p})}$, for each $f \in L^1(\nu_{F_p})$, and the extension $I_{\nu_{F_p}} : L^1(\nu_{F_p}) \to \ell^{p'}(\mathbb{Z})$ of F_p is again the map $f \mapsto \widehat{f}$, for $f \in L^1(\nu_{F_p})$, [87, Theorem 1.1(iii)]. For simplicity, we adopt the notation of [87] and, henceforth, write $\mathbf{F}^p(\mathbb{T}) := L^1(\nu_{F_p})$.

We now turn to a more concrete description of $\mathbf{F}^p(\mathbb{T})$. First,

$$\|f\|_{\mathbf{F}^p(\mathbb{T})} = \sup\left\{ \int_{\mathbb{T}} |f| \cdot |\check{\phi}|\, d\lambda : \phi \in \ell^p(\mathbb{Z}),\ \|\phi\|_p \le 1 \right\},$$

where $\check{\phi}$ is the inverse Fourier transform of $\phi \in \ell^p(\mathbb{Z}) = \ell^{p'}(\mathbb{Z})^* \subseteq \ell^2(\mathbb{Z})$. Given $1 \le p \le 2$, define a vector subspace $V^p(\mathbb{T})$ of $L^{p'}(\mathbb{T})$ by

$$V^p(\mathbb{T}) := \{h \in L^{p'}(\mathbb{T}) : h = \check{\varphi} \ \text{ for some } \varphi \in \ell^p(\mathbb{Z})\}.$$

For $p = 2$, Plancherel's theorem implies that $V^2(\mathbb{T}) = L^2(\mathbb{T})$. It is known that there exists $f \in C(\mathbb{T})$ such that $\widehat{f} \notin \ell^r(\mathbb{Z})$ for all $1 \le r < 2$ and so the containment $V^p(\mathbb{T}) \subseteq L^{p'}(\mathbb{T})$ is *proper* for $1 \le p < 2$. For each $f \in L^1(\mathbb{T})$, define a linear map $S_f : L^\infty(\mathbb{T}) \to c_0(\mathbb{Z})$ by $S_f : g \mapsto \widehat{gf}$ for $g \in L^\infty(\mathbb{T})$. Clearly $\|S_f\| \le \|f\|_1$. For each $1 \le p \le 2$ and each continuous operator $R : L^\infty(\mathbb{T}) \to \ell^{p'}(\mathbb{Z})$, let $\|R\|_{\infty,p'} := \sup_{\|g\|_\infty \le 1} \|Rg\|_{p'}$ denote its operator norm. If $f \in L^1(\mathbb{T})$ has the property that the range $S_f(L^\infty(\mathbb{T})) \subseteq \ell^{p'}(\mathbb{Z})$, then the Closed Graph Theorem implies that $\|S_f\|_{\infty,p'} < \infty$. For the next result, see [87, Theorem 1.2].

Theorem 2.20. *Let $1 \le p \le 2$. Each of the spaces*

$$\Delta^p(\mathbb{T}) = \left\{ f \in L^1(\mathbb{T}) : \int_{\mathbb{T}} |fg|\, d\lambda < \infty,\ \forall g \in V^p(\mathbb{T}) \right\},$$

$$\Phi^p(\mathbb{T}) = \left\{ f \in L^1(\mathbb{T}) : \widehat{f\chi_A} \in \ell^{p'}(\mathbb{Z}),\ \forall A \in \mathcal{B}(\mathbb{T}) \right\},$$

$$\Gamma^p(\mathbb{T}) = \left\{ f \in L^1(\mathbb{T}) : S_f(L^\infty(\mathbb{T})) \subseteq \ell^{p'}(\mathbb{Z}) \right\}, \tag{2.13}$$

coincides with the o.c. optimal lattice domain $\mathbf{F}^p(\mathbb{T})$ of the Hausdorff-Young inequality. Moreover, in the case of (2.13), *the operator norm $\|S_f\|_{\infty,p'}$ is equivalent to $\|f\|_{\mathbf{F}^p(\mathbb{T})}$, for $f \in \mathbf{F}^p(\mathbb{T})$.*

For $p = 1, 2$ it turns out that $\mathbf{F}^1(\mathbb{T}) = L^1(\mathbb{T})$ and $\mathbf{F}^2(\mathbb{T}) = L^2(\mathbb{T})$. So, both maps $F_1 : L^1(\mathbb{T}) \to \ell^\infty(\mathbb{Z})$ and $F_2 : L^2(\mathbb{T}) \to \ell^2(\mathbb{Z})$ are already defined on their o.c. optimal lattice domain; no further extension is possible. Is $\mathbf{F}^p(\mathbb{T})$ genuinely larger than $L^p(\mathbb{T})$ for $1 < p < 2$? The answer is given by the following result, [87, Theorem 1.4].

Theorem 2.21. *For $1 < p < 2$, the following inclusions are proper:*

$$L^p(\mathbb{T}) \subseteq \mathbf{F}^p(\mathbb{T}) \subseteq L^1(\mathbb{T}).$$

It is also shown in [87] that, for each $1 < r < p$, the space $\mathbf{F}^p(\mathbb{T})$ is not contained in $L^r(\mathbb{T})$ and, that $L^r(\mathbb{T})$ is not contained in $\mathbf{F}^p(\mathbb{T})$ for any $1 \leq r < p$. It is also established in [87] that $\mathbf{F}^p(\mathbb{T})$ is a translation invariant subspace of $L^1(\mathbb{T})$ and that it is a weakly sequentially complete B.f.s. with the σ-Fatou property. Moreover, the translation operators $\tau_w f(z) = f(zw^{-1})$, for $w, z \in \mathbb{T}$, are continuous in $\mathbf{F}^p(\mathbb{T})$ and τ_w converges to the identity operator (for the strong operator topology) as $w \mapsto 1$. Accordingly, $\mathbf{F}^p(\mathbb{T})$ is a *homogeneous* Banach space and so is well suited for harmonic analysis.

3. Aspects of spectral integration

A rich source of vector measures arises in spectral theory. For instance, the resolution of the identity of a normal operator in a Hilbert space is a σ-additive projection-valued measure (i.e., *spectral measure*). Such vector (= operator-valued) measures were extended to the Banach space setting by N. Dunford, [45]. The theory of spectral measures and integration in Banach and Fréchet spaces is somewhat more complicated than in Hilbert spaces (where it is more transparent because of the Mackey-Wermer theorem, [42, Proposition 8.2]). Nevertheless, there have been significant advances in this theory over the past 15 years or so. We begin with developments that occured in the general theory; the latter half of the section is devoted to applications.

Let X be a Fréchet space and $\mathcal{L}(X)$ denote the space of all continuous linear operators of X into itself. The identity operator in X is denoted by I. The strong operator topology τ_s (briefly, sot) in $\mathcal{L}(X)$ is generated by the seminorms

$$q_{x,n} : T \mapsto \|Tx\|_n, \qquad T \in \mathcal{L}(X), \tag{3.1}$$

where $x \in X$ is arbitrary and $\{\|\cdot\|_n\}_{n=1}^{\infty}$ is a sequence of continuous seminorms determining the topology of X. Then $E = \mathcal{L}_s(X)$ denotes the quasicomplete lcHs $\mathcal{L}(X)$ equipped with the continuous seminorms $Q := \{q_{x,n} : x \in X, n \in \mathbb{N}\}$ as given by (3.1).

A collection $\mathcal{M} \subseteq \mathcal{L}(X)$ of commuting projections is a *Boolean algebra* (briefly, B.a.) if $0, I \in \mathcal{M}$ and $\mathcal{M}$ is a B.a. for the partial order $\leq$ defined by $P \leq R$ iff $PR = P = RP$ (equivalently, their ranges satisfy $PX \subseteq RX$). The B.a. operations $\vee, \wedge$ and complementation in $\mathcal{M}$ are then given by $P \wedge R = PR$ and $P \vee R = P + R - PR$ and $P^c = I - P$. If $\mathcal{M}$ is equicontinuous in $\mathcal{L}(X)$, then $\mathcal{M}$ is called *equicontinuous* (or *bounded* if X is Banach). We will discuss only such B.a.'s of projections, although important examples occur in classical analysis which are not equicontinuous; see [111, 84] and [117, Ch. III Example 18]. It is remarkable that every B.a. of projections which is merely σ-complete as an abstract B.a. is already equicontinuous; see [5, Theorem 2.2] for X a Banach space, [134, Proposition 1.2] for X a Fréchet space, and [9] for some other classes of lcHs' X. Here *abstractly σ-complete* means every countable subset $A \subseteq \mathcal{M}$ has a greatest lower bound, denoted by $\wedge A$ (equivalently a least upper bound, denoted by $\vee A$).

According to a result of M.H. Stone, every B.a. of projections $\mathcal{M} \subseteq \mathcal{L}(X)$ is isomorphic to the algebra of all closed-open sets $Co(\Omega_{\mathcal{M}})$ of some totally disconnected, compact Hausdorff space $\Omega_{\mathcal{M}}$. This isomorphism is finitely additive as a set function from $Co(\Omega_{\mathcal{M}})$ into $\mathcal{L}(X)$. If $\mathcal{M}$ is abstractly σ-complete, then $\Omega_{\mathcal{M}}$ is basically disconnected and if $\mathcal{M}$ is abstractly complete (i.e., every $A \subseteq \mathcal{M}$ has a greatest lower bound $\wedge A$; equivalently, a least upper bound $\vee A$), then $\Omega_{\mathcal{M}}$ is extremely disconnected; see, e.g., [117, Ch. II] for a discussion of these classical facts. In particular, every B.a. of projections can be represented as the range $P(\Sigma) := \{P(A) : A \in \Sigma\}$ of some finitely additive *spectral measure* $P : \Sigma \to \mathcal{L}(X)$ and vice versa (with Σ an algebra of subsets of some set $\Omega \neq \emptyset$), where P satisfies $P(\emptyset) = 0$ and $P(\Omega) = I$ and P is multiplicative (i.e., $P(A \cap B) = P(A)P(B)$ for $A, B \in \Sigma$).

A B.a. of projections $\mathcal{M} \subseteq \mathcal{L}(X)$ is *Bade complete* (resp. *Bade σ-complete*) if it is abstractly complete (resp. abstractly σ-complete) and

$$(\wedge_\alpha B_\alpha)X = \cap_\alpha (B_\alpha X) \text{ and } (\vee_\alpha B_\alpha)X = \overline{\mathrm{sp}\{\cup_\alpha (B_\alpha X)\}}, \tag{3.2}$$

whenever $\{B_\alpha\}$ is a family (resp. countable family) of elements from $\mathcal{M}$, [5, 134]. The interaction between the order properties of $\mathcal{M}$ and the topology of X (via (3.2)) has some far reaching consequences.

Theorem 3.1. *Let X be a Fréchet space and $\mathcal{M} \subseteq \mathcal{L}(X)$ be a B.a. of projections.*

(a) *If $\mathcal{M}$ is Bade complete, then it is a closed (hence, also complete) subset of $\mathcal{L}_s(X)$.*

(b) *If $\mathcal{M}$ is Bade σ-complete, then its closure $\overline{\mathcal{M}}_s$ in $\mathcal{L}_s(X)$ is a Bade complete B.a. of projections.*

(c) *If $\mathcal{M}$ is Bade σ-complete (resp. Bade complete), then $\mathcal{M}$ is the range of a σ-additive $\mathcal{L}_s(X)$-valued spectral measure defined on the Baire (resp. Borel) subsets of the Stone space $\Omega_{\mathcal{M}}$ of $\mathcal{M}$.*

For Banach spaces, (a) and (b) occur in [5] and for Fréchet spaces in [40, §4]. Part (c) is folklore; for a discussion and proof see [92, §4] and [93], for example, and the references therein.

If X is separable, then every Bade σ-complete B.a. of projections $\mathcal{M}$ is Bade complete; see [5, p. 350] and [40, Proposition 4.3]. If $\mathcal{M}$ possesses a *cyclic vector* (i.e., $X = \overline{\mathrm{sp}\{Bx : B \in \mathcal{M}\}}$ for some $x \in X$), then again $\mathcal{M}$ is Bade complete; the same is true if $\mathcal{M}$ is *countably decomposable* (i.e., every pairwise disjoint family of elements from $\mathcal{M}$ is at most countable). For these claims and further sufficient conditions on $\mathcal{M}$ see [94]. If we restrict X to the class of Banach spaces, even more is known. For instance, X has the property that every bounded B.a. of projections $\mathcal{M} \subseteq \mathcal{L}_s(X)$ which is τ_s-closed is Bade complete iff X does not contain an isomorphic copy of c_0, [64]. Or, if X is separable, then a B.a. of projections in X is abstractly σ-complete iff it is abstractly complete, [116, Corollary 2.1]; this is a consequence of the fact that in any weakly compactly generated Banach space (separable spaces have this property) the τ_s-closure of any abstractly σ-complete B.a. of projections is Bade complete, [116, Theorem 2]. Recently it was shown that a Banach space X has the property that the τ_s-closure of every abstractly σ-complete B.a. of projections in X is Bade complete iff X does not contain an isomorphic copy of ℓ^∞, [35]. This is of interest because an abstractly σ-complete B.a. of projections in a Banach space not containing a copy of ℓ^∞ need not be Bade complete or even Bade σ-complete, [116, Remark 2].

In view of Theorem 3.1(b), it might be anticipated that every Bade σ-complete B.a. of projections in a Banach space X is at least a sequentially closed subset of $\mathcal{L}_s(X)$. For purely atomic B.a.'s this was known to be the case, [115], but, recently this question was answered in the *negative* in [58], even for Hilbert spaces! The paper [58] is also of interest because it exhibits a large class of *non-atomic*, Bade σ-complete (but, not Bade complete) B.a.'s of projections in the *non-separable* Banach space $ca(\Sigma)$ consisting of all σ-additive, $\mathbb{R}$-valued measures on a measurable space (Ω, Σ) equipped with the total variation norm. Such types of examples were missing in the past.

Non-trivial, concrete examples of Bade complete and σ-complete B.a.'s of projections in *Fréchet* spaces X have been somewhat scarce in the past; see [134, 135, 123, 89], for some such examples. In recent years this list of examples has been significantly extended and reveals (in certain cases) an intimate connection between properties of $\mathcal{M}$ (e.g., Dunford's boundedness criterion, boundedly σ-complete, finite τ_s-variation) and geometric properties of X (e.g., nuclear, Montel, Radon-Nikodým property); see [95, 10, 11, 13, 12, 90].

There is also a converse to Theorem 3.1.

Theorem 3.2. *Let X be a Fréchet space and $P : \Sigma \to \mathcal{L}_s(X)$ be a σ-additive spectral measure defined on a σ-algebra Σ.*

(a) *The range $P(\Sigma) \subseteq \mathcal{L}(X)$ is an equicontinuous, Bade σ-complete B.a. of projections.*

(b) *The B.a. of projections $P(\Sigma)$ is Bade complete if and only if $P(\Sigma)$ is a closed subset of $\mathcal{L}_s(X)$.*

For the proof and a discussion of this result, together with its historical origins (for X normable and non-normable), see [92, Section 3] and [93, Theorem 2, Remark 4.3] and the references therein.

Of particular interest is the case when Σ equals the Borel subsets $\mathcal{B}(K)$ of some compact set $K \subseteq \mathbb{C}$ (or $K \subseteq \mathbb{C} \cup \{\infty\}$ if X is non-normable). It is assumed that K is minimal, that is, it equals the *support* of the spectral measure P (in the sense of [42, p. 122]) and that the identity function on K belongs to $L^1(P)$, in which case the operator $I_P(z) = \int_K z \, dP(z)$ is called a *scalar-type spectral operator* and P is its (unique) *resolution of the identity*. For X a Banach space, such operators have been extensively studied in [42, 45, 117]. It turns out, for X a separable Banach space, that *every* Bade σ-complete (= Bade complete) B.a. of projections $\mathcal{M} \subseteq \mathcal{L}(X)$ coincides with the resolution of the identity of some scalar-type spectral operator [103, Proposition 2]. The proof of this result depends on the existence of *Bade functionals*, i.e., given any $x \in X$ there exists $x^* \in X^*$ with the properties that $\langle Rx, x^* \rangle \geq 0$ for all $R \in \mathcal{M}$ and $Bx = 0$ whenever $B \in \mathcal{M}$

satisfies $\langle Bx, x^* \rangle = 0$, [5, Theorem 3.1]. Unfortunately, this remarkable fact fails to hold in a general Fréchet space X. Indeed, a Fréchet space X has the property that *every* Bade σ-complete B.a. of projections $\mathcal{M} \subseteq \mathcal{L}(X)$ possesses Bade functionals (for arbitrary $x \in X$) iff X does not contain an isomorphic copy of the Fréchet sequence space $\mathbb{C}^{\mathbb{N}}$, [114, Theorem 2]. Nevertheless, it remains true that every Bade σ-complete (= Bade complete) B.a. of projections in a separable Fréchet space coincides with the resolution of the identity of some scalar-type spectral operator, [118].

For X a Fréchet space and $P : \Sigma \to \mathcal{L}_s(X)$ a σ-additive spectral measure, the integration operator $I_P : L^1(P) \to \mathcal{L}_s(X)$ is of central importance. Indeed, with respect to pointwise multiplication the lc-Riesz space $L^1(P)$ is also a commutative lc-algebra with identity and is topologically τ_s-complete iff $P(\Sigma)$ is a closed subset of $\mathcal{L}_s(X)$, [40, Proposition 1.4]. In the case when $P(\Sigma)$ is τ_s-closed, the integration operator I_P is an isomorphism of the complete lc-algebra $L^1(P)$ onto the closed operator algebra in $\mathcal{L}_s(X)$ generated by $P(\Sigma)$, [40, Proposition 1.5]. It should be pointed out that $L^1(P) = L^\infty(P)$, as vector spaces, whenever X is a Banach space, [117, Proposition V.4]. For each $x \in X$, let $P(\Sigma)[x]$ denote the closed subspace of X generated by $\{P(A)x : A \in \Sigma\}$; it is called the *cyclic space* generated by x. The X-valued vector measure defined in Σ by $A \mapsto P(A)x$ is denoted by Px. An important fact is that $L^1(P) = \bigcap_{x \in X} L^1(Px)$, [95, Lemma 2.2]. The following result, [41, Proposition 2.1], a converse to Theorem 1.3, reveals the intimate connection between B.a.'s of projections, spectral measures and certain aspects from the theory of order and positivity that arise via Banach lattices (and more general Riesz spaces); see also [39]. For the terminology of undefined notions we refer to [3].

Theorem 3.3. *Let X be a Fréchet space and $\mathcal{M} \subseteq \mathcal{L}(X)$ be a Bade complete B.a. of projections, displayed as the range of some spectral measure $P : \Sigma \to \mathcal{L}_s(X)$, that is, $\mathcal{M} = P(\Sigma)$. Then, for each $x \in X$, the integration operator $I_{Px} : L^1(Px) \to \mathcal{M}[x]$ induces on the cyclic space $\mathcal{M}[x]$ the structure of a Dedekind complete Fréchet lattice with Lebesgue topology in which x is a weak order unit. The absolute value of an element $\int_\Omega f\, dPx \in \mathcal{M}[x]$ is the element $\int_\Omega |f|\, dPx \in \mathcal{M}[x]$. Moreover, I_{Px} is a Riesz and topological isomorphism and the absolute value mapping on $\mathcal{M}[x]$ is continuous.*

Concerning Theorem 3.3, if there exists a *cyclic vector* $x_0 \in X$ for $\mathcal{M}$, then the integration operator induces on X the structure of a Dedekind complete Fréchet lattice (= Banach lattice if X is normable) with a Lebesgue topology (= order continuous norm if X is normable) which is equivalent to the original topology in X and such that x_0 is a weak order unit. Moreover, with respect to this Fréchet lattice topology, I_P is a *positive operator* and the B.a. $\mathcal{M}$ may be identified with the B.a. of all *band projections*, [41, Proposition 2.1]. For X a Banach space, this fact goes essentially back to A.I. Veksler, [133]. For X non-normable, to describe $L^1(P)$ "concretely" is, in general, rather difficult. Some illuminating and non-trivial examples occur in [10, 11, 12], where $L^1(P)$ is identified with a certain space of multiplication operators acting on X.

To decide whether a *particular* operator acting in a given Banach space is actually scalar-type spectral (briefly, scalar) can be rather difficult; see [45], for example. Here we only make some relevant comments in the direction of harmonic analysis. That certain translation and convolution operators in L^p-spaces over $\mathbb{Z}$, for $p \neq 2$, fail to be scalar operators goes back to U. Fixman, [57], and G.L. Krabbe, [69]. Translation operators in $L^p(G)$, with G an arbitrary locally compact abelian group and $p \neq 2$, were shown by T.A. Gillespie to be scalar operators iff they have finite spectrum, [63]. If one is prepared to relax the topology in $L^p(G)$, then it may happen that non-scalar translation operators in $L^p(G)$ *do* become scalar operators when extended to act in a superspace $X_p(G)$ (not necessarily Fréchet) which contains $L^p(G)$ continuously, [60, 61], but not always. As noted in Section 2, translations and convolutions are special cases of Fourier multiplier operators. For $G = \mathbb{T}$ (or any compact metrizable abelian group), those Fourier p-multiplier operators $T_\psi \in \mathcal{L}(L^p(G))$ which are scalar are characterized in [85]; see also [86]. Namely, for each λ in the countable set $\psi(\Gamma) \subseteq \mathbb{C}$ the idempotent $\chi_{\psi^{-1}(\{\lambda\})}$ should be a p-multiplier for G and the pairwise disjoint family of Fourier p-multiplier projections $\{T\chi_{\psi^{-1}(\{\lambda\})} : \lambda \in \psi(\Gamma)\} \subseteq \mathcal{L}(L^p(G))$ should be a *Littlewood-Paley p-decomposition* for G. For $G = \mathbb{R}^N$ and $\psi : \mathbb{R}^N \to \mathbb{C}$ a polynomial, (2.10) can be used to define a closed, densely defined, unbounded Fourier p-multiplier operator which is, of course, none-other than a constant coefficient linear partial differential operator in $L^p(\mathbb{R}^N)$. Such operators are rarely (unbounded) scalar operators for $p \neq 2$, [1]; see also [2, 113] for results concerning the spectrality of *matrix-valued* Fourier p-multiplier and differential operators in L^p-spaces.

We also mention some recent directions where vector and operator-valued measures occur in vector-valued harmonic analysis. Let G be a lca group and $\lambda : \mathcal{B}(G) \to [0, \infty]$ denote Haar measure. Given $1 \leq p < \infty$ and a Hilbert space $\mathcal{H}$, let $L^p(G, \mathcal{H})$ be the Banach space of (equivalence classes of) strongly λ-measurable functions $f : G \to \mathcal{H}$ such that the norm $\|f\|_p := \left(\int_G \|f(u)\|^p_{\mathcal{H}} \, d\lambda(u) \right)^{1/p} < \infty$. The Fourier transform of a function $f \in L^1(G, \mathcal{H})$ is defined in the natural way:

$$\widehat{f}(\gamma) := \int_G \overline{\langle u, \gamma \rangle} f(u) \, d\lambda(u), \qquad \gamma \in \Gamma,$$

the integral being a $\mathcal{H}$-valued Bochner integral. An operator $T \in \mathcal{L}(L^p(G, \mathcal{H}))$ is called a *Fourier p-multiplier operator* if there exists a measurable function $\Phi : \Gamma \to \mathcal{L}(\mathcal{H})$ such that Φ is essentially bounded in the operator norm of $\mathcal{L}(\mathcal{H})$ and for each $f \in L^1 \cap L^2 \cap L^p(G, \mathcal{H})$ the equality $(\widehat{Tf})(\gamma) = \Phi(\gamma)\widehat{f}(\gamma)$ holds a.e. on Γ (compare with (2.10)). Here the measurability of Φ means that the $\mathcal{H}$-valued function $\gamma \mapsto \Phi(\gamma)h$ is strongly measurable for each $h \in \mathcal{H}$. As for scalar-valued harmonic analysis, it turns out that T is a Fourier p-multiplier operator iff it commutes with each translation operator in $\mathcal{L}(L^p(G, \mathcal{H}))$, [62, Proposition 2.8]. The monograph [62] is mainly concerned with an analysis of the more special case when

$$\Phi(\gamma) = \widehat{\mu}(\gamma) := \int_G \overline{\langle u, \gamma \rangle}\, d\mu(u), \qquad \gamma \in \Gamma,$$

is the Fourier-Stieltjes transform $\widehat{\mu} : \Gamma \to \mathcal{L}(\mathcal{H})$ of a regular *operator-valued measure* $\mu : \mathcal{B}(G) \to \mathcal{L}_s(\mathcal{H})$. A collection of negative results illustrates decisively how known L^p multiplier results in the scalar setting can break down in the vector-valued setting when $p \neq 2$. For instance, because the operator-valued measure μ is not, generally, selfadjoint-valued, it can happen that $\widehat{\mu} : \Gamma \to \mathcal{L}(\mathcal{H})$ is a Fourier p-multiplier but *not* a Fourier p'-multiplier, where $\frac{1}{p} + \frac{1}{p'} = 1$. Chapter 4 of [62] is devoted to the case when $\mu : \mathcal{B}(G) \to \mathcal{L}_s(\mathcal{H})$ is a *spectral measure.* It is shown that whenever $(G, \mathcal{B}(G), \lambda)$ is a separable measure space (with G infinite), $\mathcal{H} = L^2(G)$ and μ is the canonical (selfadjoint) spectral measure acting in $L^2(G)$ via multiplication with χ_A, for $A \in \mathcal{B}(G)$, then $T_{\widehat{\mu}}$ is a Fourier p-multiplier operator iff $p = 2$, [62, Theorem 4.7]. As a consequence: according to Stone's theorem, the translation group $\Phi : \Gamma \to \mathcal{L}_s(L^2(\Gamma))$, which is bounded and τ_s-continuous, has the form $\Phi = \widehat{P}$ for some regular, selfadjoint spectral measure $P : \mathcal{B}(G) \to \mathcal{L}_s(L^2(G))$. By the previous mentioned fact it follows that $T_{\widehat{P}}$ is a Fourier p-multiplier operator (acting in $L^p(G, L^2(G))$) iff $p = 2$. Of course, [62] has further results than just the sample alluded to above; see also [131] and the references therein.

A fundamental question is to determine criteria which ensure that the sum and product of two commuting scalar operators acting in a Banach space are again scalar: an example of C.A. McCarthy shows, even in a separable reflexive Banach space, that this is not always the case, [83]. Since commutativity of the scalar operators is equivalent to the commutativity of their resolutions of the identity (a consequence of [45, XV Corollary 3.7]), the above question is intimately related to the problem of determining criteria which ensure that the B.a. generated by two commuting, bounded B.a.'s of projections is again bounded. An account of what was known up to 1970 in regard to these questions can be found in [45, pp. 2098–2101]. Much research concentrated on identifying classes of Banach spaces in which the answer is positive. For instance, C.A. McCarthy established that all L^p-spaces, for $1 \leq p < \infty$, have this property and (together with W. Littman and N. Rivière) also their complemented subspaces; see [45, pp. 2099–2100], for example. This is also the case for all Grothendieck spaces with the Dunford-Pettis property, [110], and the class of all hereditarily indecomposable Banach spaces, [112]. Also [43] is relevant to these questions. However, not so many results have appeared in this regard, since it is difficult to identify geometric conditions on a Banach space X which ensure that *all* pairs of commuting, bounded B.a.'s of projections in X automatically have a uniformly bounded product B.a. Perhaps the most recent significant result in this direction is the following one, due to T.A. Gillespie, [65].

Theorem 3.4. *Let $\mathcal{M}$ and $\mathcal{N}$ be commuting, bounded B.a.'s of projections in a Banach space X. Then the B.a. of projections $\mathcal{M} \vee \mathcal{N}$ generated by $\mathcal{M}$ and $\mathcal{N}$ is also bounded, in each of the following cases.*

(a) *X is a Banach lattice.*
(b) *X is a closed subspace of any p-concave Banach lattice (p finite).*
(c) *X is a complemented subspace of any $\mathcal{L}^\infty$-space.*
(d) *X is a closed subspace of $\mathcal{L}^p$ for any $1 \le p < \infty$.*
(e) *X has local unconditional structure (briefly, l.u.st.).*

In the recent article [104], the viewpoint was taken that the geometry of X is not the only relevant ingredient; an important property of the individual B.a.'s concerned (when available) can also play a crucial role. This is the notion of R-boundedness, introduced by E. Berkson and T.A. Gillespie in [7] (where it is called the R-property), but already explicit in earlier work of J. Bourgain, [14]. For a Banach space X, a non-empty collection $\mathcal{T} \subseteq \mathcal{L}(X)$ is called *R-bounded* if there exists a constant $M \ge 0$ such that

$$\Big(\int_0^1 \Big\|\sum_{j=1}^n r_j(t)T_j x_j\Big\|^2 dt\Big)^{1/2} \le M\Big(\int_0^1 \Big\|\sum_{j=1}^n r_j(t)x_j\Big\|^2 dt\Big)^{1/2}$$

for all $T_1, \dots, T_n \in \mathcal{T}$, all $x_1, \dots, x_n \in X$ and all $n \in \mathbb{N}$, where $\{r_j\}_{j=1}^\infty$ is the sequence of Rademacher functions on $[0,1]$. Clearly every R-bounded collection is uniformly bounded in $\mathcal{L}(X)$. An important fact is the following one, [104, Theorem 3.1].

Theorem 3.5. *Let X be a Banach space and $\mathcal{M}$ be any R-bounded B.a. of projections in X. Then the B.a. $\mathcal{M} \vee \mathcal{N}$ is bounded for every bounded B.a. of projections $\mathcal{N} \subseteq \mathcal{L}(X)$ which commutes with $\mathcal{M}$.*

The applicability of Theorem 3.5 stems from the fact that every Banach space X with property (α), a class of spaces introduced by G. Pisier, [109], has the property that *every* bounded B.a. of projections in X is automatically R-bounded, [104, Theorem 3.3]. We recall that *property* (α) holds if there exists a constant $\alpha \ge 0$ such that

$$\begin{aligned}\int_0^1\int_0^1 \Big\|\sum_{j=1}^m\sum_{k=1}^n \varepsilon_{jk} r_j(s) r_k(t) x_{jk}\Big\|^2 ds\, dt \\ \le \alpha^2 \int_0^1\int_0^1 \Big\|\sum_{j=1}^m\sum_{k=1}^n r_j(s) r_k(t) x_{jk}\Big\|^2 ds\, dt\end{aligned}$$

for every choice of $x_{jk} \in X$, $\varepsilon_{jk} \in \{-1, 1\}$ and for all $m, n \in \mathbb{N}$. It is shown in [109, Proposition 2.1] that every Banach space with l.u.st. and having finite cotype necessarily has property (α). In particular, every Banach lattice (which automatically has l.u.st., [34, Theorem 17.1]) with finite cotype has property (α). Actually, within the class of Banach spaces with l.u.st., having property (α) is equivalent to having finite cotype, [104, pp. 488–489]. Concerning some relevant examples, note that c_0 and ℓ^∞ (for instance) have l.u.st. but fail to have finite cotype (and hence, fail to have property (α)). The von Neumann-Schatten ideals $\mathfrak{S}_p$, for $1 < p < \infty$, are Banach spaces with finite cotype but, for $p \ne 2$, fail to have property (α) (hence, also fail l.u.st.); see [65, Remark 2.10] and [104, Corollary 3.4]. For every $p > 2$, it is known that there exist closed subspaces of L^p (hence, they have property (α)) which fail to have l.u.st., [109, p. 19].

For the definition and properties of Banach spaces which are *Gordon-Lewis spaces* (briefly, GL-spaces) we refer to [34, §17]. GL-spaces have property (α) iff they have finite cotype, [104, Theorem 4.4]. Every Banach space with l.u.st. is a GL-space but not conversely; see [104, p. 491] for a discussion. Accordingly, the following result, [104, Theorem 4.2], is not subsumed by Theorem 3.4 above.

Theorem 3.6. *The product B.a. generated by any pair of commuting, bounded B.a.'s of projections in a GL-space X is again bounded.*

The particular B.a. $\mathrm{Bd}(X)$ consisting of all *band projections* in a Banach lattice X has the following remarkable property, [104, Theorem 5.8].

Theorem 3.7. *For a Banach lattice X the following are equivalent.*

(a) *X is Dedekind σ-complete and* $\mathrm{Bd}(X)$ *is R-bounded.*
(b) *X has finite cotype.*
(c) *X has property (α).*

A consequence of Theorem 3.7 and the discussion immediately after Theorem 3.5 is that *every* bounded B.a. of projections in a Dedekind σ-complete Banach lattice X is R-bounded precisely when *just* the B.a. $\mathrm{Bd}(X)$ is R-bounded! The techniques used in [104] to establish these facts are of interest in their own right and have further consequences. Given any Banach space X and any bounded B.a. of projections $\mathcal{M}$ in X, it is always possible to equip the cyclic space $\mathcal{M}[x]$ with a Banach lattice structure, for each $x \in X$; see [104, §6]. This leads to the following useful fact, [104, Proposition 6.3].

Theorem 3.8. *Let $\mathcal{M}$ be a bounded B.a. of projections in a Banach space X. Then $\overline{\mathcal{M}}_s$ is a Bade complete B.a. of projections iff each Banach lattice $\mathcal{M}[x]$, for $x \in X$, has order continuous norm.*

By applying the above criteria in each Banach lattice $\mathcal{M}[x]$, for $x \in X$, it turns out, for any Banach space X, that $\overline{\mathcal{M}}_s$ is Bade complete whenever $\mathcal{M}$ is R-bounded, [104, Theorem 6.6]. A consideration of the case $X = c_0$ and $\mathcal{M} = \mathrm{Bd}(c_0)$, which even has a cyclic vector, shows that the converse is false in general.

In conclusion we briefly mention some recent results in [106] on $C(K)$-representations which are also related to spectral measures and R-boundedness; see also [105]. For the following fact see [106, Proposition 2.17 & Remark 2.18].

Theorem 3.9. *Let K be a compact Hausdorff space and X be a Banach space. Let $\Phi : C(K) \to \mathcal{L}(X)$ be a continuous representation (i.e., linear and multiplicative) which is R-bounded, that is, the image under Φ of the unit ball of $C(K)$ is R-bounded in $\mathcal{L}(X)$. Then there exists a regular spectral measure $P : \mathcal{B}(K) \to \mathcal{L}_s(X)$ which is R-bounded (i.e., $P(\mathcal{B}(K)) \subseteq \mathcal{L}(X)$ is R-bounded) and satisfies*

$$\Phi(f) = \int_K f\, dP, \qquad f \in C(K). \tag{3.3}$$

Conversely, if $P : \mathcal{B}(K) \to \mathcal{L}_s(X)$ *is any regular,* R*-bounded spectral measure, then* Φ *as defined by* (3.3) *is a continuous,* R*-bounded representation.*

Let G be a lca group and consider $L^1(G)$ as a Banach algebra under convolution. A representation $\Psi : L^1(G) \to \mathcal{L}(X)$, always assumed to be continuous, is called *essential* if $\bigcup_{f \in L^1(G)} \Psi(f)(X)$ is dense in X. The next result (see [107]) follows by applying Theorem 3.9 to the dense subalgebra $\{\widehat{f} : f \in L^1(G)\}$ of $C_0(\Gamma)$ (the Banach space of all continuous functions on Γ which vanish at ∞) and by passing to $C(\Gamma_\infty)$, where Γ_∞ is the 1-point compactification of Γ.

Theorem 3.10. *Let* $\Psi : L^1(G) \to \mathcal{L}(X)$ *be an essential representation with* $\{\Psi(f) : f \in L^1(G), \|\widehat{f}\|_\infty \leq 1\}$ *being* R*-bounded. Then there exists an* R*-bounded, regular spectral measure* $P : \mathcal{B}(\Gamma) \to \mathcal{L}_s(X)$ *with*

$$\Psi(f) = \int_\Gamma \widehat{f}\, dP, \qquad f \in L^1(G). \tag{3.4}$$

Conversely, given any R*-bounded, regular spectral measure* $P : \mathcal{B}(\Gamma) \to \mathcal{L}_s(X)$*, the map* Ψ *defined by* (3.4) *is an essential representation of* $L^1(G)$ *and* $\{\int_\Gamma \widehat{f}\, dP : f \in L^1(G), \|\widehat{f}\|_\infty \leq 1\}$ *is* R*-bounded.*

For X a Banach lattice which satisfies either a lower p-estimate for some $1 \leq p < 2$ or an upper q-estimate for some $2 < q \leq \infty$, techniques are exhibited which can be used to decide about the R-boundedness of particular representations $\Psi : L^1(G) \to \mathcal{L}(X)$ defined on particular groups G; see [105, 107] for the details.

References

[1] E. Albrecht and W.J. Ricker, Local spectral properties of constant coefficient differential operators in $L^p(\mathbb{R}^N)$ *J. Operator Theory* **24**(1990), 85–103.

[2] E. Albrecht and W.J. Ricker, Local spectral properties of certain matrix differential operators in $L^p(\mathbb{R}^N)^m$, *J. Operator Theory* **35** (1996), 3–37.

[3] C.D. Aliprantis and O. Burkinshaw, *Locally Solid Riesz Spaces*, Academic Press, New York-London, 1978.

[4] N. Aronzajn and P. Szeptycki, On general integral transformations, *Math. Ann.* **163** (1966), 127–154.

[5] W.G. Bade, On Boolean algebras of projections and algebras of operators, *Trans. Amer. Math. Soc.,* **80** (1955), 345–459.

[6] C. Bennett and R. Sharpley, *Interpolation of Operators*, Academic Press, Inc., Boston, 1988.

[7] E. Berkson and T.A. Gillespie, Spectral decompositions and harmonic analysis in UMD spaces, *Studia Math.* **112** (1994), 13–49.

[8] J. Bonet and S. Diaz-Madrigal, Ranges of vector measures in Fréchet spaces, *Indag Math. (N.S.)*, **11** (2000), 19–30.

[9] J. Bonet and W.J. Ricker, Boolean algebras of projections in (DF)- and (LF)-spaces, *Bull. Austral. Math. Soc.*, **67** (2003), 297–303.

[10] J. Bonet and W.J. Ricker, Spectral measures in classes of Fréchet spaces, *Bull. Soc. Roy. Sci. Liège*, **73** (2004), 99–117.

[11] J. Bonet and W.J. Ricker, The canonical spectral measure in Köthe echelon spaces, *Integral Equations Operator Theory*, **53** (2005), 477–496.

[12] J. Bonet, S. Okada and W.J. Ricker, The canonical spectral measure and Köthe function spaces, *Quaestiones Math.* **29** (2006), 91–116.

[13] J. Bonet and W.J. Ricker, Schauder decompositions and the Grothendieck and Dunford-Pettis properties in Köthe echelon spaces of infinite order, *Positivity*, **11** (2007), 77–93.

[14] J. Bourgain, Some remarks on Banach spaces in which martingale differences are unconditional, *Ark. Math.* **21** (1983), 163–168.

[15] A. Cianchi and L. Pick, Sobolev embeddings into BMO, VMO, and L_∞ spaces, *Ark. Mat.*, **36** (1998), 317–340.

[16] G.P. Curbera, El espacio de funciones integrables respecto de una medida vectorial, Ph.D. Thesis, Univ. of Sevilla, 1992.

[17] G.P. Curbera, Operators into L^1 of a vector measure and applications to Banach lattices, *Math. Ann.* **293** (1992), 317–330.

[18] G.P. Curbera, When L^1 of a vector measure is an AL-space, *Pacific J. Math.* **162** (1994), 287–303.

[19] G.P. Curbera, Banach space properties of L^1 of a vector measure, *Proc. Amer. Math. Soc.* **123** (1995), 3797–3806.

[20] G.P. Curbera and W.J. Ricker, Optimal domains for kernel operators via interpolation, *Math. Nachr.*, **244** (2002), 47–63.

[21] G.P. Curbera and W.J. Ricker, Optimal domains for the kernel operator associated with Sobolev's inequality, *Studia Math.*, **158** (2003), 131–152.

[22] G.P. Curbera and W.J. Ricker, Corrigenda to "Optimal domains for the kernel operator associated with Sobolev's inequality", *Studia Math.*, **170** (2005) 217–218.

[23] G.P. Curbera and W.J. Ricker, Banach lattices with the Fatou property and optimal domains of kernel operators, *Indag. Math.* (*N.S.*), **17** (2006), 187–204.

[24] G.P. Curbera and W.J. Ricker, Compactness properties of Sobolev imbeddings for rearrangement invariant norms, *Trans. Amer. Math. Soc.*, **359** (2007), 1471–1484.

[25] G.P. Curbera and W.J. Ricker, Can optimal rearrangement invariant Sobolev imbeddings be further extended?, *Indiana Univ. Math. J.*, **56** (2007), 1489–1497.

[26] G.P. Curbera and W.J. Ricker, The Fatou property in p-convex Banach lattices, *J. Math. Anal. Appl.*, **328** (2007), 287–294.

[27] O. Delgado, Banach function subspaces of L^1 of a vector measure and related Orlicz spaces, *Indag. Math.* (*N.S.*), **15** (2004), 485–495.

[28] O. Delgado, L^1-spaces for vector measures defined on δ-rings, *Archiv. Math.* (*Basel*) **84** (2005), 43–443.

[29] O. Delgado, Optimal domains for kernel operators on $[0,\infty)\times[0,\infty)$, *Studia Math.* **174** (2006), 131–145.

[30] O. Delgado and J. Soria, Optimal domains for the Hardy operator, *J. Funct. Anal.* **244** (2007), 119–133.

[31] J.C. Diaz, A. Fernández and F. Naranjo, Fréchet AL-spaces have the Dunford-Pettis property, *Bull. Austral. Math. Soc.*, **58** (1998), 383–386.

[32] J. Diestel and J.J. Uhl, Jr., *Vector Measures*, Math. Surveys 15, Amer. Math. Soc., Providence, 1977.

[33] J. Diestel and J.J. Uhl, Jr., Progress in vector measures: 1977–83, Lecture Notes Math. 1033, Springer, Berlin Heidelberg, 1984, pp. 144–192.

[34] J. Diestel, H. Jarchow and A. Tonge, *Absolutely Summing Operators*, Cambridge Studies in Advanced Mathematics 43, Cambridge University Press, 1995.

[35] J. Diestel and W.J. Ricker, The strong closure of Boolean algebras of projections in Banach spaces, *J. Austral. Math. Soc.* **77** (2004), 365–369.

[36] N. Dinculeanu, *Vector Measures*, VEB Deutscher Verlag der Wissenschaften, Berlin, 1966.

[37] N. Dinculeanu, *Integration on Locally Compact Spaces*, Noordhoff, Leyden, 1974.

[38] N. Dinculeanu, *Vector Integration and Stochastic Integration in Banach spaces*, Wiley-Interscience, New York, 2000.

[39] P.G. Dodds and B. dePagter, Orthomorphisms and Boolean algebras of projections, *Math. Z.* **187** (1984), 361–381.

[40] P.G. Dodds and W.J. Ricker, Spectral measures and the Bade reflexivity theorem, *J. Funct. Anal.*, **61** (1985), 136–163.

[41] P.G. Dodds, B. dePagter and W.J. Ricker, Reflexivity and order properties of scalar-type spectral operators in locally convex spaces, *Trans. Amer. Math. Soc.* **293** (1986), 355–380.

[42] H.R. Dowson, *Spectral Theory of Linear Operators*, Academic Press, London, 1978.

[43] H.R. Dowson, M.B. Ghaemi and P.G. Spain, Boolean algebras of projections and algebras of spectral operators, *Pacific J. Math.* **209** (2003), 1–16.

[44] N. Dunford and J.T. Schwartz, *Linear Operators I: General Theory* (2nd Ed), Wiley-Interscience, New York, 1964.

[45] N. Dunford and J.T. Schwartz, *Linear Operators III: Spectral Operators*, Wiley-Interscience, New York, 1971.

[46] D. van Dulst, *Characterizations of Banach Spaces not Containing* ℓ^1, CWI Tract 59, Centrum voor Wiskunde en Informatica, Amsterdam, 1989.

[47] D. Edmunds, R. Kerman and L. Pick, Optimal Sobolev imbeddings involving rearrangement-invariant quasinorms, *J. Funct. Anal.*, **170** (2000), 307–355.

[48] A. Fernández and F. Naranjo, Rybakov's theorem for vector measures in Fréchet spaces, *Indag. Math. (N.S.)*, **8** (1997), 33–42.

[49] A. Fernández and F. Naranjo, Operators and the space of integrable scalar functions with respect to a Fréchet-valued measure, *J. Austral. Math. Soc. (Ser. A)*, **65** (1998), 176–193.

[50] A. Fernández, F. Naranjo and W.J. Ricker, Completeness of L^1-spaces for measures with values in complex vector spaces, *J. Math. Anal. Appl.* **223** (1998), 76–87.

[51] A. Fernández, F. Mayoral, F. Naranjo and P.J. Paul, Weakly sequentially complete Fréchet spaces of integrable functions, *Arch. Math. (Basel)*, **71** (1998), 223–228.

[52] A. Fernández and F. Naranjo, Strictly positive linear functionals and representation of Fréchet lattices with the Lebesgue property, *Indag. Math. (N.S.)*, **10** (1999), 381–391.

[53] A. Fernández and F. Naranjo, AL- and AM-spaces of integrable scalar functions with respect to a Fréchet-valued measure, *Quaestiones Math.* **23** (2000), 247–258.

[54] A. Fernández and F. Naranjo, Nuclear Fréchet lattices, *J. Austral. Math. Soc.* **72** (2002), 409–417.

[55] A. Fernández, F. Mayoral, F. Naranjo, C. Sáez and E.A. Sánchez-Pérez, Vector measure Maurey-Rosenthal-type factorizations and ℓ-sums of L^1-spaces, *J. Funct. Anal.* **220** (2005), 460–485.

[56] A. Fernández, F. Mayoral, F. Naranjo, C. Sáez and E.A. Sánchez-Pérez, Spaces of p-integrable functions with respect to a vector measure, *Positivity*, **10** (2006), 1–16.

[57] U. Fixman, Problems in spectral operators, *Pacific J. Math.* **9** (1959), 1029–1051.

[58] D.H. Fremlin, B. dePagter and W.J. Ricker, Sequential closedness of Boolean algebras of projections in Banach spaces, *Studia Math.* **167** (2005), 45–62.

[59] D.H. Fremlin and D. Preiss, On a question of W.J. Ricker, *Electronic file* (December, 2006); http://www.essex.ac.uk/maths/staff/fremlin/n05403.ps.

[60] G.I. Gaudry and W.J. Ricker, Spectral properties of L_p translations, *J. Operator Theory*, **14** (1985), 87–111.

[61] G.I. Gaudry and W.J. Ricker, Spectral properties of translation operators in certain function spaces, *Illinois J. Math.*, **31** (1987), 453–468.

[62] G.I. Gaudry, B.R.F. Jefferies and W.J. Ricker, Vector-valued multipliers: convolution with operator-valued measures, *Dissertationes Math.*, **385** (2000), 1–77.

[63] T.A. Gillespie, A spectral theorem for L^p translations, *J. London Math. Soc.* **(2)11** (1975), 499–508.

[64] T.A. Gillespie, Strongly closed bounded Boolean algebras of projections, *Glasgow Math. J.*, **22** (1981), 73–75.

[65] T.A. Gillespie, Boundedness criteria for Boolean algebras of projections, *J. Funct. Anal.* **148** (1997), 70–85.

[66] B. Jefferies, *Evolution Process and the Feynman-Kac Formula*, Kluwer, Dordrecht, 1996.

[67] I.Kluvánek and G. Knowles, *Vector Measures and Control Systems*, North-Holland, Amsterdam, 1976.

[68] I. Kluvánek, Applications of vector measures, In: Integration, topology, and geometry in linear spaces, Proc. Conf. Chapel Hill/N.C. 1979, Contemp. Math. **2** (1980), 101–134.

[69] G.L. Krabbe, Convolution operators which are not of scalar type, *Math. Z.* **69** (1958), 346–350.

[70] S.G. Krein, Ju. I. Petunin and E.M. Semenov, *Interpolation of Linear Operators*, Amer. Math. Soc., Providence, 1982.

[71] I. Labuda and P. Szeptycki, Extended domains of some integral operators with rapidly oscillating kernels, *Indag. Math.* **48** (1986), 87–98.

[72] R. Larsen, *An Introduction to the Theory of Multipliers*, Springer-Verlag, Berlin Heidelberg New York, 1971.

[73] D.R. Lewis, On integrability and summability in vector spaces, *Illinois J. Math.*, **16** (1972), 294–307.

[74] J. Lindenstrauss and L. Tzafriri, *Classical Banach Spaces* vol. II, Springer-Verlag, Berlin, (1979).

[75] W.A. Luxemburg and A.C. Zaanen, Notes on Banach function spaces, I, *Nederl. Akad. Wet., Proc.*, **66** =*Indag. Math.* **25** (1963) 135–147.

[76] W.A. Luxemburg and A.C. Zaanen, Notes on Banach function spaces, II, *Nederl. Akad. Wet., Proc.*, **66** = *Indag. Math.* **25** (1963) 148–153.

[77] W.A. Luxemburg and A.C. Zaanen, Notes on Banach function spaces, III, *Nederl. Akad. Wet., Proc.*, **66** = *Indag. Math.* **25** (1963) 239–250.

[78] W.A. Luxemburg and A.C. Zaanen, Notes on Banach function spaces, IV, *Nederl. Akad. Wet., Proc.*, **66** = *Indag. Math.* **25** (1963) 251–263.

[79] W.A. Luxemburg, Spaces of measurable functions, Jeffery–Williams Lectures 1968–72 Canad. Math. Congr. (1972) 45–71.

[80] T.-W. Ma, *Banach Hilbert Spaces, Vector Measures and Group Representations,* World Scientific, Singapore, 2002.

[81] P. R. Masani and H. Niemi, The integration theory of Banach space valued measures and the Tonelli–Fubini theorems. I. Scalar-valued measures on δ-rings, *Adv. Math.* **73** (1989), 204–241.

[82] P.R. Masani and H. Niemi, The integration theory of Banach space-valued measures and the Tonelli–Fubini theorems. II. Pettis integration, *Adv. Math.* **75** (1989), 121–167.

[83] C.A. McCarthy, Commuting Boolean algebras of projections, *Pacific J. Math.* **11** (1961), 295–307.

[84] G. Mockenhaupt and W.J. Ricker, Idempotent multipliers for $L^p(\mathbb{R})$, *Arch. Math. (Basel)*, **74** (2000), 61–65.

[85] G. Mockenhaupt and W.J. Ricker, Fuglede's theorem, the bicommutant theorem and p-multiplier operators for the circle, *J. Operator Theory,* **49** (2003), 295–310.

[86] G. Mockenhaupt and W.J. Ricker, Approximation of p-multiplier operators via their spectral projections, *Positivity* (to appear).

[87] G. Mockenhaupt and W.J. Ricker, Optimal extension of the Hausdorff-Young inequality, *J. Reine Angew. Math.* (to appear).

[88] G. Muraz and P. Szeptycki, Domains of trigonometric transforms, *Rocky Mountain J. Math.* **26** (1996), 1517–1527.

[89] K.K. Oberai, Sum and product of commuting spectral operators, *Pacific J. Math.*, **25** (1968), 129–146.

[90] S. Okada, Spectrum of scalar-type spectral operators and Schauder decompositions, *Math. Nachr.* **139** (1988), 167–174.

[91] S. Okada and W.J. Ricker, Vector measures and integration in non-complete spaces, *Arch. Math. (Basel)*, **63** (1994), 344–353.

[92] S. Okada and W.J. Ricker, Boolean algebras of projections and ranges of spectral measures. *Dissertationes Math.*, **365**, 33p., 1997.

[93] S. Okada and W.J. Ricker, Representation of complete Boolean algebras of projections as ranges of spectral measures, *Acta Sci. Math.* (*Szeged*), **63** (1997), 209–227 and **63** (1997), 689–693.

[94] S. Okada and W.J. Ricker, Criteria for closedness of spectral measures and completeness of Boolean algebras of projections, *J. Math. Anal. Appl.*, **232** (1999), 197–221.

[95] S. Okada and W.J. Ricker, Integration with respect to the canonical spectral measure in sequence spaces, *Collect. Math.* **50** (1999), 95–118.

[96] S. Okada, W.J. Ricker and L. Rodríguez-Piazza, Compactness of the integration operator associated with a vector measure, *Studia Math.*, **150** (2002), 133–149.

[97] S. Okada and W.J. Ricker, Compact integration operators for Fréchet-space-valued measures, *Indag. Math.* (*New Series*), **13** (2002), 209–227.

[98] S. Okada and W.J. Ricker, Fréchet-space-valued measures and the AL-property, *Rev. R. Acad. Cien. Serie A. Mat.* RACSAM, **97** (2003), 305–314.

[99] S. Okada and W.J. Ricker, Optimal domains and integral representations of convolution operators in $L^p(G)$, *Integral Equations Operator Theory*, **48** (2004), 525–546.

[100] S. Okada and W.J. Ricker, Optimal domains and integral representations of $L^p(G)$-valued convolution operators via measures, *Math. Nachr.* **280** (2007), 423–436.

[101] S. Okada, W.J. Ricker and E.A. Sánchez-Pérez, *Optimal Domain and Integral Extension of Operators acting in Function Spaces*, Operator Theory Advances and Applications, Birkhäuser Verlag, 2008 (to appear).

[102] E. Pap (Ed.), *Handbook of Measure Theory I*, Part 2: Vector Measures, North-Holland, Amsterdam, 2002, pp. 345–502.

[103] B. dePagter and W.J. Ricker, Boolean algebras of projections and resolutions of the identity of scalar-type spectral operators, *Proc. Edinburgh Math. Soc.* **40** (1997) 425–435.

[104] B. dePagter and W.J. Ricker, Products of commuting Boolean algebras of projections and Banach space geometry, *Proc. London Math. Soc.* **(3)91** (2005), 483–508.

[105] B. dePagter and W.J. Ricker, R-boundedness of $C(K)$-representations, group homomorphisms, and Banach space geometry, In: Proceedings of the Conference "Positivity IV-Theory and Applications", July 2005, Eds. M. Weber and J. Voigt, Technische Universität Dresden, Germany, pp. 115–129 (2006).

[106] B. dePagter and W.J. Ricker, $C(K)$-representations and R-boundedness, *J. London Math. Soc.* (to appear).

[107] B. dePagter and W.J. Ricker, R-bounded representations of $L^1(G)$, *Positivity* (to appear).

[108] L. Pick, *Optimal Sobolev Embeddings*, Rudolph–Lipschitz-Vorlesungsreihe Nr. 43, SFB 256: Nichtlineare Partielle Differentialgleichungen, Univ. of Bonn, 2002.

[109] G. Pisier, Some results on Banach spaces without local unconditional structure, *Compositio Math.* **37** (1978), 3–19.

[110] W.J. Ricker, Spectral operators of scalar-type in Grothendieck spaces with the Dunford-Pettis property, *Bull. London Math. Soc.* **17**(1985), 268–270.

[111] W.J. Ricker, Spectral like multipliers in $L^p(\mathbb{R})$, *Arch. Math.* (*Basel*), **57**(1991), 395–401.

[112] W.J. Ricker, Well bounded operators of type (B) in H.I. spaces, *Acta Sci. Math. (Szeged)* **59** (1994), 475–488.

[113] W.J. Ricker, Spectrality for matrices of Fourier multiplier operators acting in L^p-spaces over lca groups, *Quaestiones Math.* **19** (1996), 237–257.

[114] W.J. Ricker, Existence of Bade functionals for complete Boolean algebras of projections in Fréchet spaces, *Proc. Amer. Math. Soc.* **125** (1997), 2401–2407.

[115] W.J. Ricker, The sequential closedness of σ-complete Boolean algebras of projections, *J. Math. Anal. Appl.*, **208** (1997), 364–371.

[116] W.J. Ricker, The strong closure of σ-complete Boolean algebras of projections, *Arch. Math. (Basel)*, **72** (1999), 282–288.

[117] W.J. Ricker, *Operator Algebras Generated by Commuting Projections: A Vector Measure Approach*, Lecture Notes Math. 1711, Springer, Berlin Heidelberg, 1999.

[118] W.J. Ricker, Resolutions of the identity in Fréchet spaces, *Integral Equations Operator Theory*, **41** (2001), 63–73.

[119] W.J. Ricker and M. Väth, Spaces of complex functions and vector measures in incomplete spaces, *J. Function Spaces Appl.*, **2** (2004), 1–16.

[120] L. Rodríguez-Piazza, Derivability, variation and range of a vector measure, *Studia Math.*, **112** (1995), 165–187.

[121] L. Rodríguez-Piazza and C. Romero-Moreno, Conical measures and properties of a vector measure determined by its range, *Studia Math.*, **125** (1997), 255–270.

[122] E.A. Sánchez-Pérez, Compactness arguments for spaces of p-integrable functions with respect to a vector measure and factorization of operators through Lebesgue-Bochner spaces, *Illinois J. Math.*, **45** (2001), 907–923.

[123] H.H. Schaefer and B. Walsh, Spectral operators in spaces of distributions, *Bull. Amer. Math. Soc.*, **68** (1962), 509–511.

[124] K.D. Schmidt, *Jordan Decompositions of Generalized Vector Measures*, Longman Scientific & Technical, Harlow, 1989.

[125] M.A. Sofi, Vector measures and nuclear operators, *Illinois J. Math.* **49** (2005), 369–383.

[126] M.A. Sofi, Absolutely p-summable sequences in Banach spaces and range of vector measures, *Rocky Mountain J. Math.*, to appear.

[127] M.A. Sofi, Fréchet-valued measures and nuclearity, *Houston J. Math.*, to appear.

[128] G. Stefansson, L^1 of a vector measure, *Le Mathematiche* **48** (1993), 219–234.

[129] P. Szeptycki, Notes on integral transformations, *Dissertationes Math.* **231** (1984), 48pp.

[130] P. Szeptycki, Extended domains of some integral operators, *Rocky Mountain J. Math.* **22** (1992), 393–404.

[131] U.B. Tewari, Vector-valued multipliers, *J. Anal.* **12** (2004), 99–105.

[132] E. Thomas, The Lebesgue-Nikodým theorem for vector valued Radon measures, *Mem. Amer. Math. Soc.* **139** (1974).

[133] A.I. Veksler, Cyclic Banach spaces and Banach lattices, *Soviet Math. Dokl.* **14** (1973), 1773–1779.

[134] B. Walsh, Structure of spectral measures on locally convex spaces, *Trans. Amer. Math. Soc.* **120** (1965), 295–326.

[135] B. Walsh, Spectral decomposition of quasi-Montel sapces, *Proc. Amer. Math. Soc.*, **(2) 17** (1966), 1267–1271.

[136] A. C. Zaanen, *Integration*, 2nd rev. ed. North-Holland, Amsterdam; Interscience, New York Berlin, 1967.

[137] A. C. Zaanen, *Riesz Spaces II*, North-Holland, Amsterdam, 1983.

G.P. Curbera
Facultad de Matemáticas
Universidad de Sevilla
Aptdo 1160
Sevilla 41080, Spain
e-mail: curbera@us.es

W.J. Ricker
Math.-Geogr. Fakultät
Katholische Universität Eichstätt-Ingolstadt
D-85072 Eichstätt, Germany
e-mail: werner.ricker@ku-eichstaett.de

Positivity
Trends in Mathematics, 161–195

The Role of Frames in the Development of Lattice-ordered Groups: A Personal Account

Jorge Martínez

Abstract. A frame is a complete lattice in which finite meets distribute over arbitrary joins.

Frames have only recently made a formal entry into the development of lattice-ordered groups. On the other hand, the work of Paul Conrad and some of his students of the sixties and seventies, analyzing a lattice-ordered group through its lattice of convex ℓ-subgroups, is frame theory in disguise. In more recent work, pure frame theory has found application to problems in ℓ-groups, producing, in several cases, theorems which had not been possible with more traditional techniques. And now this turning of the tables has been taken a step further: proving theorems from the theory of ℓ-groups in frame-theoretic settings, without invoking the Axiom of Choice or other axioms which imply the existence of points in spectra.

This article aims to inform and convince the reader: inform, in broad terms, and convince that the phenomena discussed in the preceding paragraph constitute an honorable research activity. This is a survey article of modest length: selectivity is a must – with the choices of illustrations being left, for good or ill, to the taste and prejudices of the author.

The exposition is in three parts, following the three (chronological) aspects of the role of frame theory in the development of ℓ-groups. First up is the famous theorem of Conrad on finite-valued ℓ-groups. This is followed by an account of dimension theory, particularly as it applies to the z-dimension of rings of continuous functions. Finally, there is an account of the recent and ongoing work on the epicompletion in a category of regular frames, and related issues concerning archimedean frames.

1. Part One: Preliminaries

1.1. Introduction

An important strand in the development of lattice-ordered groups has involved structure theory in terms of the lattice of convex ℓ-subgroups of the group in question. There has evolved a kind of synergy, which this exposition will illustrate with the selection a few examples which, in my opinion, best witness this phenomenon. The choices are somewhat personal, reflecting the impact on my own work. But these illlustrations are chosen principally because they witness rather well these three important elements of the progress of the influence of frame theory in the structure theory of lattice-ordered groups:

- *The intuitive reliance on frame-theoretic principles, guided by formal algebraic or topological or analytical motivation, or a combination of the three.*
- *The reliance on spectra, while seeking formulations of structure theory that are pointfree.*
- *The evolution towards approaches to structure theory that are pointfree and* Choice-free – which means, using only Zermelo-Fraenkel principles.

In making these selections, no apologies are added for the bias they reflect. Moreover, although this exposition is being given some amplitude, this kind of circumstance seems to call for selectivity and impact. Without further ado then, here is the first assumption which will be carried throughout: that all groups are abelian. To open, here are the definitions of a lattice-ordered group and of a frame.

Definition & Remarks 1.1.1. A *lattice-ordered group* (henceforth, *ℓ-group*) G is a group, which is simultaneously a lattice, such that, for each $a, b, c \in G$,

$$a + (b \vee c) = (a + b) \vee (a + c). \tag{1.1.1.1}$$

We will assume familiarity with the arithmetic of ℓ-groups. In particular, the reader should know:

- that the dual of (1.1.1.1), with respect to meets, also holds;
- that the underlying group is torsion-free, and the underlying lattice is distributive.

Any unexplained terminology is surely to be found in either [BKW77] or [D95], and we shall often refer to these.

Definition & Remarks 1.1.2. Let L be a complete lattice; denote its top and bottom by 1 and 0, respectively. L is a *frame* if for each $a \in L$ and each subset S of L,

$$a \wedge \left(\bigvee S\right) = \bigvee \left\{ a \wedge x \, : \, x \in S \right\}. \tag{1.1.2.1}$$

The next section will consist of a short dictionary of frame-theoretic terms, along with some categorical observations which ought to help the reader further on. The standard reference for the general theory of frames remains [J82], although [PT01, Chapter 2] is a personal favorite.

To conclude this introduction, it seems fair to formulate a sort of thesis in this exposition; namely, that there are important aspects of the theory of ℓ-groups which are entirely frame-theoretic, and that their presentation frequently benefits from a pointfree approach.

1.2. The language of frames and categories

As advertised, this section assembles the necessary frame-theoretic resources. The reader who is familiar with the fundamental terms and concepts from frame theory should be able to skip this section; in any event, it is here for easy reference.

We begin with a catalogue or dictionary of terms; each item is suitably illustrated.

Definition & Remarks 1.2.1. L is a complete lattice. For $x \in L$, denote the set of elements of L less than or equal to (resp. greater than or equal to) x by $\downarrow x$ (resp. $\uparrow x$).

- $c \in L$ is *compact*: if $c \leq \bigvee S$, then $c \leq \bigvee F$, for some finite subset F of S. We say that L is *compact* if the top, 1, is compact.
- L is an *algebraic* lattice: every element is a supremum of compact elements. Most examples of algebraic lattices arise as lattices of subobjects of some algebraic structure. In such a context, the compact elements coincide with the finitely generated subobjects.

  The subset of compact elements is denoted $\mathfrak{k}(L)$.

  It is well known that if L is an algebraic lattice, then it is a frame if and only it satisfies the (finite) distributive law.
- L has the *finite-intersection property* (abbr. *FIP*):

  $$a, b \in \mathfrak{k}(L) \implies a \wedge b \in \mathfrak{k}(L).$$

  Note that a finite supremum of compact elements is always compact.
- L is *coherent*: it is compact and satisfies the FIP.
- In a frame L, $y^{\perp} \equiv \vee\{\, x \in L \,:\, x \wedge y = 0\,\}$. $p \in L$ a *polar*: it is of the form $p = y^{\perp}$, for some $y \in L$. $x^{\perp\perp}$ signifies $(x^{\perp})^{\perp}$. We record for later use the closure operator p defined by $p(x) = x^{\perp\perp}$. It is well known that the set $\mathcal{P}L$ of all polars forms a complete boolean algebra, in which infima agree with those in L.

  Many authors use the term "pseudo-complemented" in place of "polar". A polar x is *complemented* if $x \vee x^{\perp} = 1$.
- $a \preceq b$: (in a frame) $b \vee a^{\perp} = 1$. $x \in L$ is *regular*: $x = \vee\{\, a \in L \,:\, a \preceq x\,\}$.

  A frame L *regular*: each element of L is regular.
- A frame L is *normal*: whenever $x \vee y = 1$, there exist disjoint $u \wedge v = 0$ in L, such that $u \leq x$ and $v \leq y$, and $1 = x \vee v = u \vee y$.
- An algebraic frame L has *disjointification*: for each pair $a, b \in \mathfrak{k}(L)$, there exist disjoint c, d, both compact, such that $c \leq a$, $d \leq b$, and

  $$a \vee b = c \vee b = a \vee d.$$

Banaschewski calls this property "coherent normality", while in [ST93] it is called "relative normality". In spite of the obvious connection to normality, "disjointification" seems to better cut to the quick.

Remark 1.2.2. Frame theory was appropriated by topologists, who reverse all the arrows and call the objects of the resulting category *locales*. Not being a topologist, I prefer frames for their own sake. That having been said, much of the motivation and terminology comes from topology, and we should address the spatial side of things. In this remark, we outline the basic adjointness (and resulting duality) between spaces and frames; said outline will be informal and light on the category theory.

(a) **Frames from spaces.** If X is any topological space, then $\mathfrak{O}(X)$ denotes the frame of open sets. Arbitrary meets in $\mathfrak{O}(X)$ are defined

$$\bigwedge \mathcal{S} \equiv \mathrm{int}_X\Big(\bigcap \mathcal{S}\Big).$$

If $g : X \longrightarrow Y$ is a continuous function between spaces, then $\mathfrak{O}(g) : \mathfrak{O}(Y) \longrightarrow \mathfrak{O}(X)$ denotes the map $U \mapsto g^{-1}(U)$.

(b) **Spaces from frames: Spectra.** Let L be a frame. $p < 1$ in L is said to be *prime* if $x \wedge y \le p$ implies that either $x \le p$ or $y \le p$. The set of all primes of L is denoted $\mathrm{Spec}(L)$ and referred to as the *(prime) spectrum* of L. It becomes a topological space under the *hull-kernel* topology, whose open sets are the subsets

$$\mathrm{Coz}(x) \equiv \{\, p \in \mathrm{Spec}(L) \,:\, x \not\le p \,\} \quad \text{over all} \quad x \in L.$$

(c) **Frame homomorphisms.** A map $h : L \longrightarrow M$ between frames is a *frame homomorphism* (or simply a *frame map*) if it preserves all suprema and all finite infima. Taking infima and suprema of empty families, a frame homomorphism also preserves top and bottom, respectively.

If $g : X \longrightarrow Y$ is a continuous map between spaces, then it is easily seen that $\mathfrak{O}(g)$ is a frame map, and that $\mathfrak{O}$ is a contravariant functor between the category $\mathfrak{Top}$ of all topological spaces with all continuous maps, and $\mathfrak{Frm}$, the category of all frames with all frame maps.

Conversely, if $h : L \longrightarrow M$ is a frame homomorphism, then, for each prime p of M, the map

$$\mathrm{Spec}(h)(p) \equiv \vee\{\, x \in L \,:\, h(x) \le p \,\}$$

defines a continuous function $\mathrm{Spec}(h) : \mathrm{Spec}(M) \longrightarrow \mathrm{Spec}(L)$. It is routine to check that Spec defines a contravariant functor from $\mathfrak{Frm}$ to $\mathfrak{Top}$.

(d) **Back-and-forth: "Adjointness-Lite".** We have a surjective function $\mathrm{Coz} : L \longrightarrow \mathfrak{O}(\mathrm{Spec}(L))$, which, in view of the identities:

- $\mathrm{Coz}(0) = \emptyset$ and $\mathrm{Coz}(1) = \mathrm{Spec}(L)$,
- $\mathrm{Coz}(x \wedge y) = \mathrm{Coz}(x) \cap \mathrm{Coz}(y)$,
- $\mathrm{Coz}(\bigvee S) = \cup\{\, \mathrm{Coz}(x) \,:\, x \in S \,\}$,

is a frame map. A frame L is *spatial* if Coz is one-to-one, and thereby an isomorphism. Thus, as is easy to verify, L is spatial if and only if each $x \in L$ is an infimum of primes.

Dually, for each space X one has, for each $p \in X$, the open set $U_p \equiv X \setminus \mathrm{cl}_X\{p\}$, and it is easy to verify that U_p is prime in $\mathfrak{O}(X)$. The map $\varepsilon : p \mapsto U_p$ then defines a continuous function $\varepsilon : X \longrightarrow \mathrm{Spec}(\mathfrak{O}(X))$. When ε is a homeomorphism the space X is said to be *sober*.

Let $\mathfrak{SpFrm}$ denote the full subcategory of spatial frames, and $\mathfrak{Sob}$ denote the full subcategory of sober spaces. The upshot of the above discussion is that the two functors

$$\mathrm{Spec} : \mathfrak{SpFrm} \longleftrightarrow \mathfrak{Sob} : \mathfrak{O}$$

define a duality of categories.

(e) **Frames without points.** Any complete boolean algebra B is a frame; this is an interesting exercise in itself. It is quite easy indeed to establish that $p \in B$ is prime if and only if p is a co-atom. Since the passage $x \mapsto x^{\perp}$ in B carries atoms to co-atoms, one realizes that $\mathrm{Spec}(B)$ is empty precisely when B is atomless. On the other hand B is spatial if and only if it is atomic; if this is the case, then $\mathrm{Spec}(B)$ is a discrete space.

2. Part Two: Dual Frames

2.1. Conrad's theorem for finite-valued ℓ-groups

Conrad's theorem on finite-valued ℓ-groups characterizes, in a number of ways, the ℓ-groups in which every positive element can be decomposed into a finite supremum of so-called "special" elements. This theorem first appeared in [C65]. The result was then abstracted and generalized for partially ordered sets, in [M72], and reprised in [ST93]. The interest in Conrad's theorem here is due to its characterization of a class of ℓ-groups in terms of the frame of convex ℓ-subgroups of its members. From one point of view Conrad's theorem characterizes those ℓ-groups G for which the frame $\mathcal{C}(G)$ of all convex ℓ-subgroups is completely distributive. Another viewpoint regards this theorem as capturing those ℓ-groups G for which $\mathcal{C}(G)$ is dually a frame.

After studying Conrad's theorem for a period of time, one is more than tempted to ask: *What is going on here?* So let us have a "vertical" look at the situation, beginning with the original formulation of Conrad's theorem and a brief discussion of its proof, and then proceed to later, more abstract versions of the theorem, to attempt to decipher what – in Conrad's own words – makes his theorem tick.

We recite the theorem without further preface and supply clarifications afterwards. The reader will find, in each item in the following theorem, a specific reference to explanatory commentary in 2.1.2. We follow [M06b] closely, referring the reader there for the details of proofs.

Theorem 2.1.1 (Conrad's theorem). *Suppose that G is an ℓ-group. The following are equivalent statements.*

1. *$\mathcal{C}(G)$ is freely generated by its values.* (2.1.2(e))
2. *$\mathcal{C}(G)$ is a completely distributive frame* (2.1.2(a)).
3. *$\mathcal{C}(G)$ satisfies the dual frame law*

$$a \vee \left(\bigwedge S\right) = \bigwedge \left\{ a \vee x \,:\, x \in S \right\}. \quad (2.1.2(a)) \tag{2.1.1.1}$$

4. *Each $0 < g \in G$ is a disjoint supremum of special elements: $g = g_1 \vee \cdots \vee g_n$* (2.1.2(b)).
5. *Each nonzero $g \in G$ has at most finitely many values* (2.1.2(b)).
6. *Each value of G is special* (2.1.2(b)).

Definition & Remarks 2.1.2. G stands for an arbitrary ℓ-group. An *ℓ-subgroup* is a subgroup which is simultaneously a sublattice. $C \subseteq G$ is a *convex* subgroup if it is a subgroup and $a \leq g \leq b$, with $a, b \in C$, implies that $g \in G$. The set of all convex ℓ-subgroups of G is throughout denoted by $\mathcal{C}(G)$.

Throughout this commentary L will denote a complete lattice.

(a) $\mathcal{C}(G)$ is a complete sublattice of the algebraic lattice of all subgroups of G ([D95, Theorem 7.5]). This fact, plus an application of the Riesz Interpolation Property ([D95, Theorem 3.11]), enables one to prove Birkhoff's Theorem ([D95, Proposition 7.10], 1942), stating that $\mathcal{C}(G)$ is a frame.

Complete distributivity in L is simply the validity of the most general possible distributive law:

$$\bigwedge_{i \in I} \bigvee_{j \in J} x_{ij} = \bigvee_{f \in J^I} \bigwedge_{i \in I} x_{if(i)}.$$

This clearly implies both the frame law and its dual.

(b) In an algebraic frame L the element v is a *value* if it is maximal with respect to $a \not\leq v$, for a suitable $a \in \mathfrak{k}(L)$. If this is the case, then we also say that v is a *value of* a. It is well known – and the argument is not *Choice-dependent* – that values are prime. $\mathrm{Val}(L)$ denotes the set of values of L. (We abbreviate, when dealing with ℓ-groups, $\mathrm{Spec}(G) \equiv \mathrm{Spec}(\mathcal{C}(G))$ and $\mathrm{Val}(G) \equiv \mathrm{Val}(\mathcal{C}(G))$.)

The compact $a \in L$ is said to be *special* if it has only one value. Any $v \in \mathrm{Val}(L)$ which is a value of some special element is also called special. L is *finite-valued* if each compact element of L has finitely many values.

(c) The compact elements of $\mathcal{C}(G)$ are precisely the principal ones ([D95, Proposition 7.14]). The convex ℓ-subgroup generated by $a \in G$ is

$$G(a) = \{\, x \in G \,:\, |x| \leq n|a|, \quad \text{for suitable} \quad n \in \mathbb{N} \,\}.$$

Note that $|x| \equiv x \vee -x$.

This observation prompts the meta-statement, which the reader needs to apply with caution: *any affirmation about compact elements and their values in an algebraic frame has a corresponding interpretatation in $\mathcal{C}(G)$ for the principal convex ℓ-subgroups, and, consequently, for the elements of G themselves.*

(d) As has already been noted, if one assumes the Axiom of Choice, then each algebraic frame L is spatial, and, in fact, each $x \in L$ is an infimum of values of L. Thus, with *Choice*, L is (meet) generated by its values. Further, let us observe that in most arguments involving statements about elements of an ℓ-groups *vis-à-vis* their values, require the property that for any prime $p \in \mathrm{Spec}(L)$ such that p does not exceed the compact element $a \in L$, there be a value v of a such that $p \leq v$. For this to work the Axiom of Choice is typically invoked in the guise of Zorn's lemma.

Following Conrad's lead, define L to be *freely generated* by $\mathrm{Val}(L)$ if for any two upsets S_1 and S_2 of values,

$$\bigwedge S_1 = \bigwedge S_2 \quad \Longrightarrow \quad S_1 = S_2. \tag{2.1.2.1}$$

(If P is any poset and $S \subseteq P$, we say that S is an *upset* if $y \geq x \in S$ implies that $y \in S$. A *downset* is defined dually.)

With the preceding comments in mind, let us analyze the proof of Conrad's theorem.

Remark 2.1.3. Conrad makes straightforward assumptions. He begins with a lattice L.

A *meet-irreducible* element $m \in L$ is one for which, given $S \subseteq L$ such that $m < x \in S$ (for each $x \in S$) implies that $m < \bigwedge S$, provided $\bigwedge S$ exists. Observe that this means that if L has a top, then $m < 1$. Now let $\mathcal{M}(L)$ denote the set of meet-irreducible elements of L. Conrad assumes that $\mathcal{M}(L)$ *generates* L: that is, each $x \in L$ is a meet of members of $\mathcal{M}(L)$. The first thing he gets from this assumption is that L *is* complete ([C65, 2.4]). The second is that L is necessarily a frame ([C65, 2.5]).

One amazing aspect of Conrad's work in [C65] is that he obtains, as a preliminary to Theorem 2.1.1, the equivalence of 1, 2, and 3 in the theorem:

Theorem A. *Suppose L is a lattice that is generated by $\mathcal{M}(L)$. Then the following are equivalent:*

(i) *$\mathcal{M}(L)$ freely generate L* (2.1.2.1).
(ii) *L is completely distributive.*
(iii) *The dual frame law holds for members of $\mathcal{M}(L)$.*

The above facts are established within ZF, quite simply because Choice is already incorporated into the assumption in Theorem A.

In [M72] Conrad's theorem is generalized to posets P and their frame of ideals $\mathrm{Idl}(P)$. In that context an ideal J of P is a subset which is a downset, and closed under all existing finite suprema. $\mathrm{Idl}(P)$ is an algebraic frame, in which the finitely generated ideals are the compact elements.

The contribution in [M72] amounts to tying the three conditions of Theorem A to obtain the following generalization of Conrad's theorem, after realizing that the meet-irreducibles generate $\mathrm{Idl}(P)$ ([M72, Theorem 3.1]), with an application of Zorn's lemma, of course:

Theorem B. ([M72, Corollary 3.1.1]) *Let P be a poset. Then the following are equivalent:*

(i) $\mathcal{M}(\mathrm{Idl}(P))$ *freely generates* $\mathrm{Idl}(P)$.
(ii) $\mathrm{Idl}(P)$ *is completely distributive.*
(iii) *The dual frame law holds for members of* $\mathcal{M}(\mathrm{Idl}(P))$.
(iv) *Each meet-irreducible ideal of P is special.*
(v) *Each $x \in P$ can be decomposed, uniquely, as a finite supremum of pairwise incomparable strongly finitely join-irreducible elements.*

To amplify (iv), one observes that $\mathcal{M}(\mathrm{Idl}(P)) = \mathrm{Val}(\mathrm{Idl}(P))$. As to (v), to say that $a \in P$ is a "strongly finitely join-irreducible element" means the following:

$$a \leq \bigvee F, \quad F \text{ finite}, \quad \Longrightarrow \quad a \leq y, \quad \text{for some} \quad y \in F.$$

These take the place of Conrad's special elements. They are the elements having exactly one value. If P is a distributive lattice then $a \in P$ is strongly finitely join-irreducible precisely when it is finitely join-irreducible ([M72, Proposition 2.5]).

2.2. Conrad's theorem for frames

By now it should be evident that Conrad's theorem is a theorem about algebraic frames. Let us briefly reflect upon what properties of $\mathcal{C}(G)$ might not play a role in the proof of the theorem.

Remark 2.2.1. The frame of convex ℓ-subgroups $\mathcal{C}(G)$ of an ℓ-group G has the disjointification! (See 1.2.1.) However, any finite distributive lattice has all the properties of the theorem, without necessarily having disjointification. So one ought to be able to drop the disjointification in a reasonable generalization of Conrad's theorem to frames. This seems to be the message of Theorem B as well.

Yet, the best we can do is prove in ZF that the conditions of Theorem A can be attached to most of the other desired equivalences, but that the disjointification seems to play a crucial role anyway.

As a first step in a proof of Conrad's theorem, within ZF, we have a lemma on decomposition into indecomposable elements. We say that the compact element c in the frame L is *indecomposable* if $c = a \vee b$ implies that a or b is c.

Lemma 2.2.2. *Suppose that L is an algebraic frame satisfying the dual frame law. Then every compact element of L may be expressed, uniquely, as a finite supremum of pairwise disjoint indecomposable compact elements.*

Proof. Suppose that $a \in L$ is compact; if $a = 0$ there is nothing to prove, so assume that $a > 0$. Furthermore, the uniqueness follows routinely, so we proceed to sketch the proof of existence, in ZF.

If a is indecomposable, we're done. Assume, therefore, that it is decomposable and write $a = a_1 \vee b_1$, with $a_1 \wedge b_1 = 0$ and nontrivial. The compactness of a implies that a_1 and b_1 are also compact. If either one of these components is decomposable – without loss of generality a_1 – write $a_1 = a_2 \vee b_2$, with $a_2 \wedge b_2 = 0$ and nontrivial. Induct and get two sequences of nontrivial compact elements $a, a_1, \ldots, a_n, \ldots$ and

$b_1, \ldots, b_n, \ldots$, such that, for each n, $a_{n+1} \wedge b_{n+1} = o$, and $a_n = a_{n+1} \vee b_{n+1}$. The reader may refer to the picture (2.2.2.1) below.

(2.2.2.1)

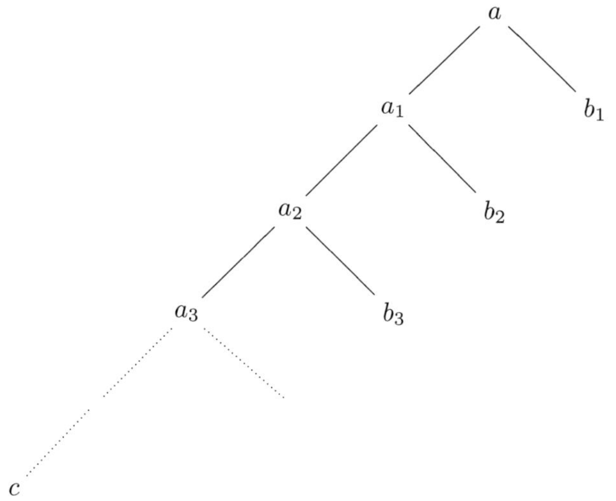

Put $c \equiv \bigwedge_{n=1}^{\infty} a_n$, and observe that, since L is a frame, c is disjoint to $\bigvee_{n=1}^{\infty} b_n$. Now apply the dual frame law: since each a_i exceeds all b_j, with $j > i$, we have

$$c \vee \Big(\bigvee_{n=1}^{\infty} b_n \Big) = \Big(a_1 \vee \Big(\bigvee_{n=1}^{\infty} b_n \Big) \Big) \wedge \Big(a_2 \vee \Big(\bigvee_{n=1}^{\infty} b_n \Big) \Big) \wedge \cdots = a.$$

Since a is compact, there exist indices $j_1 < j_2 < \ldots < j_k$ such that $a = c \vee (b_{j_1} \vee \cdots \vee b_{j_k})$. On the other hand,

$$c \vee (b_{j_1} \vee \cdots \vee b_{j_k}) \leq a_{j_k+1} \vee (b_{j_1} \vee \cdots \vee b_{j_k}),$$

a contradiction. □

The next lemma is also proved within ZF. Note that the use of the dual frame law in the proof of Lemma 2.2.2 is rather mild: what is needed is the distributivity of joins over countable meets. However, the next proof uses the full force of the dual frame law.

Lemma 2.2.3. *Suppose that L is an algebraic frame with disjointification, satisfying the dual frame law. Then every indecomposable element is finitely join-irreducible, and hence special.*

Proof. Suppose $a \in \mathfrak{k}(L)$ and $b, c \in L$, both $< a$, such that $a = b \vee c$. Using the dual frame law, we may assume b and c are the smallest such elements. By the disjointification, there exist disjoint b_1 and c_1 such that $a = b \vee c_1 = b_1 \vee c$. The minimal features of b and c in the decomposition $a = b \vee c$ then imply that $b \wedge c = 0$.

Thus, if a is indecomposable it must also be finitely join-irreducible. Finally, a compactness argument then also shows that there is a largest element beneath a. As is well known, this is equivalent to a being special. □

One then has the following version of Conrad's theorem, in ZF. We refer the reader to [M06b] for the particulars of the proof. We simply note here that the relevant argument in the proof of [M72, Theorem 3.1] shows that 5 implies 1, and does not involve *Choice issues.*

Theorem 2.2.4. *For an algebraic frame L with the disjointification, the following are equivalent:*

1. *$\mathcal{M}(L)$ freely generate L.*
2. *L is completely distributive.*
3. *The dual frame law holds L.*
4. *Each $a \in \mathfrak{k}(L)$ can be decomposed, uniquely, as a finite supremum of pairwise disjoint finitely join-irreducible elements.*
5. *$\mathcal{M}(L)$ generates L, and each meet-irreducible element of L is special.*

Remark 2.2.5. (a) On the proof that $4 \implies 5$ in Theorem 2.2.4: first, it is easy to see that each $a \in \mathfrak{k}(L)$ has n values, corresponding to the n special components in the disjoint supremum of 4, and each value of a is special. This is more than enough to show that the infimum of all values is 0. Furthermore, condition 4 also holds in any quotient frame $\uparrow x$, which shows that $\mathcal{M}(L)$ generates L. Finally, as a meet-irreducible m must be a value of some compact element, it must be special.

(b) In Theorem 2.2.4 the condition *"L is finite-valued,"* is conspicuous by its absence. The proof that $4 \implies 5$ shows that this condition follows from the theorem. We are unable to show the converse in ZF, even when it is accompanied by the assumption that $\mathcal{M}(L)$ generates L.

Remark 2.2.6. Notably absent in the formulation of Theorem 2.2.4 is any mention of the FIP; that is, the property that the meet of two compact elements be compact. In fact, the FIP is a consequence of the theorem ([M06c]).

To conclude this part of the exposition, let us underscore that it has been intended as an illustration of an almost accidental or implicit early manifestation of frame theory in a theorem about ℓ-groups, a manifestation which should now be sufficiently explicit.

3. Part Three: Dimensions of Algebraic Frames

3.1. On d-elements and z-elements

In the early eighties, C. B. Huijsmans and B. de Pagter produced a trio of papers, [HdP80a, HdP80b, Pa81], on the subject of d-ideals and z-ideals of Riesz spaces. The concept of a d-ideal had been around for a number of years, albeit under a different name. But it was their introduction of the "abstract" z-ideal and the richness of the content of these articles that made such an impact. They managed to weave these special convex ℓ-subgroups into a thorough analysis of uniformly complete Riesz spaces. The thing to stress here is the connection with the theme of this exposition: these authors proved structure theorems for uniformly complete Riesz spaces through conditions on the frame of convex ℓ-subgroups.

Much later Mel Henriksen, Grant Woods and I became interested in rings of continuous functions $C(X)$ with the property that every prime z-ideal is either maximal or minimal. The fruits of our efforts appear in [HMW03]; we caled the Tychonoff spaces characterized by this condition *quasi-P spaces.* As is well known, P-spaces have this feature. In [HMW03], we achieved many nice results, but the answer to one question eluded us: *Is the topological sum of quasi P-spaces again quasi P?*

One could not help being struck by the dimensional angle to this study: we were, after all, looking at archimedean f-rings for which the "Krull" dimension of the lattice of z-ideals was at most one! And Huijsmans and de Pagter had given us z-ideals, although they had not explicitly considered the abstract concept in the context of f-rings.

Then in [MZ03], Eric Zenk and I stumbled onto the frame-theoretic gateway to systematically dealing with the phenomenon of d-ideals and z-ideals. That is the subject of this section. A bit later on we went further, to develop the dimension theory of z-ideals, and to apply it to study $C(X)$ with finite dimension. An account of these results, in [MZ05], will be given in §3.2.

To put things in the proper perspective, we need to devote some attention to closure operators, and, in particular, to nuclei on frames. In short, it is time to expand our vocabulary.

Definition & Remarks 3.1.1. Throughout L denotes a frame. The reader is urged to review the vocabulary in 1.2.1.

- A *closure operator* $j : L \longrightarrow L$: an order-preserving map such that $x \leq j(x) = j(j(x))$, for each $x \in L$. We define $jL \equiv \{\, x \in L \,:\, j(x) = x \,\}$; its members are called *j-elements.*
  j is *dense*: $j(0) = 0$. Note that j is dense if and only if $0 \in jL$.
- A *nucleus* j: a closure operator such that $j(a \wedge b) = j(a) \wedge j(b)$.
- L is *j-regular* if jL is a regular frame.
- For closure operators j and j': $j \leq j'$ means that $j(x) \leq j'(x)$, for each $x \in L$. This condition is equivalent to $j'L \subseteq jL$.

For the remainder of this commentary j stands for a nucleus on an algebraic frame L. We consider the process of manufacturing a new nucleus which is in some sense algebraic. This process can be carried out more generally, but goes through more smoothly by assuming L has the FIP, so we assume that as well throughout these remarks. In any event, the applications we have in mind are to ℓ-groups G and their convex ℓ-subgroups, and $\mathcal{C}(G)$ has the FIP.

1. j is *inductive*: for each $x \in L$,

$$j(x) = \bigvee \left\{\, j(a) \,:\, a \leq x,\, a \in \mathfrak{k}(L) \,\right\}.$$

This identity is equivalent to the condition: *jL is closed under the supremum of updirected sets.* ([MZ03, Proposition 4.3]) It implies that jL is algebraic, but the converse is false. ([MZ03, Example 4.4])

2. Define

$$\widehat{j}(x) = \bigvee \left\{ j(a) \,:\, a \le x,\, a \in \mathfrak{k}(L) \right\}.$$

It is shown in [MZ03, §4] that

- $\widehat{j}$ is an inductive nucleus, and $\widehat{j} \le j$. Furthermore, $\widehat{j}$ is the largest inductive nucleus $\le j$.
- $x \in \widehat{j}L$ if and only if $a \le x$ with a compact, implies that $j(a) \le x$.
- $\mathfrak{k}(\widehat{j}L) = \{ j(a) \,:\, a \in \mathfrak{k}(a) \}$

$\widehat{j}$ is called the *inductivization* of j.

By way of examples, we now give an account of the two inductivizations we are most interested in. Our discussion of d-elements will be brief, as our main interest lies with z-elements.

Recall from 1.2.1 the definition $p(x) = x^{\perp\perp}$; it is well known that this defines a nucleus on any frame L. Recall as well that the preferred notation for pL is $\mathcal{P}L$, which is a complete boolean algebra.

Definition & Remarks 3.1.2. Let L be an algebraic frame with the FIP, and let $d \equiv \widehat{p}$. Eplicitly,

$$d(x) = \bigvee \left\{ c^{\perp\perp} \,:\, c \le x, \quad c \in \mathfrak{k}(L) \right\},$$

for each $x \in L$. The d-elements are the members of dL. Let us summarize the principal features of d:

(i) d is an inductive nucleus, and dL is an algebraic frame with the FIP.
(ii) For each $c \in \mathfrak{k}(L)$, $d(c) = c^{\perp\perp}$.
(iii) The term "prime d-element" is unambiguous. Note that a minimal prime element of L is a d-element.
(iv) d is the largest among the inductive nuclei j, for which $j(0) = 0$.

Let us interpret the foregoing in the frame $\mathcal{C}(G)$, for a typical ℓ-group G.

Remark 3.1.3. Let G be an ℓ-group and denote $d\mathcal{C}(G) \equiv \mathcal{C}_d(G)$.

(a) We shall refer to the elements of $\mathcal{C}_d(G)$ as *d-subgroups*, following the usage of [HdP80a] and [HdP80b]. Early in the literature these were called "z-subgroups", but this conflicted with more established terminology from general topology. The nomenclature introduced here has become common.
(b) We say that G is *d-regular* if $\mathcal{C}(G)$ is d-regular. Now, $\mathcal{C}_d(G)$ is compact (and therefore coherent) if and only if G has a *weak order unit*; that is, there is in G an element $u > 0$ such that $u \wedge g = 0$ implies that $g = 0$. If G has a weak order unit then it is easily seen to be d-regular if and only if for each element $0 < g \in G$ there is an $0 < h \in G$ such that $g \wedge h = 0$ and $g \vee h$ is a weak order unit. An ℓ-group with this feature is said to be *complemented.*

Complemented ℓ-groups were introduced in [CM90]. For Riesz spaces see [HdP80b, Theorem 9.8(i)].

What follows is a frame-theoretic formalization of the concept of an archimedean ℓ-group. Recall that the ℓ-group G is *archimedean* if $0 \le a, b \in G$ and $na \le b$, for each $n \in \mathbb{N}$ together imply that $a = 0$. We introduce upper-archimedean elements of an algebraic frame, and use them to define the abstract notion of a z-element. Throughout the remarks that follow, L denotes an algebraic frame with the FIP. These concepts were first discussed in [M73].

Definition & Remarks 3.1.4. $\mathrm{Max}(L)$ denotes the (possibly empty) set of maximal elements of L. It is a routine matter to show that a maximal element is necessarily prime.

(a) With $x \in L$ we say that $m < x$ is *maximal under* x if $m \in \mathrm{Max}(\downarrow x)$. Note that $\mathrm{Max}(\downarrow x) \neq \emptyset$, if x is compact and one assumes (for example) Zorn's lemma.
The set $\mathrm{Max}(\downarrow x)$ is in one-to-one correspondence with

$$\{\, p \in \mathrm{Spec}(L) \,:\, x \not\le p \quad \text{and } p \text{ is maximal with this property}\}.$$

Thus, if x is compact, then the preceding set of primes is simply the set of values of x.

(b) L is said to be an *archimedean lattice* if, for each $c \in \mathfrak{k}(L)$, $\bigwedge \mathrm{Max}(\downarrow c) = 0$. We shall say that $x \in L$ is *upper-archimedean* if $\uparrow x$ is archimedean. Denote the set of all upper-archimedean elements of L by $\mathbf{a}^{\uparrow}(L)$.

(c) Observe that if L is compact then L is archimedean if and only if $\bigwedge \mathrm{Max}(L) = 0$. Thus, if L is compact then $x \in \mathbf{a}^{\uparrow}(L)$ precisely when x is an infimum of maximal elements of L.

The following lemma is a paraphrase of [MZ03, Lemma 6.2].

Lemma 3.1.5. *Suppose that L is an algebraic frame with the FIP. Then $\mathbf{a}^{\uparrow}(L)$ is closed under arbitrary infima, and if*

$$ar(x) \equiv \bigwedge \left\{ y \in \mathbf{a}^{\uparrow}(L) \,:\, y \ge x \right\},$$

then ar defines a nucleus for which $arL = \mathbf{a}^{\uparrow}(L)$.

Definition & Remarks 3.1.6. Suppose, throughout this commentary, that L is an algebraic frame with the FIP.

(a) The proof of Lemma 3.1.5 underscores that if L is archimedean then $\mathbf{a}^{\uparrow}(L)$ contains all the polars of L.
If L is a compact algebraic frame then, in view of the comments in 3.1.4(c),

$$ar(x) = \bigwedge \left\{ m \in \mathrm{Max}(L) \,:\, x \le m \right\}.$$

(b) Now define $z = \widehat{ar}$. We summarize the properties of the operator z.

(i) For each $x \in L$,

$$z(x) = \bigvee \left\{ ar(c) \,:\, c \leq x, \quad c \in \mathfrak{k}(L) \right\},$$

(ii) Assume L is archimedean. Then $ar(x) \leq x^{\perp\perp}$, for each $x \in L$, whence $z \leq d$. Thus, every d-element is necessarily a z-element.

(iii) z is the largest inductive nucleus under ar.

In the context of an ℓ-group G, the frame $z\mathcal{C}(G)$ of z-elements is denoted $\mathcal{C}_z(G)$.

3.2. Krull dimension without primes

The subject of a Krull-style dimension for either distributive lattices with top and bottom, or algebraic frames with the FIP, has received considerable attention in recent years. The subject has been investigated by researchers in real algebra, frequently employing the techniques and terminology of logic. In [M06a], and together with Zenk in [MZ05], I approached the subject from a frame-theoretic point of view. In [CL02] and [CLR03], the authors investigated the subject in distributive lattices, and established a pointfree criterion for such lattices to have dimension not exceeding n. In [M06a], a similar principle was developed for algebraic frames with the FIP and disjointification, which allows dimension to be computed using certain finite sets of compact elements of the frame. The criterion, Theorem 3.2.6 below, offers certain advantages over the one in [M06a, Theorem 3.8], the principal one being that it lends itself to inductive arguments.

As we have already signalled in the introduction to §3.2, the original motivating force behind our interest in dimension in the setting of frames, was the desire to have a vehicle by means of which one could study the frame of z-ideals of a ring $C(X)$ of continuous real-valued functions on a Tychonoff space X. The approach via Theorem 3.2.6, together with the observation that the lattice of cozerosets and the sublattice of principal z-ideals are isomorphic, permits us to compute z-dimension purely spatially.

In advance of Theorem 3.2.6 we need three preliminaries. The first two are part of the stock of frame-theoretic observations. The third, Lemma 3.2.5, will be an inductive estimation of dimension. This is also the place where boundary quotients make their appearance.

The proof of the following lemma involves ultrafilters of compact elements. By a *filter F of compact elements* we mean a subset of $\mathfrak{k}(L) \setminus \{0\}$, closed under finite meets and such that $c \leq d$ in $\mathfrak{k}(L)$ with $c \in F$ implies that $d \in F$. An *ultrafilter of compact elements* is a maximal filter of compact elements.

Lemma 3.2.1 ([M73, Corollary 2.5.1]). *Suppose L is an algebraic frame. Then $p \in \mathrm{Spec}(L)$ is minimal if and only if $F_p = \{\, c \in \mathfrak{k}(L) \,:\, c \not\leq p \,\}$ is an ultrafilter on $\mathfrak{k}(L)$. If this is the case, then $p = \vee \{\, c^{\perp} \,:\, c \in F_p \,\}$.*

Remark 3.2.2. Let $\mathrm{Min}(L)$ denote the set of all minimal prime elements of L. Zorn's lemma easily shows that in any frame each prime element exceeds a minimal prime. It is also a routine matter to verify that, in any algebraic frame, each polar is an infimum of minimal primes.

Lemma 3.2.1 implies the following; this corollary amounts to half the proof of Lemma 3.2.5.

Corollary 3.2.3. *Let L be an algebraic frame. For each $a \in \mathfrak{k}(L)$ and each $p \in$ $\mathrm{Min}(L)$, we have $a \vee a^{\perp} \not\leq p$.*

Lemma 3.2.4. *Let L be an algebraic frame. For each $y \in L$, the map $j^y(x) = x \vee y$ is an inductive nucleus and $j^y L = \uparrow y$. Thus,* $\mathrm{Spec}(\uparrow y)$ *consists of the primes of L that exceed y.*

In an algebraic frame L, and for a compact $a \in L$, denote $L^a \equiv \uparrow (a \vee a^{\perp})$, and call L^a the *boundary quotient over a*.

Lemma 3.2.5. *Suppose that L is an algebraic frame. Then* $\dim(L) \leq k$ *if and only if, for each $a \in \mathfrak{k}(L)$, the dimension of the boundary quotient L^a over a is $\leq k-1$.*

The proof of Theorem 3.2.6 is now a relatively easy induction argument. We refer the reader to [MZ05, Theorem 2.7].

Theorem 3.2.6. [The Coquand–Lombardi–Roy theorem.] *Let L be an algebraic frame. Then* $\dim(L) \leq k$ *if and only if*

$$1 = x_k \vee (x_k \rightarrow (\cdots (x_1 \vee (x_1 \rightarrow (x_0 \vee x_0^{\perp}))) \cdots)),$$

for all $x_0, x_1, \ldots, x_k \in \mathfrak{k}(L)$.

Theorem 3.2.7. *Let L be an algebraic frame. Then* $\dim(L) \leq k$ *if and only if for each set of compact elements $a_0, a_1, \ldots, a_k, a_{k+1}$ there exist compact elements $b_0, b_1, \ldots, b_k$ such that*

$$a_{k+1} \leq a_k \vee b_k,\ a_k \wedge b_k \leq a_{k-1} \vee b_{k-1}, \ldots, a_1 \wedge b_1 \leq a_0 \vee b_0, \quad \textit{and} \quad a_0 \wedge b_0 = 0.$$

Proof. Apply Theorem 3.2.6, iterating the observation that, for any compact elements a and b in L,

$$a \leq b \vee (b \rightarrow y) \quad \text{iff} \quad \exists c \in \mathfrak{k}(L), \quad \text{with} \quad a \leq b \vee c \quad \text{and} \quad b \wedge c \leq y.$$

□

Remark 3.2.8. Briefly, we make note of the fact that the condition in Theorem 3.2.6 coincides with the one obtained in [AB91]. The context in that article is that of clopen downsets in a Priestley space.

3.3. The z-dimension of a Tychonoff space

We explore a very particular application of Theorem 3.2.6, namely, the z-dimension of a Tychonoff space. The foregoing two sections having set us to calculate the dimension of any algebraic frame with FIP, we go one step further: by establishing that there is an isomorphism between the frame of ideals of $\mathrm{Coz}(X)$, the lattice of cozerosets of the space X and $\mathcal{C}_z(X)$, the frame of all z-ideals of $C(X)$, we are able to phrase the concept of dimension in purely topological terms.

Throughout this section, X denotes a *Tychonoff space*, which is to say, a Hausdorff topological space such that for any point $p \in X$ and closed set K in X, not containing p, there is a real-valued continuous function f such that $f(p) = 1$ and $f(K) = \{0\}$. $C(X)$ stands, as is customary, for the ring of all real-valued continuous functions defined on X. $C(X)$ is also an ℓ-group and an f-ring. All the pertinent operations are to be taken pointwise. If $f \in C(X)$, we denote

$$\mathrm{coz}(f) \equiv \{\, x \in :\ f(x) \neq 0 \,\},$$

the *cozeroset of* f, and $Z(f)$ stands for the set-theoretic complement of $\mathrm{coz}(f)$, and is the *zeroset* of f. It is well known that a Hausdorff space is Tychonoff if and only if $\mathrm{Coz}(X)$, the collection of all cozerosets of X, forms a base of the open sets of X. Observe that $\mathrm{Coz}(X)$ is a distributive lattice with respect to ordinary set-theoretic union and intersection, with top X and bottom $\emptyset$.

Typically, we shall spell out only the terminology from general topology which is strictly necessary for this narrative, referring te reader to [GJ76] for all unexplained issues.

Definition & Remarks 3.3.1. A *z-ideal* $\mathfrak{r}$ of $C(X)$ is a subgroup for which $f \in \mathfrak{r}$ and $\mathrm{coz}(g) \subseteq \mathrm{coz}(f)$ (with $g \in C(X)$) together imply that $g \in \mathfrak{r}$. It is easy to see that a z-ideal is indeed a ring ideal as well as a convex ℓ-subgroup of $C(X)$. Moreover, it emerges from [HdP80a] that the z-ideals are precisely the z-elements of $\mathcal{C}(C(X))$.

Thus, $\mathcal{C}_z(X)$ is a frame algebraic frame with FIP under inclusion. It is easy to check directly that $\mathfrak{k}(\mathcal{C}_z(X))$ consists of the *principal* z-ideals; that is, the z-ideals of the form, for each $f \in C(X)$,

$$\langle f \rangle_z = \{\, g \in C(X) \ :\ \mathrm{coz}(g) \subseteq \mathrm{coz}(f) \,\}.$$

For example, if $\mathfrak{r}$ is a z-ideal which is generated by $f_1, f_2, \ldots, f_m$, we may assume without loss of generality – by passing from f_i to $|f_i|$ – that each of the generators is positive. It is then clear that $\mathfrak{r} = \langle (f_1 + f_2 + \cdots + f_m) \rangle_z$.

Note, finally, that $\mathcal{C}_z(X)$ is compact, and therefore coherent; the top is $\langle 1 \rangle_z$.

The following lemma establishes the crucial link between the algebraic and the purely spatial.

Lemma 3.3.2. ([MZ05, Lemma 4.2]) *Let X be a space. The map*

$$\eta^z_X(\mathrm{coz}(f)) = \langle f \rangle_z,$$

is a lattice isomorphism from $\mathrm{Coz}(X)$ *onto* $\mathfrak{k}(\mathcal{C}_z(X))$.

Definition & Remarks 3.3.3. The *z-dimension* of $C(X)$ is

$$\dim_z(C(X)) \equiv \dim(\mathcal{C}_z(X)) = dim(\mathcal{I}(\mathrm{Coz}(X))).$$

We shall also speak of the z-dimension of X itself, and write it $\dim_z(X)$.

The translation of Theorem 3.2.7 reads as follows.

Theorem 3.3.4 ([MZ05, Theorem 4.5]). *Let X be a space. Then $\dim_z(X) \leq k$ if and only if for each sequence of cozerosets $U_0, U_1, \ldots, U_k$ there exist cozerosets $V_0, V_1, \ldots, V_k$ such that*

$$X = U_k \cup V_k, U_k \cap V_k \subseteq U_{k-1} \cup V_{k-1}, \ldots, U_1 \cap V_1 \subseteq U_0 \cup V_0, \quad \textit{and} \quad U_0 \cap V_0 = \emptyset.$$

Remarks 3.3.5.
(a) It is easy to see from Theorem 3.3.4 that $\dim_z(X) = 0$ precisely when each cozeroset is closed. This is one of the many equivalent definitions of a P-space; see [GJ76, Theorem 14.29].

(b) As was already advertised in the introduction to §3.1, in [HMW03] we studied the spaces for which $\dim_z(X) \leq 1$ (without using any of the machinery or terminology introduced here, and without mentioning dimension). These are quasi P-spaces; reciting Theorem 3.3.4 for $k = 1$, yiels the following characterization of quasi P-spaces: *X is quasi P if and only if for each cozero sets U_0 and U_1 there exist cozerosets V_0 and V_1 such that $X = U_1 \cup V_1$, $U_1 \cap V_1 \subseteq U_0 \cup V_0$, with $U_0 \cap V_0 = \emptyset$.*

(c) We recall the open question from [HMW03], which can now be answered with ease. Recall that if $\{ X_i : i \in I \}$ is a family of spaces, and X denotes the disjoint union of the X_i, then X is called the *topological union* if its topology is defined as follows: $V \in \mathfrak{O}(X)$ if and only if each $V \cap X_i \in \mathfrak{O}(X_i)$. Note that if X is the topological union of the X_i, then $V \in \mathrm{Coz}(X)$ precisely when $V \cap X_i \in \mathrm{Coz}(X_i)$; then also $C(X)$ is canonically isomorphic – as a ring and as an ℓ-group – to the direct product $\prod_{i \in I} C(X_i)$.

In [HMW03] it was asked whether the topological union of any number of quasi P spaces is quasi P. Several affirmative partial results were obtained, but the general question remained unresolved.

Theorem 3.3.4 immediately implies the following, which also answers the question affirmatively, without exception.

Proposition 3.3.6. *Suppose that X is the topological union of the spaces X_i ($i \in I$). Then*

$$\dim_z(X) = \sup_{i \in I} \dim_z(X_i).$$

The preceding application is straightforward. To obtain meaningful topological structure theory from Theorem 3.2.6, one has to dig a little deeper. In [MZ05] we discussed the notion of a "natural typing of open sets", which sets up the connection between boundary quotients as defined prior to Lemma 3.2.5, and topological *boundaries* in the following sense: for any cozeroset U of the space X, we consider $bU \equiv \mathrm{cl}_X U \setminus U$. The reader who is interested in the details is referred to [MZ05, §3]. What follows is our best result for z-dimension. Recall that a space is *Lindelöf* when every cover by open sets has a countable subcover.

Theorem 3.3.7. *Suppose that X is a Lindelöf space. Then for each nonnegative integer k, $\dim_z(X) \leq k$ if and only if $\dim_z(bU) \leq k - 1$, for each boundary bU ($U \in \mathrm{Coz}(X)$).*

For compact spaces there is a complete characterization of spaces whose z-dimension is finite. To understand Theorem 3.3.10, a brief primer on scattered spaces seems necessary, and, in particular, one should highlight the so-called Cantor-Bendixson derivatives of a space.

Definition & Remarks 3.3.8. In this general commentary Y is an arbitrary Tychonoff space.

(a) Y is said to be *scattered* if each nonvoid subspace S has an isolated point of S. Many properties of scattered spaces are summarized in Z. Semadeni's memoir [Se59]; we also refer the reader to his book [Se71]. It is easy to see that if each nonempty closed subspace of Y has an isolated point, then Y is scattered.

(b) If Y is a space let $\mathrm{Is}(Y)$ denote its set of isolated points, and let: $Y^{(0)} = Y$, $Y^{(1)} = Y \setminus \mathrm{Is}(Y)$. For any ordinal η, let $Y^{(\eta+1)} = (Y^{(\eta)})^{(1)}$, and if η is a limit ordinal, let

$$Y^{(\eta)} = \cap\{\, Y^{(\xi)} : \xi < \eta \,\}.$$

The spaces $Y^{(\eta)}$ are called *Cantor-Bendixson derivatives of* Y. The reader will note that these derivatives form a decreasing transfinite sequence of closed subspaces of Y. From cardinality considerations there is an ordinal α such that $Y^{(\alpha)} = Y^{(\alpha+1)}$; then, in fact, $Y^{(\alpha)} = Y^{(\beta)}$, for each $\beta > \alpha$. Let $\mathrm{CB}(Y)$ denote the smallest ordinal for which $Y^{(\alpha)} = Y^{(\alpha+1)}$; this is the *CB-index* of a space Y.

Now, it is easily seen that Y is scattered if and only if $Y^{(\alpha)} = \emptyset$, for suitable α. If Y is scattered and $\mathrm{CB}(Y) = \alpha$, then α is also the least ordinal for which $Y^{(\alpha)} = \emptyset$. In particular, $\mathrm{CB}(Y) = 1$, with Y scattered, simply means that Y is a nontrivial discrete space.

Obviously, if Y is scattered, then any subspace S is also scattered, and $\mathrm{CB}(S) \leq \mathrm{CB}(Y)$.

If Y is compact, scattered, and $\alpha = \mathrm{CB}(Y)$, then it is clear that $\bigcap_{\eta<\alpha} Y^{(\eta)}$ is nonempty. It follows that α has a predecessor γ such that $Y^{(\gamma)}$ is finite and, hence, the last nonempty Cantor-Bendixson derivative. (To illustrate, $\mathrm{CB}(Y) = 1$ means that Y itself is finite and nonempty; $\mathrm{CB}(Y) = 2$ means that $Y \setminus \mathrm{Is}(Y)$ is finite, but nonvoid; and so on.)

Note that if Y is compact and scattered, then $\mathrm{CB}(Y) = 2$ if and only if Y is a finite topological sum of one-point compactifications of discrete spaces (of which at least one is infinite).

(c) If X is scattered, with finite CB-index, then an easy induction argument establishes that each nonisolated point $p \in X$ is the limit of a sequence $p_1, p_2, \ldots$; moreover, if p is isolated in $X^{(i)}$, then p_n may be chosen so that it is isolated in $X^{(i_n)}$, with $i_n \leq i$.

To prove Theorem 3.3.10, applying Theorem 3.3.7, the following lemma is very handy. The nontrivial implication – (c) $\Rightarrow$ (a) – in the lemma is, probably, part of the folklore of scattered spaces. A proof, due to Martínez and McGovern, appears in [MZ05].

Lemma 3.3.9. *For any (compact) space X space the following are equivalent.*

(a) *X is scattered.*
(b) *For each open set O, bO is scattered.*
(c) *For each cozeroset U, bU is scattered.*

If X is scattered, then, for each nonnegative integer k, $\mathrm{CB}(X) \leq k$ *if and only if* $\mathrm{CB}(bU) \leq k-1$, *for each cozeroset U of X.*

It should be noted that Theorem 3.3.10 generalizes [HMW03, Theorem 4.1(II)]. We sketch the proof.

To emphasize, though the Cantor-Bendixson apparatus is a purely topological concept, apart from the case for quasi P-spaces, in [HMW03], we believe that no topological proof exists of the theorem.

Theorem 3.3.10. *Suppose X is a space. Then* $\dim_z(X) \leq k$ *if and only if X is scattered and* $\mathrm{CB}(X) \leq k+1$; *($k \geq -1$ is an integer).*

Proof. For $k = -1$, both $\dim_z(X) \leq k$ and $\mathrm{CB}(X) \leq k+1$ are true precisely when the space $X = \emptyset$. Now suppose that $k \geq -1$, and the theorem holds for all compact spaces of z-dimension $\leq k$. Observe that $\dim_z(X) \leq k+1$ if and only if $\dim_z(bU) \leq k$, for each cozeroset U of X, which, by induction, is true if and only if each cozeroset boundary bU is scattered of CB-index $\leq k+1$. Finally, applying Lemma 3.3.9, the latter holds if and only if X itself is scattered and $\mathrm{CB}(X) \leq k+2$. □

Remark 3.3.11. Note that $\beta\mathbb{N}$, the Stone-Čech compactification of the discrete natural numbers, has infinite z-dimension. Thus, any space containing a copy of $\beta\mathbb{N}$ also has infinite z-dimension. This includes all the compact F-spaces ([GJ76, Theorem 14.25]).

Briefly, we comment on what is known about d-dimension; there is very little to report. We confine our remarks to $\mathcal{C}_d X(X)$.

Remark 3.3.12. Let X be a space. The *d-dimension* of X (or of $C(X)$), denoted $\dim_d(X)$ (resp. $\dim_d((C(X)))$ is the dimension of $\mathcal{C}_d(X)$.

As every d-element is a z-element, it follows that $\dim_d(X) \leq \dim_z(X)$, for every space X. There are interesting situations when these dimensions agree; let us refer the reader to [MZ05, §6].

It is easily seen that $\dim_d(X) = 0$ precisely when, for each cozeroset U of X, there is a cozeroset V of X such that $U \cap V = \emptyset$ and $U \cup V$ is dense. Such spaces are called *cozerocomplemented*; they have been extensively studied, and most recently by Henriksen and Woods ([HW04]). This class includes all metric spaces. However, $\beta\mathbb{N}$ is cozerocomplemented, but its z-dimension is infinite, one quickly realizes that z-dimension and d-dimension can be quite different.

There is a characterization of the the condition $\dim_d(X) \leq k$, as counterpart to Theorem 3.3.4, in terms of closures of cozerosets; see [MZ05, Theorem 6.6]. Likewise, one has the analogue of Proposition 3.3.6:

Suppose that X is the topological union of the spaces X_i ($i \in I$). Then

$$\dim_d(X) = \sup_{i \in I} \dim_d(X_i).$$

4. Part Four: Epicompletion in Frames

Here we present an account of work with Eric Zenk, some aspects of which is very much in progress, in [M06c, MZ07]. The goal of this research is to understand, in frame-theoretic terms, the "construction" of the essential closure of an archimedean ℓ-group, and then to view it functorially. Alternatively, from the topological point of view, one seeks a frame-theoretic approach to the absolute of a compact space, and terms under which the construction is reflective. The reference for this is, primarily, [MZ06b].

In the interest of economy, we make several assumptions in this part of the exposition. First, we shall suppose that the reader is familiar with the elements of category theory – with terms such as functor, whether covariant or contravariant; in the first, short section to follow, we will provide the basic tools which the reader will need to understand the rest of the narrative. Second, and for background purposes, we hope that the reader will be able to work around the few references to Stone duality. And, third, we suppose that if the reader has made it this far into the exposition, then this attempt of ours to ratchet up the discourse will not discourage.

4.1. Categorical preliminaries

This section consists of a brief introduction to the theory of reflections and coreflections of categories. We shall endeavor to be as plain-spoken as possible, but category theory being category theory, the reader may reasonably expect a blitz of terminology. Our main reference is [HS79].

4.1.1. Monos and Epis. The seasoned navigator around issues related to monomorphisms and epimorphisms will know that in most *reasonable* categories "monomorphism" means "one-to-one", whereas "epimorphism" frequently does not imply "surjective". For the record, we note that a morphism $g : A \longrightarrow B$ in a category $\mathfrak{C}$ is a *monomorphism* (resp. *epimorphism*) if $g \cdot h_1 = g \cdot h_2$ (resp. $h_1 \cdot g = h_2 \cdot g$) implies that $h_1 = h_2$ (whenever the compositions make sense).

4.1.2. Reflections and Coreflections. In this commentary we consider only covariant functors. Suppose that $\mathfrak{B}$ is a category having a *full* subcategory $\mathfrak{A}$; (that is to say, between two $\mathfrak{A}$-objects, all $\mathfrak{B}$-morphisms are in $\mathfrak{A}$.)

A *reflection of* $\mathfrak{B}$ *in* $\mathfrak{A}$ is an assignment which associates with each $\mathfrak{B}$-object B an $\mathfrak{A}$-object ρB, as well as a morphism $\rho_B : B \longrightarrow \rho B$, having the following universal property: for each $\mathfrak{A}$-object X and each morphism $g : B \longrightarrow X$, there is a unique morphism $\widehat{g} : \rho B \longrightarrow X$ such that $\widehat{g} \cdot \rho_B = g$; that is, the diagram below commutes:

(4.1.2.ref)

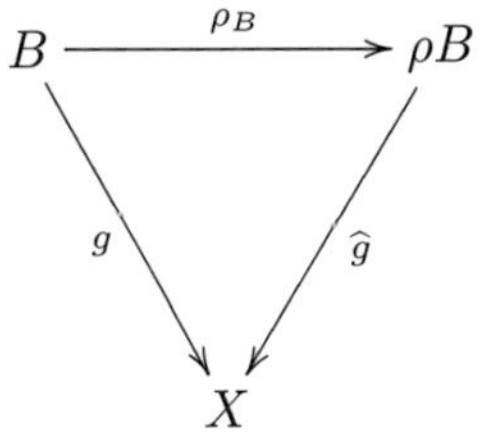

It then follows that $\rho : \mathfrak{B} \longrightarrow \mathfrak{A}$ is a covariant functor ([HS79]), and the knowledgeable reader will recognize ρ as the left (or "front") adjoint of the inclusion functor U of $\mathfrak{A}$ in $\mathfrak{B}$. Also worth a mention is the fact that ρ defines a natural transformation between the identity functor $1_{\mathfrak{B}}$ and the composite $U \cdot \rho$.

Examples of reflections abound in mathematics. Likely to be familiar to many mathematicians are: the reflection of groups in abelian groups, by factoring out the commutator subgroup; the reflection of torsion free abelian groups in divisible abelian groups, by the formation of the divisible hull; the reflection of the category of Tychonoff spaces in the subcategory of compact spaces, by way of the Stone-Čech compactification.

For additional examples of reflections, as well as amplification of the above ones, the reader is referred to [HS79].

A *coreflection* is the dual concept, with all arrows reversed, associated with the following commutative diagram:

(4.1.2.coref)

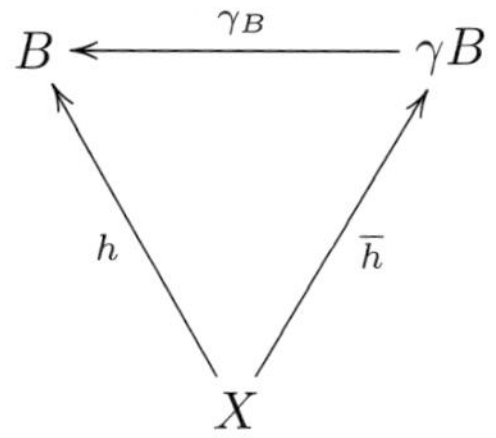

It is an exercise worth the reader's while to actually formulate the dual statement. The standard example in algebra of a coreflection is that of abelian groups in torsion abelian groups, by passing from a groups G to its subgroups of torsion elements.

The reader should also reflect upon the following (purposefully) imprecise statement: that in most situations involving a pair of contravariant adjoint functors between categories $\mathfrak{A}$ and $\mathfrak{B}$, a reflection of $\mathfrak{A}$ in a subcategory has a counterpart in a coreflection of $\mathfrak{B}$, and vice-versa. Applying this to the situation involving the functors $\mathfrak{O}$ and Spec of 1.2.2, between the categories $\mathfrak{Frm}$ of all frames and $\mathfrak{Top}$ of all topological spaces, the reader might reasonably expect some "new" examples of coreflections of frames to arise from reflective counterparts in $\mathfrak{Top}$.

We comment, explicitly, upon two of these, next.

4.1.3. The Regular Coreflection. Recall from 1.2.1 that a frame L is *regular* if each $x \in L$ is the supremum of the elements of L which are well below it. Indeed, $x \in L$ is regular if and only if

$$x = \vee \{ a \in L : a \preceq x \}.$$

Let us denote the subset of L of all regular elements by $\mathrm{Reg}(L)$; it is well known and easily verified that $\mathrm{Reg}(L)$ is a subframe of L. The following are also readily checked; we illustrate in one, typical instance:

(a) Any regular subframe of L is contained in $\mathrm{Reg}(L)$.
(b) The subframe generated by any collection of regular subframes of L is regular; thus, the supremum ϱL of all the regular subframes of L is regular and $\varrho L \subseteq \mathrm{Reg}(L)$. We note without comment that ϱL and $\mathrm{Reg}(L)$ are, in general different; $\mathrm{Reg}(L)$ may fail to be regular.
(c) We say that $\preceq$ *interpolates* if $a \preceq b$ implies that $c \in L$ exists such that $a \preceq c \preceq b$. If $\preceq$ interpolates then $\mathrm{Reg}(L)$ is regular, and hence equal to ϱL.
(For, if $\overline{a}$ denotes the supremum of all elements of L which are well below a, and note that $\overline{a} \in \mathrm{Reg}(L)$. Next, if x is regular and $c \preceq x$, then there is a $y \in L$ such that $c \preceq y \preceq x$, and so $c \preceq \overline{y} \leq y \preceq x$, and therefore x is the supremum over the $\overline{a}$, with $a \preceq x$.)
(d) If L is normal (see 1.2.1) then $\preceq$ interpolates.
(e) The image under any frame homomorphism of a regular frame is regular.

As a consequence of (e) above, the inclusion $\varrho L \subseteq L$ defines a coreflection of the category $\mathfrak{Frm}$ in the subcategory $\mathfrak{RegFrm}$ of all regular frames.

4.1.4. The Stone-Čech Coreflection. Now assume that L is regular. Consider the ideal frame $\mathrm{Idl}(L)$. (Recall: $J \subseteq L$ is an ideal of L if it is nonempty, closed under finite suprema, and $a \in J$ implies that $\downarrow a \subseteq J$.) Consider the join map $\bigvee_L : \mathrm{Idl}(L) \longrightarrow L$, given by $\bigvee_L(J) = \bigvee J$. Note that $\mathrm{Idl}(L)$ is compact. Now suppose that $h : A \longrightarrow L$ is any frame map out of a compact regular frame A. Define $\bar{f} : A \longrightarrow \mathrm{Idl}(L)$ by

$$\bar{f}(a) \equiv \langle\, f(x) \,:\, x \preceq a \,\rangle,$$

where $\langle S \rangle$ indicates the ideal generated by S. One has to verify that $\bar{f}$ is a frame homomorphism. By the regularity of A one gets that $\bigvee_L \cdot \bar{f} = f$.

Apply the regular coreflection ϱ, to obtain the passage from the commutativity on the left of the diagram below to the one on the right; $\beta L \equiv \varrho(\mathrm{Idl}(L))$, and is the *Stone-Čech compactification* of L:

(4.1.4.1)

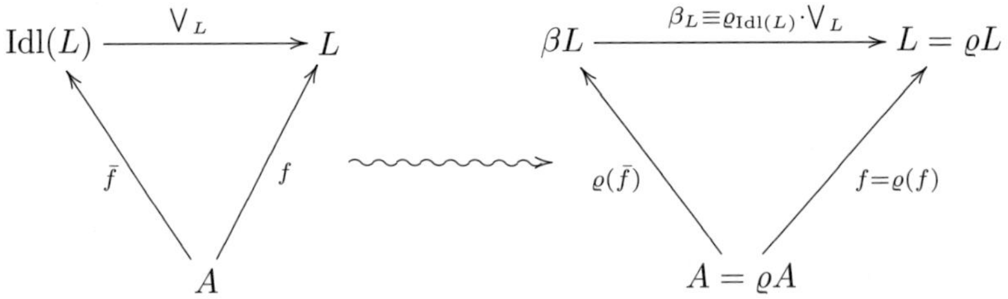

It is a routine exercise to show that $\varrho(\bar{f})$ is actually uniquely determined by f. That shows that β is a coreflection.

Finally in this section, we introduce the morphisms to be used in §4.4, which will render the frame-theoretic absolute to be discussed there functorial.

4.1.5. Skeletal maps. The reader familiar with skeletal maps in topology, in the sense of [HS68] and [DPR81], will find the frame-theoretic counterpart natural enough.

The frame homomorphism $h : L \longrightarrow M$ is *skeletal* if $x^{\perp\perp} = 1$ in L implies that $h(x)^{\perp\perp} = 1$. It is easy to verify that h is skeletal if and only if

$$x_1^{\perp} = x_2^{\perp} \quad \Longrightarrow \quad h(x_1)^{\perp} = h(x_2)^{\perp}.$$

Then it is also easy to check that h is skeletal precisely when there is a (unique) frame homomorphism $\mathcal{P}(h) : \mathcal{P}L \longrightarrow \mathcal{P}M$ making the diagram below commute:

(4.1.5.1)
$$\begin{array}{ccc} L & \xrightarrow{\;h\;} & M \\ \downarrow{\scriptstyle p_L} & & \downarrow{\scriptstyle p_M} \\ \mathcal{P}L & \xrightarrow{\;\mathcal{P}(h)\;} & \mathcal{P}M \end{array}$$

In figure (4.1.5.1), p_L denotes the nucleus defined by $p_L(x) = x^{\perp\perp}$. (We do not decorate the $\perp$s to indicate which frame the complements are taken in.)

As noted in [BaP96], if one considers the subcategory $\mathfrak{FrmS}$ of frames with skeletal maps, then $\mathcal{P}$ turns into a functor – which is, evidently, a reflection ([HS79]).

The concept of a skeletal map is also discussed in [BaP94].

4.2. Gleason and Conrad

We summarize the work of A. Gleason on the absolute of a compact Hausdorff space, and the dual concept of the essential closure of an archimedean ℓ-group. We refer the reader to [Wa74] for an account of the former, and to [C71] for one of the latter.

It should be emphasized that this section is meant to give basic background information. Technically speaking the reader should be able to proceed to §4.4 and pick up the "point-free" thread of this narrative.

4.2.1. The Absolute of a Compact Space. In this commentary all topological spaces are assumed to be compact and Hausdorff. If X is a space then $\mathfrak{RO}(X)$ denotes the collection of regular open sets; we remind the reader that an open set U is *regular open* if it is the interior of a closed set. Equivalently, $U \in \mathfrak{O}(X)$ is regular open if and only if $U = \mathrm{int}_X(\mathrm{cl}_X U)$, if and only if U is a polar of the frame $\mathfrak{O}(X)$.

Thus, $\mathfrak{RO}(X)$ is itself a boolean frame, and, in particular, it is complete. The Stone dual of $\mathfrak{RO}(X)$, is denoted EX; applying the machinery of Stone duality, one obtains that EX is an *extremally disconnected space* – that is, every open set has an open closure. One may view EX as the space of maximal ideals of $\mathfrak{RO}(X)$ – the latter viewed as a boolean ring, while EX bears the hull-kernel topology;

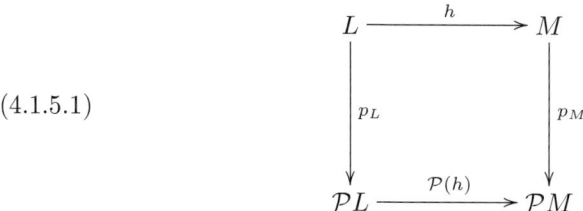

each maximal ideal $\mathfrak{m}$ is *fixed*, in the sense that there is a (unique) point $p \in X$ such that

$$\mathfrak{m} = \mathfrak{m}_p \equiv \{\, U \in \mathfrak{m} \,:\, p \notin U \,\}.$$

The existence of p uses the compactness of X, while the uniqueness follows from the Hausdorff separation.

Define $e_X(\mathfrak{m}) = p$; this defines a continuous map of EX onto X, which is *irreducible*, meaning that no proper closed subset maps onto X.

If Y is any extremally disconnected space and $g : Y \longrightarrow X$ is an irreducible continuous surjection, then there is a homeomorphism $h : Y \longrightarrow EX$ such that $e_X \cdot h = g$. One then says that EX is unique *up to a homeomorphism over* X; EX is the *absolute of* X.

Whether one obtains the absolute of a compact Hausdorff space X as a space of maximal ideals of the Stone dual $\mathfrak{RO}(X)$, or, as other approaches do, by identifying it as the maximum in a poset – see, for example [H89] – the existence of the absolute involves, somewhere in the process, an application of Choice. The frame-theoretic approach to be outlined in §4.4 is Choice-free: it invokes the Stone-Čech coreflection, which is Choice-free, as was pointed out in 4.1.4.

4.2.2. D(X). We review the basic facts about this construct, the set of all "almost real-valued" continuous functions defined on the space X. Spaces are here assumed to be Tychonoff, and frequently will have strong disconnectivity properties as well.

$D(X)$ stands for the set of all continuous functions f defined of X with values in the extended reals $\mathbb{R} \cup \{\pm\infty\}$, such that $f^{-1}(\mathbb{R})$ is dense in X. $D(X)$ is easily seen to be a lattice under pointwise supremum and infimum. On the other hand, pointwise addition and multiplication yield a group and ring structure, respectively, in only very particular circumstances, which ought not concern us here. Suffice it to observe that (a) for each $f \in D(X)$, $f^{-1}(\mathbb{R})$ is a dense cozeroset, and (b) dense cozerosets have to be C^*-embedded. (We refer the interested reader to [BKW77, Chapter 13] and [PW89, 8.4] for further discussion.)

4.2.3. Essential Extensions. Suppose that G is an ℓ-subgroup of the ℓ-group H. Call H an *essential extension of* G if each nontrivial convex ℓ-subgroup of H intersects G nontrivially. One also uses the following phrases, synonymously: G *is an essential* ℓ*-subgroup of* H, and G *is a large* ℓ*-subgroup of* H.

In the context of archimedean ℓ-groups, H is an essential extension of G if and only if the trace map $P \mapsto G \cap P$ is a boolean isomorphism of $\mathcal{P}H$ onto $\mathcal{P}G$ ([BKW77, Theorem 11.1.15]); this due to Conrad. For archimedean ℓ-groups also one may consider the concept of an "essential closure": first, H is *essentially closed* if there are no proper (archimedean) essential extensions of H. Further, H is an *essential closure* of G if G is essential in H and H is essentially closed.

Note that one is forced to confine oneself to archimedean ℓ-groups if the concept of an essentially closed object is to amount to anything. Else, for any G one may lexicographically construct $G \times \mathbb{Z}$, making the elements of G infinitesimals to the pairs (g, n) (with $n \in \mathbb{N}$), and this is an essential extension.

4.2.4. The Essential Closure. *From this point onward we assume all ℓ-groups entering the conversation are archimedean, unless the contrary is explicitly stated.* The reader is reminded that all archimedean ℓ-groups are abelian ([BKW77, Theorem 11.1.3]).

For these results on $D(X)$ the reader may go directly to [C71], where the existence and uniqueness of the essential closure was first proved by Conrad; alternatively, one may refer to [BKW77, §13.4].

If X is a compact extremally disconnected space, then $D(X)$ is an ℓ-group, since in such a space every dense open set is C^*-embedded ([GJ76, 1H]), and $D(X)$ is also essentially closed. Moreover, every essentially closed ℓ-group is a $D(X)$, for a suitable compact extremally disconnected space X. Conrad also characterizes such a $D(X)$ is terms of completeness properties ([BKW77, Corollary 13.4.3]); we attend to that in brief in the comments of 4.3.4, below.

Indeed, the facts concerning the essential closure – at least, Conrad's approach to them – seem to depend on the following theorem. Many authors can claim some version of it; however, in its full generality, the theorem is most often credited to Simon Bernau ([Be65], or [BKW77, Theorem 13.4.1]). We state it here in a manner convenient to our purposes.

> *Each archimedean ℓ-group G may be essentially embedded as an ℓ-subgroup of $D(X)$, for a suitable compact extremally disconnected space X. Moreover, X is the Stone dual of the boolean algebra $\mathcal{P}G$ of polars of G.*

This representation theorem makes it is clear that each ℓ-group G has an essential closure, which is unique up to an isomorphism over G.

4.3. The category $\mathfrak{W}$ and its relations

It is virtually impossible anymore, to say nothing of how perilous it would be, to raise the subject matter of archimedean ℓ-groups, without highlighting the Yosida Embedding Theorem. And since we are about to shift the discussion to categorical questions, the appearance of the category $\mathfrak{W}$ is downright inevitable.

It is at this point too that the narrative of this article begins to acquire the personal color announced in the title. Meanwhile, the reader should be made aware of what is missing in this narrative.

The reader should know that there is a frame-theoretic perspective of the category $\mathfrak{W}$ and the Yosida Embedding Theorem which is different from the one in the pages ahead. In broad terms, one may define concepts such as *completely regular* and *Lindelöf* frames, as pointfree companions of their topological counterparts. The Yosida Embedding Theorem then has a frame-theoretic formulation; the reader will find fine expositions of it in [BH91] and [MaV90]. Both these references discuss the characterization of regular Lindelöf frames in terms of the ℓ-group $C(L)$ of frame homomorphisms of $\mathfrak{O}(\mathbb{R})$ into L.

Madden obtained a constructive proof of the aforementioned frame-theoretic Yosida Embedding Theorem in [Ma90]. (For practical purposes the reader may interpret the term *constructive* as indicating a process which is free of Choice or other axioms used to produce the existence of points or cardinal bounds. In this regard we recommend [Ba94], which contains a purely constructive account of the frame of real numbers.)

We record the definition of $\mathfrak{W}$, followed by a statement of the (classical!) Yosida Embedding Theorem. The most comprehensive reference for this version of the theorem remains [HR77].

Definition & Remarks 4.3.1. The category $\mathfrak{W}$ consists of all the archimedean ℓ-groups G together with a designated unit $u > 0$, while the morphisms are the ℓ-homomorphisms that preserve the designated unit. (Recall that $0 < u \in G$ is a *unit* of G if $u^\perp = \{0\}$.)

Associated with each $\mathfrak{W}$-object G with designated unit u, one has the *Yosida space* $Y(G)$; namely the space of all values of u, endowed with the hull-kernel topology. $Y(G)$ is a compact Hausdorff space.

For each $\mathfrak{W}$-morphism $\phi : G \longrightarrow H$, there is an induced continuous map

$$Y(\phi) : Y(H) \longrightarrow Y(G),$$

defined by

$$Y(\phi)(V) = \phi^{-1}(V),$$

for each $V \in Y(H)$.

Three additional comments are in order:

1. As defined, Y becomes a contravariant functor from $\mathfrak{W}$ to the category $\mathfrak{KT}_2$ of compact Hausdorff spaces.
2. If ϕ stands for the inclusion of the ℓ-group in the extension H, then $G \subseteq H$ is essential if and only if $Y(\phi)$ is an irreducible surjection. This fact, effectively, makes the work of Conrad and Gleason duals of one another.
3. $f : G \longrightarrow H$ preserves all existing suprema precisely when $Y(\phi)$ is a skeletal map ([BH90, Lemma 9.4]).

Without further ado we formulate the Yosida Embedding Theorem, followed by a few clarifying remarks.

Theorem 4.3.2 (Yosida Embedding Theorem)**.** *Suppose that G is an archimedean ℓ-group with designated unit u. Then there is a compact Hausdorff space Y and an ℓ-group G' in $D(Y)$, along with an ℓ-isomorphism $\Theta : G \longrightarrow G'$ onto G' which separates the points of Y and such that $\Theta(u) = 1$.*

The space Y and the map Θ are unique, in the sense that, if Y is a compact Hausdorff space and Φ is an ℓ-isomorphism onto an ℓ-group $\Phi(G)$ in $D(X)$ which separates the points of X, and such that $\Phi(u) = 1$, then there is a homeomorphism

$\tau : X \longrightarrow Y$, *such that the diagram below commutes; that is, for each* $g \in G$,

$$D(\tau)(\Theta(g)) = \Phi(g).$$

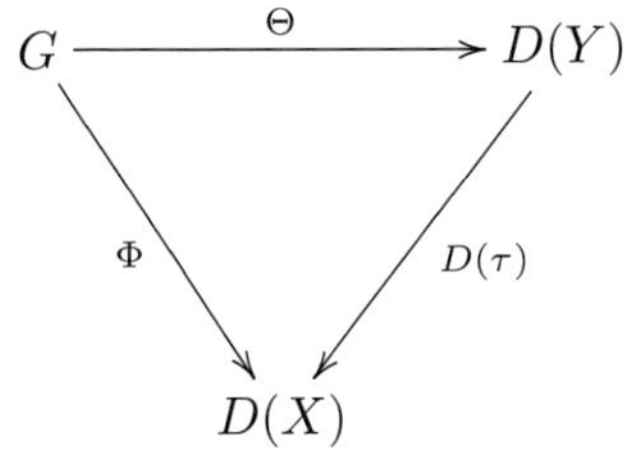

Remarks 4.3.3. (a) Let X be a Tychonoff space, and $G \subseteq D(X)$. We say that G is *an ℓ-group in* $D(X)$ if G is a lattice-ordered group, and the group operation is pointwise wherever possible, that is, for each $f, g \in G$,

$$(f+g)(x) = f(x) + g(x), \text{ for all } x \in f^{-1}(\mathbb{R}) \cap g^{-1}(\mathbb{R}).$$

In the formulation of Theorem 4.3.2 it is assumed that the ℓ-groups of extended real-valued functions have the constant function 1 as their designated units.

(b) $A \subseteq D(X)$ *separates the points of* X if for each pair $x, y \in X$ of distinct points, there is an $f \in A$ such that $f(x) \neq f(y)$. Note that if A is an ℓ-group in $D(X)$, and it separates the points, then, with $x \neq y$, we may pick $f \in A$ such that $f(x) = 0$ and $0 < f(y) < \infty$.

(c) Finally, the reader should realize that if $\tau : X \longrightarrow Y$ is a homeomorphism then the map $D(\tau) : D(Y) \longrightarrow D(X)$ defined by $D(\tau)(g) = g{\cdot}\tau$ is properly defined, and a lattice isomorphism, since $(D(\tau)(g))^{-1}(\mathbb{R}) = \tau^{-1}(g^{-1}(\mathbb{R}))$.

Now it is well known that the essential closure is not functorial in $\mathfrak{W}$, and, in particular, not a reflection in the subcategory of essentially closed objects. To make this closure a functor, one must restrict the maps. The category of choice is $\mathfrak{W}_\infty$, which has the same objects as $\mathfrak{W}$, but uses only the morphisms of $\mathfrak{W}$ which preserve all existing suprema. The dual category of compact Hausdorff spaces then, likewise, employs only the skeletal maps, in view of 4.3.1.3.

That the essential closure is a reflection of $\mathfrak{W}_\infty$ in its full subcategory of essentially closed groups is conjectured in [BH90, 9.12]. A proof, making heavy use of Yosida representation, appears in Ricardo Carrera's dissertation ([Cr04, §2.2]). Let us give a more detailed account, in view of the goal of ultimately formulating a frame-theoretic version. We quote from [Cr04], and leave it to the enterprising reader to look up the relevant categorical terms in [HS79].

Remarks 4.3.4. All categorical references in this commentary are to $\mathfrak{W}_\infty$. We follow the development of results in [Cr04].

(a) ([Cr04, Theorem 2.2.10]) *The essential closure is the maximum essential monoreflection.*

(b) ([Cr04, Proposition 4.1.1]) *The extension H of G (in $\mathfrak{W}_\infty$) is epic if and only if it is essential.* Together with (a), this immediately implies that the essential closure is the maximum monoreflection.

Carrera goes on to show that various other classes of ℓ-groups, which are not reflective in $\mathfrak{W}$, are epireflective in $\mathfrak{W}_\infty$. We summarize this next. Let α denote an uncountable cardinal or the symbol ∞. Define three classes of ℓ-groups (for each α) as follows:

- $\mathbf{P}(\alpha)$: ℓ-groups in which every polar which is generated by fewer than α elements is complemented. These are the *α-projectable groups.*
- $\mathbf{C}(\alpha)$: ℓ-groups in which every bounded set S of positive elements, with $|S| < \alpha$, has a supremum. These are the *conditionally α-complete groups.*
- $\mathbf{L}(\alpha)$: ℓ-groups in which every set S of pairwise disjoint elements, with $|S| < \alpha$, has a supremum. These are the *laterally α-complete groups.*

(In the preceding definitions, the case $\alpha = \infty$ is interpreted as placing no cardinality bounds whatsoever.) For each α, $\mathbf{L}(\alpha)$ and $\mathbf{C}(\alpha)$ are epireflective in $\mathfrak{W}_\infty$ ([Cr04, Theorem 4.2.15]), while it is known that only $\mathbf{L}(\omega_1)$ is epireflective in $\mathfrak{W}$ ([HM97, §7]). $\mathbf{P}(\alpha)$ too is epireflective in $\mathfrak{W}_\infty$ ([Cr04, Corollary 5.1.11]), whereas it not reflective in $\mathfrak{W}$, for any α.

Finally, a formulation of Conrad's characterization of the essentially closed ℓ-groups: *G is essentially closed if and only it is divisible, laterally and conditionally (∞)-complete* ([BKW77, Corollary 13.4.3]).

Remark 4.3.5. There are a variety of successful attempts in the literature to make Gleason's absolute functorial and coreflective. We refer the reader to two of these. In [DPR81] the situation is more general, since compactness is dropped; the maps of choice are the skeletal maps. In [Wo89] Woods proves that in the category of compact Hausdorff spaces with continuous "li-maps" the absolute is coreflective. However, while it is clear that the li-maps are skeletal, it is not at all obvious that the reverse is true.

4.4. Absolutes, framewise

We have taken up the spatial and algebraic preliminaries, and the time is nigh to elevate the subject to a frame-theoretic context. The task consists of "translating" Gleason's absolute to compact regular frames. The construction of the absolute may be found in [Ba88]. This section summarizes the results of [MZ06b], where the reflective properties of [Cr04] are lifted to the category $\mathfrak{KRegS}$ of all compact regular frames with all skeletal maps.

Definition & Remarks 4.4.1. Throughout this commentary A stands for an object in $\mathfrak{KRegS}$.

Suppose that D is a distributive lattice with bottom 0. $\mathrm{Idl}(D)$ denotes the frame of all ideals of D. Recall that $\mathrm{Idl}(D)$ is an algebraic frame, in which the ideals of the form $\downarrow x$ are the compact elements.

Recall that a A is a *strongly projectable* frame if every polar is complemented. If X is a compact Hausdorff space, then $\mathfrak{O}(X)$ is strongly projectable if and only if X is extremally disonnected. $\mathfrak{SPRegS}$ stands for the full subcategory of $\mathfrak{KRegS}$ whose objects are the strongly projectable frames.

We record some basic features of the ideal frame of a boolean algebra B; see [MZ06b, §3]. Define $\bigvee_B : \mathrm{Idl}(B) \longrightarrow B$ by $\bigvee_B(J) = \bigvee J$.

1. $\mathrm{Idl}(B)$ *is a compact regular algebraic frame in which each compact element is complemented.*
2. $B \cong \mathfrak{k}(\mathrm{Idl}(B))$, *via the isomorphism* $\bigvee_B$.
3. *Conversely, if L is a compact regular frame which is generated by $\mathfrak{k}(L)$, then each compact element is complemented, and $\mathfrak{k}(L)$ is a boolean algebra. Moreover, the map*
$$x \mapsto \{\, a \le x : a \in \mathfrak{k}(L) \,\},$$
for $x \in L$, is an isomorphism of frames.
4. $\mathrm{Idl}(B)$ *is strongly projectable if and only if B is complete. If this is the case, then* $\mathcal{P}(\mathrm{Idl}(B)) = \mathfrak{k}(\mathrm{Idl}(B))$.

Define εA to be $\mathrm{Idl}(\mathcal{P}A)$; this is the *absolute* of A. We record [MZ06b, Proposition 3.4].

Suppose that A is a compact regular frame. Then there is a skeletal frame embedding $\varepsilon_A : A \longrightarrow \varepsilon A$, so that the following diagram commutes:

(4.4.1.1)

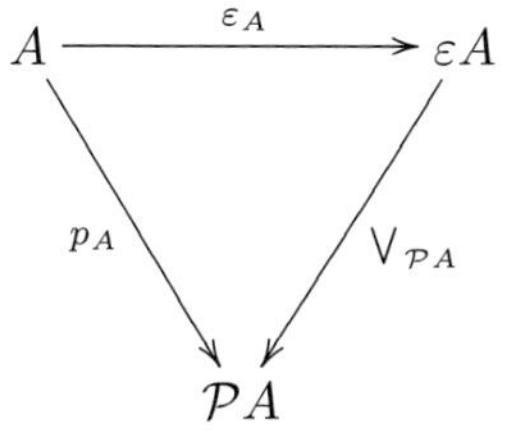

In fact, $\varepsilon_A(x) = \{\, a \in \mathcal{P}A : a \preceq x^{\perp\perp} \,\}$.

We recite the main results of [MZ06b].

Theorem 4.4.2 ([MZ06b, Theorem 4.3]). *Let $h : A \longrightarrow B$ be a $\mathfrak{KRegS}$-morphism. Then there is a unique skeletal frame map $\varepsilon(h) : \varepsilon A \longrightarrow \varepsilon B$ making the rectangles in which it lies, in the prism below, commutative:*

(4.4.2.1)

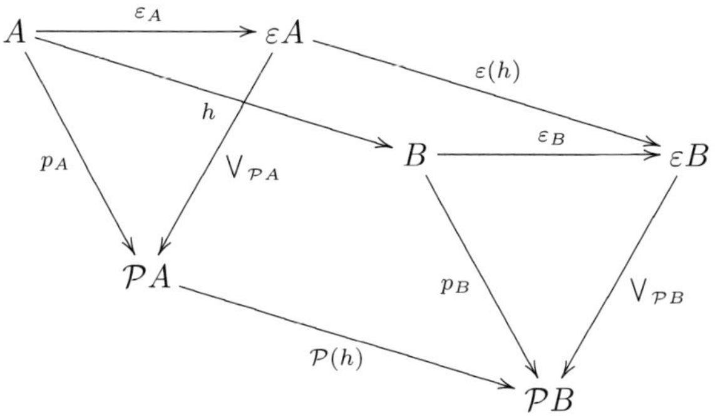

Thus, $\varepsilon : \mathfrak{KRegS} \longrightarrow \mathfrak{SPRegS}$ *defines a monoreflection.*

Definition & Remarks 4.4.3. A morphism $e : A \longrightarrow B$ which is both monic and epic will be called an *epic extension*. An object E is *epicomplete* if every epic extension $m : E \longrightarrow F$ is an isomorphism. $E(\mathfrak{B})$ will stand for the full subcategory of all epicomplete objects of the category $\mathfrak{B}$.

Evidently, $E(\mathfrak{B})$ is contained in each monoreflective subcategory of $\mathfrak{B}$. Frequently, as is the case in $\mathfrak{KRegS}$, $E(\mathfrak{B})$ is the least monoreflective subcategory. However, there are examples in which $E(\mathfrak{B})$ is not even epireflective.

Theorem 4.4.4 ([MZ06b, Theorem 5.5]). $E(\mathfrak{KRegS}) = \mathfrak{SPRegS}$.

Remark 4.4.5. Spurred on by the results quoted in 4.3.4, (a) and (b), one would reasonably conjecture that ε ought to be an *essential* reflection, in the sense that if $h : A \longrightarrow B$ is monic in $\mathfrak{KRegS}$, then $\varepsilon(h)$ is also monic. The trouble is that one cannot seem to decide what the monomorphisms are in this category. It is unknown whether monics in $\mathfrak{KRegS}$ are dense; if this is true, then, without too much trouble, one can proceed to show that monics are, in fact, one-to-one, and that ε is essential.

What is known is the following; the reader is referred to [Ba88]. Now, ε has the ostensibly weaker property that if $h : A \longrightarrow B$ is one-to-one, then the extension $\varepsilon(h)$ is also one-to-one. More generally, let us say, for purposes of this remark, that a frame embedding $m : A \longrightarrow B$ is *large* if for each $0 < b \in B$, there exists an $a > 0$ in A such that $m(a) \leq b$. Banaschewski shows in [Ba88] that an embedding $m : A \longrightarrow B$ of compact regular frames is large if and only if g is one-to-one whenever $g \cdot m$ is one-to-one, for each frame map $g : B \longrightarrow C$ of compact regular frames. Further, a compact regular frame A is strongly projectable if and only if A has no proper large extensions ([Ba88, Proposition 2]).

All of which yields the following observation, which parallels 4.3.4(b).

Proposition 4.4.6. *Suppose that $m : A \longrightarrow B$ is a one-to-one $\mathfrak{KRegS}$-morphism. Then m is an epic extension if and only if it is large.*

Proof. Observe that a composition of frame extensions $m_2 \cdot m_1$ is large if and only if both m_1 and m_2 are large. Thus, since $\varepsilon_A : A \longrightarrow \varepsilon A$ is large, we have that that if m is large then so is $\varepsilon(m)$. Then, by the foregoing remarks, $\varepsilon(m)$ is an isomorphism. Therefore m may be viewed as a first factor. Since ε_A is epic, by a routine argument one is able to show that m too is epic.

Conversely, suppose m is an epic extension. Then, as a second factor of an epimorphism, $\varepsilon(m)$ is also an epic extension. Then, by [MZ06b, Lemma 5.4], $\varepsilon(m)$ is an isomorphism, whence m is large. □

4.5. Postscript

The classical work of Conrad described in §4.2 begs to be carried over to the level of frames, and along the lines suggested by Carrera's contribution. The goal in this closing section is to give the reader an idea of the progress that has been made; the references we give are to ongoing work, in [M06c] and [MZ07].

The reader will perhaps already have guessed that the category where the proper discourse will eventually take place should consist of archimedean frames and frame maps which are both skeletal and coherent. We refer the reader back to 3.1.4(b) for the notion of an archimedean frame, and recall that the notion is abstracted from archimedean ℓ-groups. The coherent frame homomorphisms between algebraic frames are the ones mapping compact elements to compact elements. Moreover, since the motivation for this work comes from ℓ-groups and f-rings, one might find the hypothesis of normality desirable.

This research has, in fact, prompted a second look at archimedean frames, *per sé*, which is the major focus of [M06c]. We give an account of the elementary ideas, in part because it will increase the desirability of the assumption of normality.

Definition 4.5.1. Let A be an arbitrary frame. If for each $0 < a \leq b \in A$, there exists a $c \in A$, with $c < b$, such that $b = a \vee c$, we say that A is *joinfit*. If A is algebraic and this condition holds with a, b and c compact, then we say that A is *finitely* joinfit.

It is straightforward that a joinfit algebraic frame is finitely joinfit. The distributive law also allows for the following simplification: for A to be joinfit, it suffices that the above definition be satisfied for $b = 1$. Then it is clear that if A is compact and finitely joinfit, it is also joinfit.

The following proposition summarizes the relationship between joinfitness and the archimedean condition; it will appear in [M06c]. We remind the reader that ϱ stands for the regular coreflection.

Proposition 4.5.2. *Suppose A is an algebraic frame. Regarding the conditions below,*

- (a) *implies* (b), *and the reverse is true with the Axiom of Choice.*
- (b) *if and only if* (d), *if A is normal.*
- *If A is a compact algebraic normal frame they are all equivalent, with Choice.*

(a) *A is archimedean.*
(b) *A is joinfit.*
(c) *A is finitely joinfit.*
(d) *ϱA is large in A.*

Condition (d) in the proposition appears to be key. Note that there are examples of joinfit frames which are not normal, for which this condition fails.

Of note and consequence is the following; the reader will observe the role of normality.

Proposition 4.5.3. *Suppose that A is a compact normal frame. Then $(\varrho_A)_*$ is a frame homomorphism.*

Using the absolute of section §4.4 one is able to construct an epireflection $\widehat{\varepsilon}$ of $\mathfrak{KNArS}$, the category of compact normal joinfit frames in the full subcategory $\mathfrak{SPArS}$ of all strongly projectable frames. We give an account of what is known about $\widehat{\varepsilon}$ in the following commentary ([MZ07]).

Remarks 4.5.4. In the sequel A denotes a $\mathfrak{KNArS}$-object. $\widehat{\varepsilon}A$ is constructed via the pushout in $\mathfrak{Frm}$, the category of all frames and frame maps:

(4.5.4.1)
$$\begin{array}{ccc} \varrho A & \xrightarrow{\varepsilon_{\varrho A}} & \varepsilon(\varrho A) \\ {\scriptstyle \varrho_A}\downarrow & & \downarrow{\scriptstyle \theta} \\ A & \xrightarrow{\widehat{\varepsilon}_A} & \widehat{\varepsilon}A \end{array}$$

We observe the following about the pushout:

1. The maps in the pushout are skeletal; in fact,
$$\varrho(\widehat{\varepsilon}A) \cong \varepsilon(\varrho A),$$
canonically, and $\theta \cong \varrho_{\widehat{\varepsilon}A}$.
2. The functor $\mathcal{P}$ – as discussed in 4.1.5 – induces isomorphisms throughout in Figure (4.5.4.1).
3. By Proposition 4.5.3, ϱ_A, and therefore also θ, by properties of pushouts, are sections ([HS79]).
4. By properties of pushouts, and because ϱ and ε are coreflections and reflections, respectively, if $h : A \longrightarrow B$ is skeletal, then there is a unique $\widehat{\varepsilon}(h) : \widehat{\varepsilon}A \longrightarrow \widehat{\varepsilon}B$ making the diagram below commute – and $\widehat{\varepsilon}(h)$ too is skeletal.

(4.5.4.2)
$$\begin{array}{ccc} A & \xrightarrow{\widehat{\varepsilon}_A} & \widehat{\varepsilon}A \\ {\scriptstyle h}\downarrow & & \downarrow{\scriptstyle \widehat{\varepsilon}(h)} \\ B & \xrightarrow{\widehat{\varepsilon}_B} & \widehat{\varepsilon}B \end{array}$$

   In particular, $\widehat{\varepsilon}$ defines a covariant functor.
5. If A is strongly projectable then so is ϱA, and it follows easily that $\widehat{\varepsilon}_A$ is an isomorphism. In general, it is shown that $\widehat{\varepsilon}A$ is strongly projectable, and it follows that $\widehat{\varepsilon}$ is an epireflection.

This almost suffices to show that $\widehat{\varepsilon}$ epireflects $\mathfrak{KNArS}$ in $\mathfrak{SPArS}$. The missing details are the compactess, normality, and joinfitness of $\widehat{\varepsilon}A$, all of which stem from the fact that the pushout is a coequalizer of a coproduct. And the relevance of these observations is:

- that the coequalizer taken is of a pair of frame maps having a regular source (ϱA) – see [PT01, Chapter 2, Theorem 4.6], and
- that the coproduct involved, that of A and $\varepsilon(\varrho A)$, has compact, normal and joinfit factors, and that these properties are preserved by coproducts, and by a coequalizer of a pair of skeletal frame maps having a regular source.

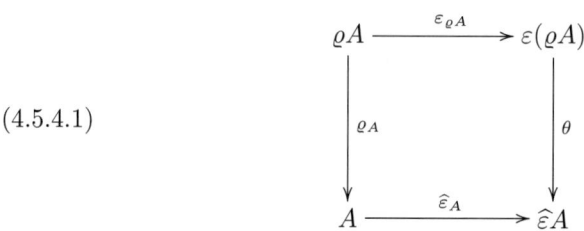

Similar arguments apply to prove the coherence of $\widehat{\varepsilon}A$ from the assumption of that property on A. The coherence of the maps $\widehat{\varepsilon}_A$ and θ, likewise, follows from properties of the pushout.

And that is, more or less, where things stand. All that's left to do is summarize what we don't know – yet:

- Is $\widehat{\varepsilon}$ a monoreflection? Or, what is an equivalent question: *Does every compact, normal joinfit frame embed skeletally in one which is also strongly projectable?*
  This question is made more mysterious by the fact that we do not know what the skeletal monomorphisms are between normal joinfit frames. Unlike the issue with monics in $\mathfrak{KRegS}$ – see 4.4.5 – in $\mathfrak{KNArS}$ monics are dense, but dense maps need not be one-to-one in this category.
- It is not difficult to show that $\mathfrak{SPRegS} \subseteq E(\mathfrak{KNArS})$, the class of epicomplete $\mathfrak{KNArS}$-objects (4.4.3); but we do not know whether equality holds. It is known that $\mathfrak{SPRegS}$ is epi- but not monoreflective.

References

[AB91] M.E. Adams & R. Beazer, *Congruence properties of distributive double p-algebras.* Czech. Math. Jour. **41** (1991), 395–404.

[BH90] R.N. Ball & A.W. Hager, *Epicompletion of archimedean ℓ-groups and vector lattices with weak unit.* Jour. Austral. Math. Soc. (Series A) **48** (1990), 25–56.

[BH91] R.N. Ball & A.W. Hager, *On the localic Yosida representation of an archimedean lattice-ordered group with weak unit.* Jour. of Pure & Appl. Alg. **70** (1991), 17–43.

[Ba88] B. Banaschewski, *Compact regular frames and the Sikorski theorem.* Kyungpook Math. Jour. **28** No. 1 (1988), 1–14.

[Ba94] B. Banaschewski, *The real numbers in pointfree topology.* Textos de Matematica (Série B) **No. 12** (1994) Univ. de Coimbra.

[BaP94] B. Banaschewski & A. Pultr, *Variants of openness.* Appl. Cat. Struct. **2** (1994), 331–350.

[BaP96] B. Banaschewski & A. Pultr, *Booleanization.* Cah. de Top. et Géom. Diff. Cat., **XXXVII-1** (1996), 41–60.

[Be65] S.J. Bernau, *Unique representation of archimedean lattice groups and normal archimedean lattice rings.* Proc. London Math. Soc. **(3) 15** (1965), 599–631.

[BKW77] A. Bigard, K. Keimel & S. Wolfenstein, *Groupes et Anneaux Réticulés.* Lecture Notes in Math. **608** (1977), Springer Verlag, Berlin-Heidelberg-New York.

[Cr04] R.E. Carrera, *Operators on the Category of Archimedean Lattice-Ordered Groups with Designated Weak Unit.* Doctoral Dissertation (2004), Univ. of Florida, Gainesville.

[C65] P.F. Conrad, *The lattice of all convex ℓ-subgroups of a lattice-ordered group.* Czech. Math. Jour. **15 (90)** (1965), 101–123.

[C71] P.F. Conrad, *The essential closure of an archimedean lattice-ordered group.* Duke Math. Jour. **38** (1971), 151–160.

[CM90] P. Conrad & J. Martinez, *Complemented lattice-ordered groups.* Indag. Math. (N.S.) **1** (1990), 281–297.

[CL02] Th. Coquand & H. Lombardi, *Hidden constructions in abstract algebra (3): Krull dimension of distributive lattices and commutative rings.* In *Commutative Ring Theory and Applications* (M. Fontana, S.-E. Kabbaj & S. Wiegand, Eds.) **131** M. Dekker (2002), 477–499.

[CLR03] Th. Coquand, H. Lombardi & M.-F. Roy, *Une caractérisation élémentaire de la dimension de Krull.* Preprint.

[D95] M.R. Darnel, *The Theory of Lattice-Ordered Groups.* Pure and Applied Math. **187**; Marcel Dekker, Basel-Hong Kong-New York.

[DPR81] R.F. Dickman, J.R. Porter & L.R. Rubin, *Completely regular absolutes and projective objects.* Pacific Jour. Math. **94**, No. 2 (1981), 277–295.

[GJ76] L. Gillman & M. Jerison, *Rings of Continuous Functions.* Grad. Texts Math. **43** Springer Verlag (1976), Berlin-Heidelberg-New York.

[H89] A.W. Hager, *Minimal covers of topological spaces.* In Papers on General Topology and Related Category Theory and Topological Algebra; Annals of the N.Y. Acad. Sci. **552** March, 15, 1989, 44–59.

[HM97] A.W. Hager & J. Martínez, *The laterally σ-complete reflection of an archimedean lattice-ordered group.* In *Ordered Algebraic Structures,* Curaçao (1995), W.C. Holland & J. Martínez, Eds.; (1997) Kluwer, Dordrecht, 217–236.

[HR77] A.W. Hager & L.C. Robertson, *Representing and ringifying a Riesz space.* Symp. Math. **27** (1977), 411–431.

[HMW03] M. Henriksen, J. Martínez & R.G. Woods, *Spaces X in which all prime z-ideals of $C(X)$ are either minimal or maximal.* Comm. Math. Univ. Carol. **44** (2) (2003), 261–294.

[HW04] M. Henriksen & R.G. Woods, *Cozero complemented spaces: when the space of minimal prime ideals of a $C(X)$ is compact.* To appear, Topol. & Appl.

[HS68] H. Herrlich & G. Strecker, *H-closed spaces and reflective subcategories.* Math. Annalen **177** (1968), 302–309.

[HS79] H. Herrlich & G. Strecker, *Category Theory.* Sigma Series Pure Math. **1** (1979), Heldermann Verlag, Berlin.

[HdP80a] C.B. Huijsmans & B. de Pagter, *On z-ideals and d-ideals in Riesz spaces, I.* Indag. Math. **42** (1980), 183–195.

[HdP80b] C.B. Huijsmans & B. de Pagter, *On z-ideals and d-ideals in Riesz spaces, II.* Indag. Math. **42** (1980), 391–408.

[J82] P.J. Johnstone, *Stone Spaces.* Cambridge Studies in Adv. Math, **3** (1982), Cambridge Univ. Press.

[Ma90] J.J. Madden, *Frames associated with an abelian ℓ-group.* Trans. AMS **331** (No. 1) (1992), 265–279.

[MaV90] J.J. Madden & J. Vermeer, *Epicomplete archimedean ℓ-groups via a localic Yosida theorem.* Jour. of Pure & Appl. Alg. **68** (1990), 243–252.

[M72] J. Martínez, *Unique factorization in partially ordered sets.* Proc. AMS **33** (No. 2), (June 1972), 213–220.

[M73] J. Martínez, *Archimedean lattices.* Alg. Univ. **3** (fasc. 2) (1973), 247–260.

[M06a] J. Martínez, *Dimension in algebraic frames.* Czech. Jour. Math. **50 (131)** (2006), 437–474.

[M06b] J. Martínez, *A theorem of Conrad, framed, without Choice.* Submitted.

[M06c] J. Martínez, *Disjointifiable ℓ-groups.* Submitted.

[M06c] J. Martínez, *Archimedean frames, revisited.* Submitted.

[MZ03] J. Martínez & E.R. Zenk, *When an algebraic frame is regular.* Alg. Univ. **50** (2003), 231–257.

[MZ05] J. Martínez & E.R. Zenk, *Dimension in algebraic frames, II: applications to frames of ideals in $C(X)$.* Comm. Math. Univ. Carol. **46, 4** (2005), 607–636.

[MZ06a] J. Martínez & E.R. Zenk, *Regularity in algebraic frames.* To appear, Jour. of Pure & Appl. Alg.

[MZ06b] J. Martínez & E.R. Zenk, *Epicompletion in frames with skeletal maps, I: compact reular frames.* Submitted.

[MZ07] J. Martínez & E.R. Zenk, *Epicompletion in frames with skeletal maps, II: compact normal archimedean frames.* Work in progress.

[Pa81] B. de Pagter, *On z-ideals and d-ideals in Riesz spaces, III.* Indag. Math. **43**, Fasc. 4 (1981), 409–422.

[PT01] M.C. Pedicchio & W. Tholen, *Special Topics in Order, Topology, Algebra and Sheaf Theory.* Cambridge Univ. Press (2001), Cambrdge, UK.

[PW89] J.R. Porter & R.G. Woods, *Estensions and Absolutes of Hausdorff Spaces.* Springer (1989), Berlin-Heidelberg-New York.

[Se59] Z. Semadeni, *Sur les ensembles clairsemés.* Rozprawy Matem. **19** (1959), Warsaw.

[Se71] Z. Semadeni, *Banach Spaces of Continuous Functions.* Polish Scient. Publ. (1971), Warsaw.

[ST93] J.T. Snodgrass & C. Tsinakis, *Finite-valued algebraic lattices.* Alg. Univ. **30** (1993), 311–318.

[Wa74] R.C. Walker, *The Stone-Čech Compactification.* Ergeb. der Math. und ihrer Grenzgeb. **83** (1974), Springer, Berlin-Heidelberg-New York.

[Wo89] R.G. Woods, *Covering properties and coreflective subcategories.* Papers on Gen. Topology and Rel. Category Th. and Top. Alg. **552** Annals NY Acad. of Sci. (1989).

Jorge Martínez
Department of Mathematics
University of Florida
Gainesville, Fl 32611-8105, USA
e-mail: jmartine@math.ufl.edu

Positivity
Trends in Mathematics, 197–227

Non-commutative Banach Function Spaces

Ben de Pagter

1. Introduction

In this paper we survey some aspects of the theory of *non-commutative Banach function spaces*, that is, spaces of measurable operators associated with a semi-finite von Neumann algebra. These spaces are also known as *non-commutative symmetric spaces.* The theory of such spaces emerged as a common generalization of the theory of classical ("commutative") rearrangement invariant Banach function spaces (in the sense of W.A.J. Luxemburg and A.C. Zaanen) and of the theory of symmetrically normed ideals of bounded linear operators in Hilbert space (in the sense of I.C. Gohberg and M.G. Krein). These two cases may be considered as the two extremes of the theory: in the first case the underlying von Neumann algebra is the commutative algebra L_∞ on some measure space (with integration as trace); in the second case the underlying von Neumann algebra is $B(\mathcal{H})$, the algebra of all bounded linear operators on a Hilbert space $\mathcal{H}$ (with standard trace). Important special cases of these non-commutative spaces are the non-commutative L_p-spaces, which correspond in the commutative case with the usual L_p-spaces on a measure space, and in the setting of symmetrically normed operator ideals they correspond to the Schatten p-classes $\mathfrak{S}_p$.

In the present paper we take the Banach function spaces as our point of departure. As will become clear, there are many results in the general theory which are direct analogues of the corresponding results in the classical theory. But, we hasten to say that the proofs are quite different in most cases (partly due to the lack of lattice structure in the non-commutative situation). However, there are also many instances where the non-commutative situation essentially differs from the commutative setting (this is in particular illustrated by some of the results in Section 8, concerning the continuity of so-called operator functions).

Definitions and results are stated in detail, but most of the proofs are omitted (with references to the relevant literature). Some proofs have been included, in particular of results which have not yet appeared in print, as well as some relatively short arguments.

In Section 2 we review some of the basic features of the classical (that is, commutative) Banach function spaces associated with a measure space. We assume that the reader is familiar with the terminology of the theory of Riesz spaces and Banach lattices (as may be found in, e.g., [1], [40]). In particular we discuss rearrangement invariant spaces and the so-called Köthe duality. In Section 3 we review some basic facts concerning von Neumann algebras and in Section 4 we discuss in some detail the τ-measurable operators (associated with a semi-finite von Neumann algebra $\mathcal{M}$ equipped with trace τ). Particular attention will be given to the properties of the order structure of the (self-adjoint part of the) space $S(\tau)$ of all such τ-measurable operators. The measure topology on the space $S(\tau)$ is introduced and its properties are discussed in Section 5. Again we will digress somewhat on the interplay between the order structure and the topology.

An important role in the theory of non-commutative spaces is played by the *generalized singular value function* (of a τ-measurable operator). In the commutative theory this corresponds to the so-called decreasing rearrangement of a function and, in the setting of compact operators in Hilbert space, to the sequence of singular values of an operator. The properties of the generalized singular value function are discussed in Section 6.

Non-commutative Banach function spaces are defined in Section 7 and some of their basic properties are reviewed and some aspects of the duality theory are discussed (in particular, trace duality and the Köthe dual). We end the paper with a brief introduction to the study of so-called operator functions in Section 8.

2. Banach function spaces

Let (X, Σ, ν) be a measure space. We always assume that (X, Σ, ν) is *Maharam*, that is, it has the *finite subset property* (i.e., for every $A \in \Sigma$ with $\nu(A) > 0$ there exists $B \in \Sigma$ such that $B \subseteq A$ and $0 < \nu(B) < \infty$) and is *localizable* (i.e., the measure algebra is a complete Boolean algebra; recall that the *measure algebra* of (X, Σ, ν) is obtained from Σ by identifying sets which are ν-almost equal). Note that any σ-finite measure space is a Maharam measure space.

The complex Riesz space of all complex-valued measurable Σ-measurable functions on X (with identification of ν-a.e. equal functions) is denoted by $L_0(\nu)$. Since we assume (X, Σ, ν) to be Maharam, $L_0(\nu)$ is Dedekind complete.

Definition 2.1. *A Banach function space on (X, Σ, ν) is an ideal $E \subseteq L_0(\nu)$ (that is, E is a linear subspace of $L_0(\nu)$ with the additional property that $f \in L_0(\nu)$, $g \in E$ and $|f| \leq |g|$ imply $f \in E$) equipped with a norm $\|\cdot\|_E$ such that $(E, \|\cdot\|_E)$ is a Banach lattice.*

Evidently, any Banach function space is Dedekind complete. We may, and shall, always assume that the *carrier* of E is equal to X (that is, for every $A \in \Sigma$ with $\nu(A) > 0$ there exists $B \in \Sigma$ such that $B \subseteq A$, $\nu(B) > 0$ and $\chi_B \in E$). Examples of Banach function spaces are the L_p-spaces ($1 \leq p \leq \infty$), *Orlicz spaces, Lorentz spaces*, and *Marcinkiewicz spaces.* A concise introduction in the theory of

Banach function spaces can be found in Chapter 15 of the book [39] (as in most of the literature on Banach function spaces, the underlying measure space is assumed to be σ-finite; for a treatment in the setting of more general measure spaces, we refer the reader to [17]). In this paper we will be interested mainly in a special class of Banach function spaces, the so-called rearrangement invariant Banach function spaces.

2.1. Rearrangements

For $f \in L_0(\nu)$ its *distribution function* $d_f : [0,\infty) \to [0,\infty]$ is defined by

$$d_f(\lambda) = \nu(\{x \in X : |f(x)| > \lambda\}), \quad \lambda \geq 0.$$

Note that d_f is decreasing and right-continuous. We are interested only in those functions $f \in L_0(\nu)$ for which there exists $\lambda_0 \geq 0$ such that $d_f(\lambda_0) < \infty$, which implies that $\lim_{\lambda\to\infty} d_f(\lambda) = 0$. We define

$$S(\nu) = \{f \in L_0(\nu) : \exists\, \lambda_0 \geq 0 \text{ s.t. } d_f(\lambda_0) < \infty\}. \tag{1}$$

If $f \in L_0(\nu)$, then $f \in S(\nu)$ if and only if f is bounded except on a set of finite measure. Evidently, $S(\nu)$ is an ideal in $L_0(\nu)$.

For $f \in S(\nu)$ the *decreasing rearrangement* $\mu(f) : [0,\infty) \to [0,\infty]$ of $|f|$ is defined by

$$\mu(f;t) = \inf\{\lambda \geq 0 : d_f(\lambda) \leq t\}, \quad t \geq 0.$$

Observe the following simple properties of the function $\mu(f)$.

Lemma 2.2. *If $f \in S(\nu)$, then*

(i) *$\mu(f;t) < \infty$ for all $t > 0$;*
(ii) *$\mu(f)$ is decreasing and right-continuous;*
(iii) *$\mu(f;0) < \infty$ if and only if $f \in L_\infty(\nu)$ and in this case $\mu(f;0) = \|f\|_\infty$;*
(iv) *the functions f and $\mu(f)$ have the same distribution function, that is, $d_{\mu(f)} = d_f$ on $[0,\infty)$ (where $d_{\mu(f)}$ is computed with respect to Lebesgue measure on $[0,\infty)$).*

For a detailed account of the properties of decreasing rearrangements of functions we refer the reader to the books [3] and [24]. The decreasing rearrangement of $|f|$ is frequently denoted by f^*. However, in the setting of the present paper we prefer the notation $\mu(f)$, in particular since the $*$ will be used later on to indicate the adjoints of Hilbert space operators.

Furthermore, we note that

$$\int_X |f|\, d\nu = \int_0^\infty \mu(f;t)\, dt$$

for all $f \in S(\nu)$. If $f \in S(\nu)$ and if $\varphi : [0,\infty) \to [0,\infty)$ is *continuous and increasing*, then $\mu(\varphi \circ |f|) = \varphi \circ \mu(|f|)$, which implies in particular that $\mu(|f|^p) = \mu(f)^p$, $1 \leq p < \infty$. Consequently,

$$\int_X |f|^p\, d\nu = \int_0^\infty \mu(f;t)^p\, dt, \quad 1 \leq p < \infty, \quad \text{for all } f \in S(\nu). \tag{2}$$

2.2. Rearrangement invariant Banach function spaces

Let E be a Banach function space on the Maharam measure space (X, Σ, ν).

Definition 2.3. *The Banach function space* $E \subseteq S(\nu)$ *is called rearrangement invariant if* $f \in E$, $g \in S(\nu)$ *and* $\mu(g) = \mu(f)$ *imply that* $g \in E$ *and* $\|g\|_E = \|f\|_E$.

Rearrangement invariant spaces are discussed in, e.g., [3], [24] or [27] (however, the results in [3] should be handled with some care, as the class of Banach function spaces considered is more restrictive: the so-called Fatou property is included in their definition of a Banach function space!). For rearrangement invariant function spaces on non-σ-finite measure spaces, see [17].

It follows from (2) and Lemma 2.2, (iii) that L_p-spaces are rearrangement invariant Banach function spaces. Other examples are Orlicz spaces, Lorentz spaces and Marcinkiewicz spaces.

The following two examples are of particular interest. The space

$$(L_1 \cap L_\infty)(\nu) = L_1(\nu) \cap L_\infty(\nu)$$

equipped with the norm given by

$$\|f\|_{L_1 \cap L_\infty} = \max\left(\|f\|_1, \|f\|_\infty\right)$$

is a rearrangement Banach function space. An alternative expression for the norm is given by

$$\|f\|_{L_1 \cap L_\infty} = \sup_{t>0} \frac{1}{\min(t,1)} \int_0^t \mu(f;s)\, ds.$$

The other example is the space

$$(L_1 + L_\infty)(\nu) = L_1(\nu) + L_\infty(\nu),$$

where the norm is defined by

$$\|f\|_{L_1 + L_\infty} = \inf\left\{\|g\|_1 + \|h\|_\infty : f = g + h, g \in L_1(\nu), h \in L_\infty(\nu)\right\}.$$

This norm is also given by

$$\|f\|_{L_1 + L_\infty} = \int_0^1 \mu(f;s)\, ds, \quad f \in (L_1 + L_\infty)(\nu).$$

If E is an rearrangement invariant Banach function space on $(0, \infty)$ (with respect to Lebesgue measure), then

$$(L_1 \cap L_\infty)(0,\infty) \subseteq E \subseteq (L_1 + L_\infty)(0,\infty), \tag{3}$$

with continuous embeddings (see, e.g., [24], Theorem II.4.1). Actually, these inclusions hold whenever the measure space (X, Σ, ν) is non-atomic or, is atomic with all atoms having equal measure.

2.3. Submajorization

Most of the classical rearrangement invariant Banach function spaces satisfy a stronger condition than just being rearrangement invariant. To discuss this stronger property we introduce the following notion. As before, (X, Σ, ν) is a Maharam measure space.

Definition 2.4. *Given $f, g \in S(\nu)$ we say that f is submajorized by g (in the sense of Hardy, Littlewood and Polya), denoted by*

$$f \prec\prec g,$$

if

$$\int_0^t \mu(f; s)\, ds \leq \int_0^t \mu(g; s)\, ds, \quad t \geq 0.$$

Definition 2.5. *A Banach function space $E \subseteq S(\nu)$ is called symmetric if it satisfies the following three conditions:*

(a) *E is rearrangement invariant;*
(b) *$L_1 \cap L_\infty(\nu) \subseteq E \subseteq (L_1 + L_\infty)(\nu)$ with continuous embeddings;*
(c) *if $f, g \in E$ and $f \prec\prec g$ then $\|f\|_E \leq \|g\|_E$.*

As we have already observed before, any rearrangement invariant Banach function space on the interval $(0, \infty)$ automatically satisfies condition (b) of the above definition. The following theorem exhibits a large class of symmetric Banach function spaces. Recall that the norm $\|\cdot\|_E$ on a Banach function space E is called a *Fatou norm* if $0 \leq f_\alpha \uparrow f \in E$ implies that $\|f_\alpha\|_E \uparrow \|f\|_E$.

Theorem 2.6 (Luxemburg). ([27]) *If E is a rearrangement invariant Banach function space on $(0, \infty)$ with a Fatou norm, then E is a symmetric Banach function space.*

For simplicity, we have formulated the above result only for the measure space $(0, \infty)$. Actually, this result holds for any measure space which is either non-atomic or is atomic with all atoms having equal measure. However, the result of the theorem is not valid for any measure space, as is illustrated by the following simple example.

Example 2.7. *Let $X = \{1, 2\}$ and define the measure ν by $\nu(\{1\}) = 2$ and $\nu(\{2\}) = 1$. For E we take $\mathbb{C}^X = \mathbb{C}^2$, equipped with the norm given by $\|(f_1, f_2)\|_E = |f_1| + |f_2|$.*

The importance of the class of symmetric Banach function spaces is already indicated by the following result: any symmetric Banach function space on $(0, \infty)$ has offspring on every Maharam measure space. For convenience, we denote a Banach function space E on the interval $(0, \infty)$ explicitly by $E(0, \infty)$. The following theorem has been obtained by W.A.J. Luxemburg ([27]) under the assumption that the norm on $E(0, \infty)$ is Fatou.

Theorem 2.8. *Let $E(0,\infty)$ be a symmetric Banach function space on $(0,\infty)$ and let (X,Σ,ν) be a Maharam measure space. If we define*

$$E(\nu) = \{f \in S(\nu) : \mu(f) \in E(0,\infty)\}$$

and

$$\|f\|_{E(\nu)} = \|\mu(f)\|_{E(0,\infty)}, \quad f \in E(\nu),$$

then $\left(E(\nu), \|\cdot\|_{E(\nu)}\right)$ is a symmetric Banach function space on (X,Σ,ν).

Another important property which is stronger than symmetry is presented in the next definition.

Definition 2.9. *A Banach function space $E \subseteq S(\nu)$ is called fully symmetric if it follows from $f \in S(\nu)$, $g \in E$ and $f \prec\prec g$ that $f \in E$ and $\|f\|_E \leq \|g\|_E$.*

It is easily verified that any fully symmetric Banach function space is symmetric in the sense of Definition 2.5, but not conversely. As was shown by A.P. Calderón ([6]), the fully symmetric Banach function spaces are precisely the *exact* (L_1, L_∞)*-interpolation spaces* (cf. also [3], Chapter 5). In connection with Theorem 2.6, we mention that any rearrangement invariant Banach function space on $(0,\infty)$ with the *Fatou property* (that is, $0 \leq f_\alpha \uparrow f$ in $L_0(0,\infty)$, $f_\alpha \in E$ and $\sup_\alpha \|f_\alpha\|_E < \infty$ imply that $f \in E$ and $\|f_\alpha\|_E \uparrow \|f\|_E$) is fully symmetric, as was shown by Luxemburg ([27]).

2.4. Köthe duality

Next we discuss some aspects of the duality theory for Banach function spaces. Given a Banach function space E on a Maharam measure space (X,Σ,ν), the *Köthe dual space* $E^\times$ of E is defined by

$$E^\times = \left\{ g \in L_0(\nu) : \int_X |fg|\, d\nu < \infty \ \forall f \in E \right\}.$$

Evidently, $E^\times$ is an ideal in $L_0(\nu)$ and it can be shown that the carrier of $E^\times$ is equal to X. For $g \in E^\times$ we define the linear functional $\varphi_g : E \to \mathbb{C}$ by

$$\varphi_g(f) = \int_X fg d\nu, \quad f \in E.$$

The functional φ_g is bounded, that is, $\varphi_g \in E^*$ and the map $g \longmapsto \varphi_g$ is linear and injective. Hence, we may identify $E^\times$ with a subspace of E^*. If we define

$$\|g\|_{E^\times} = \|\varphi_g\|_{E^*} = \sup\left\{ \left| \int_X fg d\nu \right| : f \in E, \|f\|_E \leq 1 \right\}$$

for all $g \in E^\times$, then $(E^\times, \|\cdot\|_{E^\times})$ is a Banach function space on (X,Σ,ν).

Denoting by E_n^* the band in E^* consisting of all *order continuous* (or, *normal*) functionals on E, the following result shows the importance of the Köthe dual space (see, e.g., [39], Chapter 15).

Theorem 2.10. *If E is a Banach function space, then*

$$E_n^* = \{\varphi_g : g \in E^\times\}$$

In particular, the norm on E is order continuous if and only if

$$E^* = \{\varphi_g : g \in E^\times\}.$$

If E is a rearrangement invariant Banach function space and if the measure space (X, Σ, ν) is either non-atomic or is atomic with all atoms having equal measure, then it can be shown that $(E^\times, \|\cdot\|_{E^\times})$ is also a rearrangement invariant (and, actually, fully symmetric) Banach function space (cf. [3], Section 2.4). For general measure spaces the following result may be obtained.

Theorem 2.11. *If E is a symmetric Banach function space, then $E^\times$ is a fully symmetric Banach function space.*

3. Von Neumann algebras

In this section we review some relevant notions related to von Neumann algebras. For the details we refer the reader to any of the books [9], [21], [22] or [34]. Given a complex Hilbert space $(\mathcal{H}, \langle \cdot, \cdot \rangle)$, we denote by $B(\mathcal{H})$ the algebra of all bounded linear operators on $\mathcal{H}$ equipped with the operator norm. The identity operator on $\mathcal{H}$ is denoted by $\mathbf{1}$. For any operator $x \in B(\mathcal{H})$ we denote by x^* its *adjoint*. Recall that an operator $a \in B(\mathcal{H})$ satisfying $a^* = a$ is called *self-adjoint* (or, *hermitian*); the real subspace of $B(\mathcal{H})$ consisting of all self-adjoint operators is denoted by $B(\mathcal{H})_h$. An operator $a \in B(\mathcal{H})_h$ is said to be *positive* if $\langle a\xi, \xi \rangle \geq 0$ for all $\xi \in \mathcal{H}$. The collection of all positive operators on $\mathcal{H}$ is denoted by $B(\mathcal{H})^+$, which is a proper closed generating cone in $B(\mathcal{H})_h$.

Definition 3.1. *A von Neumann algebra $\mathcal{M}$ on $\mathcal{H}$ is a subalgebra of $B(\mathcal{H})$ such that:*

(i) *$\mathcal{M}$ is $*$-closed (that is, $x \in \mathcal{M}$ implies that $x^* \in \mathcal{M}$) and $\mathbf{1} \in \mathcal{M}$;*
(ii) *$\mathcal{M}$ is closed in $B(\mathcal{H})$ for the weak operator topology.*

For any non-empty subset $\mathcal{A} \subseteq B(\mathcal{H})$ we denote by $\mathcal{A}'$ the commutant of $\mathcal{A}$, that is,

$$\mathcal{A}' = \{y \in B(\mathcal{H}) : xy = yx \ \ \forall x \in \mathcal{A}\}.$$

If $\mathcal{A}$ is $*$-closed, then $\mathcal{A}'$ is a von Neumann algebra. We denote $\mathcal{A}'' = (\mathcal{A}')'$, the *double commutant* of $\mathcal{A}$. The following fundamental result provides an alternative definition of von Neumann algebras.

Theorem 3.2 (Von Neumann's Double Commutant Theorem).
A $*$-subalgebra $\mathcal{M}$ of $B(\mathcal{H})$ is a von Neumann algebra if and only if $\mathcal{M} = \mathcal{M}''$.

Evidently, $\mathcal{M} = B(\mathcal{H})$ is a von Neumann algebra. Suppose that (X, Σ, ν) is a Maharam measure space and consider the Hilbert space $\mathcal{H} = L_2(\nu)$. For $f \in L_\infty(\nu)$ define the multiplication operator

$$M_f : L_2(\nu) \to L_2(\nu), \quad M_f(g) = fg, \; g \in L_2(\nu).$$

Then $M_f \in B(L_2(\nu))$ and $\|M_f\| = \|f\|_\infty$. The mapping $f \longmapsto M_f$ is an algebraic isomorphism and isometry from $L_\infty(\nu)$ into $B(L_2(\nu))$. Moreover, $M_f^* = M_{\bar{f}}$, where $\bar{f}$ is the complex conjugate of f.

Proposition 3.3. *Defining*

$$\mathcal{M} = \{M_f : f \in L_\infty(\nu)\},$$

$\mathcal{M}$ is a commutative von Neumann algebra on the Hilbert space $\mathcal{H} = L_2(\nu)$.

Actually, every *commutative* von Neumann algebra is of this form (see, e.g., [9], Chapter I.7). Frequently, the von Neumann algebra $\mathcal{M} = \{M_f : f \in L_\infty(\nu)\}$ is identified with the algebra $L_\infty(\nu)$.

Given a von Neumann algebra $\mathcal{M} \subseteq B(\mathcal{H})$ we define $\mathcal{M}_h = \mathcal{M} \cap B(\mathcal{H})_h$, which is a real linear subspace of $\mathcal{M}$, and $\mathcal{M}^+ = \mathcal{M} \cap B(\mathcal{H})^+$, which is a proper closed and generating cone in $\mathcal{M}_h$. We consider $\mathcal{M}_h$ as an ordered vector space with $\mathcal{M}^+$ as its positive cone.

Definition 3.4. *A trace τ on $\mathcal{M}$ is a map $\tau : \mathcal{M}^+ \to [0, \infty]$ which is additive, positive homogeneous and unitarily invariant, that is,*

$$\tau(uau^*) = \tau(a)$$

for all $a \in \mathcal{M}^+$ and all unitary $u \in \mathcal{M}$.

Definition 3.5. *A trace $\tau : \mathcal{M}^+ \to [0, \infty]$ is called:*

(i) *faithful if $\tau(a) > 0$ whenever $0 < a \in \mathcal{M}$;*
(ii) *semi-finite if for every $a \in \mathcal{M}^+$ with $\tau(a) > 0$ there exists $0 \le b \le a$ such that $0 < \tau(b) < \infty$;*
(iii) *normal if $\tau(a_\beta) \uparrow \tau(a)$ whenever $a_\beta \uparrow a$ in $\mathcal{M}^+$.*

A von Neumann algebra equipped with a semi-finite faithful normal trace is called a *semi-finite von Neumann algebra.*

Example 3.6.

(i) *Let $\mathcal{H}$ be a Hilbert space and $\mathcal{M} = B(\mathcal{H})$. Given a maximal orthonormal system $\{e_\alpha\}$ in $\mathcal{H}$ we define*

$$\tau(a) = \sum_\alpha \langle ae_\alpha, e_\alpha \rangle, \quad a \in B(\mathcal{H})^+.$$

The value of $\tau(a)$ does not depend on the particular choice of the maximal orthonormal system in $\mathcal{H}$ and $\tau : B(\mathcal{H})^+ \to [0, \infty]$ is a semi-finite faithful normal trace on $B(\mathcal{H})$. This is called the standard trace on $B(\mathcal{H})$.

(ii) *Let* $\mathcal{H} = L_2(\nu)$*, where* (X, Σ, ν) *is a Maharam measure space. On* $L_2(\nu)$ *we consider the von Neumann algebra* $\mathcal{M} = L_\infty(\nu)$ *(see Proposition* 3.3*). If we define* $\tau : L_\infty(\nu)^+ \to [0, \infty]$ *by*

$$\tau(f) = \int_X f d\nu, \quad 0 \leq f \in L_\infty(\nu),$$

then τ *is a semi-finite faithful normal trace on* $L_\infty(\nu)$.

An important object in the study of von Neumann algebras is the collection of all orthogonal projections in $\mathcal{M}$, which is denoted by $P(\mathcal{M})$. It is the analogue in non-commutative integration theory of the underlying σ-algebra in classical integration theory. The partial ordering in $\mathcal{M}_h$ induces a partial order in $P(\mathcal{M})$. If $p, q \in P(\mathcal{M})$, then $p \leq q$ if and only if $\operatorname{Ran}(p) \subseteq \operatorname{Ran}(q)$. For any $p, q \in P(\mathcal{M})$ the infimum $p \wedge q \in P(\mathcal{M})$ and supremum $p \vee q \in P(\mathcal{M})$ exists (and are given by the orthogonal projections onto $\operatorname{Ran}(p) \cap \operatorname{Ran}(q)$ and $\overline{\operatorname{Ran}(p) + \operatorname{Ran}(q)}$, respectively). Actually, $P(\mathcal{M})$ is a *complete lattice*, that is, for each collection $\{p_\alpha\}$ in $P(\mathcal{M})$, the supremum $\bigvee_\alpha p_\alpha$ and infimum $\bigwedge_\alpha p_\alpha$ exist (and are given by the projections onto $\overline{\operatorname{span}}_\alpha \{\operatorname{Ran}(p_\alpha)\}$ and $\bigcap_\alpha \operatorname{Ran}(p_\alpha)$, respectively). Every $p \in P(\mathcal{M})$ has a *complement*, given by $p^\perp = \mathbf{1} - p$, which satisfies $p \wedge p^\perp = 0$ and $p \vee p^\perp = \mathbf{1}$. Two projections $p, q \in P(\mathcal{M})$ are called *equivalent* (with respect to $\mathcal{M}$), denoted by $p \sim q$, if there exist a partial isometry $v \in \mathcal{M}$ such that $p = v^*v$ and $q = vv^*$ (that is, p and q are the initial and final projection of v, respectively). If τ is a trace on $\mathcal{M}$, then $p \sim q$ implies that $\tau(p) = \tau(q)$. Furthermore, p is said to be *majorized* by q (relative to $\mathcal{M}$), denoted by $p \precsim q$, if there exists $r \in P(\mathcal{M})$ such that $r \leq q$ and $p \sim r$. Note that $p \precsim q$ implies that $\tau(p) \leq \tau(q)$. A detailed account of this so-called *comparison of projections* can be found, e.g., in [9], Chapter III.1 or [22], Chapter 6). An important fact is that $p - p \wedge q \sim p \vee q - q$ for all $p, q \in P(\mathcal{M})$, which implies in particular that $p \precsim q^\perp$ whenever $p \wedge q = 0$.

4. Measurable operators

As is clear from the definitions, the space of all measurable functions on a measure space provides the general framework for the theory of Banach function spaces. Analogously, the space of all measurable operators is the setting for theory of non-commutative Banach function spaces and non-commutative integration. These measurable operators are in general unbounded linear operators (think of unbounded measurable functions acting via multiplication on the space $L_2(\nu)$). Therefore we first recall some facts about unbounded linear operators in Hilbert space (see, e.g., [4] or [21], [22]).

A linear operator in a Hilbert space $\mathcal{H}$ is a linear map $x : \mathcal{D}(x) \to \mathcal{H}$, where the domain $\mathcal{D}(x)$ is a linear subspace of $\mathcal{H}$. If $\mathcal{D}(x)$ is dense in $\mathcal{H}$, then we say that x is *densely defined*. The operator x is called *closed* whenever its graph is a closed subspace of $\mathcal{H} \times \mathcal{H}$. Any closed and densely defined linear operator has a closed and densely defined *adjoint* $x^*: \mathcal{D}(x^*) \to \mathcal{H}$, which is uniquely determined by the relation $\langle x\xi, \eta\rangle = \langle \xi, x^*\eta\rangle$, $\xi \in \mathcal{D}(x)$, $\eta \in \mathcal{D}(x^*)$. Note that $x^{**} = x$.

A closed densely defined linear operator $a : \mathcal{D}(a) \to \mathcal{H}$ is called self-adjoint if $a^* = a$ (meaning that also the domains coincide). If in addition $\langle a\xi, \xi\rangle \geq 0$ for all $\xi \in \mathcal{D}(\mathcal{H})$, then a is said to be *positive* (which is also denoted by $a \geq 0$). For every self-adjoint operator a there exists a unique *spectral measure* $e^a : \mathcal{B}(\mathbb{R}) \to B(\mathcal{H})$ (that is, e^a takes its values in the orthogonal projections and is σ-additive with respect to the strong operator topology) such that

$$a = \int_{\mathbb{R}} \lambda de^a(\lambda) \tag{4}$$

as a *spectral integral.* Here, $\mathcal{B}(\mathbb{R})$ is the Borel σ-algebra of $\mathbb{R}$. The spectral measure e^a is actually supported on the spectrum $\sigma(a)$ of a. In particular, if $a \geq 0$, then e^a is supported on $[0, \infty)$. Using the spectral measure of a we may define the *Borel functional calculus* for a: for any Borel function $f : \sigma(a) \to \mathbb{C}$ the operator $f(a)$ is defined by

$$f(a) = \int_{\sigma(a)} f(\lambda)\, de^a(\lambda), \tag{5}$$

which is normal operator on $\mathcal{H}$ (recall that the closed and densely defined operator x is called normal whenever $xx^* = x^*x$, with equality of domains). In particular, if $a \geq 0$, then the (positive) square root of a is given by $a^{1/2} = \int_{[0,\infty)} \lambda^{1/2} de^a(\lambda)$. It can be shown that $a^{1/2}$ is the unique positive operator satisfying $\left(a^{1/2}\right)^2 = a$.

If $x : \mathcal{D}(x) \to \mathcal{H}$ is a closed densely defined linear operator, then it can be shown that the operator x^*x is self-adjoint and actually, positive. The *modulus* $|x|$ of x is defined by

$$|x| = (x^*x)^{1/2},$$

that is,

$$|x| = \int_{[0,\infty)} \sqrt{\lambda} de^{x^*x}(\lambda).$$

The operator x can be written as

$$x = v\,|x|,$$

where v is a *partial isometry.* This is called the *polar decomposition* of x.

Now we are ready to introduce the notion of measurable operator (the details may be found in, e.g., [36] or [35], Chapter IX). We assume that $(\mathcal{M}, \tau)$ is a semi-finite von Neumann algebra on the Hilbert space $\mathcal{H}$, with a fixed faithful normal semi-finite trace τ. A linear operator $x : \mathcal{D}(x) \to \mathcal{H}$ is called *affiliated with* $\mathcal{M}$, if $ux = xu$ for all *unitary* $u \in \mathcal{M}'$. This is denoted by $x\eta\mathcal{M}$. Note that the equality $ux = xu$ involves in particular equality of the domains of the operators ux and xu, that is, $\mathcal{D}(x) = u^{-1}(\mathcal{D}(x))$. If $x \in B(\mathcal{H})$, then x is affiliated with $\mathcal{M}$ if and only if $x \in \mathcal{M}$ (as follows from Von Neumann's Double Commutant Theorem; see Theorem 3.2). A useful characterization of affiliated operators is presented in the next proposition.

Proposition 4.1. *If $x : \mathcal{D}(x) \to \mathcal{H}$ is a closed and densely defined linear operator with polar decomposition $x = v|x|$, then x is affiliated with $\mathcal{M}$ if and only if:*

(i) $e^{|x|}(B) \in \mathcal{M}$ *for all* $B \in \mathcal{B}(\mathbb{R})$;
(ii) $v \in \mathcal{M}$.

If $\mathcal{M} = B(\mathcal{H})$, then it is clear that every closed and densely defined linear operator x in $\mathcal{H}$ is affiliated with $B(\mathcal{H})$. Hence, the affiliated operators do not have any reasonable algebraic structure in general. To obtain this we further restrict the class of operators to be considered.

Definition 4.2. *A closed and densely defined linear operator $x : \mathcal{D}(x) \to \mathcal{H}$ is called τ-measurable if:*

(a) $x\eta\mathcal{M}$;
(b) *there exists* $\lambda \geq 0$ *such that* $\tau\left(e^{|x|}(\lambda, \infty)\right) < \infty$.

Condition (b) in the above definition guarantees that the domain of the operator x is "reasonably large" (with respect to the trace τ). In fact, if a closed operator $x : \mathcal{D}(x) \to \mathcal{H}$ is affiliated with $\mathcal{M}$, then x is τ-measurable if and only if its domain $\mathcal{D}(x)$ is τ-*dense* in $\mathcal{H}$ (that is, there exists a sequence $\{p_n\}_{n=1}^{\infty}$ of orthogonal projections in $\mathcal{M}$ such that $p_n(\mathcal{H}) \subseteq \mathcal{D}(x)$ for all n, $p_n \uparrow \mathbf{1}$ and $\tau(\mathbf{1} - p_n) \downarrow 0$ as $n \to \infty$).

The collection of all τ-measurable operators is denoted by $S(\tau)$. If $x, y \in S(\tau)$, then the *algebraic* sum $x + y$ and product xy need not be τ-measurable: these may fail to be closed. However, it can be shown that the operators $x + y$ and xy are closable and that there closures, $x\hat{+}y$ and $x\hat{\cdot}y$ (called the *strong sum and strong product*, respectively) are τ-measurable. Moreover, if $x \in S(\tau)$, then $x^* \in S(\tau)$. All this leads to the following result.

Theorem 4.3. *The set $S(\tau)$ is a complex $*$-algebra with unit element $\mathbf{1}$, with respect to the operations of strong sum and strong product and the $*$-operation of taking adjoints. The von Neumann algebra $\mathcal{M}$ is a $*$-subalgebra of $S(\tau)$.*

From now on we denote the strong sum $x\hat{+}y$ and product $x\hat{\cdot}y$ of two elements $x, y \in S(\tau)$ simply by $x + y$ and xy, respectively.

Example 4.4.

(i) *If $\mathcal{M} = B(\mathcal{H})$ with standard trace τ (see Example* 3.6 (i)), *then $S(\tau) = B(\mathcal{H})$.*
(ii) *If $\mathcal{H} = L_2(\nu)$, $\mathcal{M} = L_\infty(\nu)$ and $\tau(f) = \int_X f d\nu$, $0 \leq f \in L_\infty(\nu)$ (see Example* 3.6 (ii)), *then $S(\tau) = S(\nu)$ (see* (1)), *where the functions in $S(\nu)$ are identified with (in general unbounded) multiplication operators on $L_2(\nu)$.*

The real subspace of $S(\tau)$ consisting of all self-adjoint elements is denoted by $S_h(\tau)$. Note that $S(\tau) = S_h(\tau) \oplus iS_h(\tau)$. Indeed, any $x \in S(\tau)$ can be written as $x = \mathrm{Re}(x) + i\mathrm{Im}(x)$, where $\mathrm{Re}(x) = 1/2(x + x^*)$ and $\mathrm{Im}(x) = 1/2i(x - x^*)$. The set of all positive elements in $S_h(\tau)$, denoted by $S_h(\tau)^+$, is a proper cone in $S_h(\tau)$. For $a, b \in S_h(\tau)$ we define $a \leq b$ whenever $b - a \in S_h(\tau)^+$. With respect to this ordering, $S_h(\tau)$ is a partially ordered vector space. Evidently, this partial ordering is an extension of the ordering in $\mathcal{M}_h$. For every $a \in S_h(\tau)$ the operators $a^+ = \int_{\mathbb{R}} \lambda^+ de^a(\lambda)$ and $a^- = \int_{\mathbb{R}} \lambda^- de^a(\lambda)$ belong to $S_h(\tau)^+$ and satisfy $a = a^+ - a^-$. Consequently, the positive cone $S_h(\tau)^+$ is generating in $S_h(\tau)$. In the next proposition we collect some simple properties of the partial ordering in $S_h(\tau)$ (cf. [12]).

Proposition 4.5.

(i) *If $a, b \in S_h(\tau)^+$, then $a \leq b$ if and only if $\mathcal{D}(b^{1/2}) \subseteq \mathcal{D}(a^{1/2})$ and $\|a^{1/2}\xi\|_{\mathcal{H}} \leq \|b^{1/2}\xi\|_{\mathcal{H}}$ for all $\xi \in \mathcal{D}(b^{1/2})$.*

(ii) *If $a, b \in S_h(\tau)^+$, then $a \leq b$ if and only if there exists $x \in \mathcal{M}$ such that $a^{1/2} = xb^{1/2}$ and $\|x\|_{B(\mathcal{H})} \leq 1$.*

(iii) *If $a \leq b$ in $S_h(\tau)$ and $x \in S(\tau)$, then $x^*ax \leq x^*bx$.*

(iv) *If $a \in S_h(\tau)^+$ is invertible in $S(\tau)$, then $a^{-1} \geq 0$.*

(v) *If $0 \leq a \leq b$ in $S_h(\tau)$ and a is invertible in $S(\tau)$, then b is invertible in $S(\tau)$ and $0 \leq b^{-1} \leq a^{-1}$.*

As (i) of the above proposition shows, on $S_h(\tau)^+$ the partial ordering in $S_h(\tau)$ coincides with the usual quadratic form ordering of positive operators (see, e.g., [23], Section VI.2.5). Statement (ii) follows almost immediately from (i) and (iii), (iv) are more or less evident. Let us indicate a proof of (v). Since $a^{-1} \geq 0$, it follows that $a^{-1/2} \in S_h(\tau)^+$ and so, $\mathbf{1} \leq a^{-1/2}ba^{-1/2}$. By (i), there exists $x \in \mathcal{M}$ such that $\mathbf{1} = x\left(a^{-1/2}ba^{-1/2}\right)^{1/2} = \left(a^{-1/2}ba^{-1/2}\right)^{1/2} x^*$. This shows that $\left(a^{-1/2}ba^{-1/2}\right)^{1/2}$, and hence, b is invertible in $S(\tau)$ with $b^{-1} \geq 0$. Now, it follows from $0 \leq a \leq b$ that $0 \leq b^{-1/2}ab^{-1/2} \leq \mathbf{1}$. Since $b^{-1/2}ab^{-1/2} = \left(a^{1/2}b^{-1/2}\right)^* \left(a^{1/2}b^{-1/2}\right)$, this implies that $\|a^{1/2}b^{-1/2}\|_{B(\mathcal{H})} \leq 1$ and so, $\left\|\left(a^{1/2}b^{-1/2}\right)^*\right\|_{B(\mathcal{H})} \leq 1$, which implies that $0 \leq a^{1/2}b^{-1}a^{1/2} \leq \mathbf{1}$. Using (iii) once again (with $x = a^{-1/2}$), we may conclude that $0 \leq b^{-1} \leq a^{-1}$.

Using (i) of Proposition 4.5, one may prove that $S_h(\tau)$ is Dedekind complete in the following sense (see Proposition 1.1 in [12]).

Proposition 4.6. *If $\{a_\beta\}$ is an increasing net in $S_h(\tau)$ and there exists $b \in S_h(\tau)$ such that $a_\beta \leq b$ for all β, then $\sup_\beta a_\beta$ exists in $S_h(\tau)$.*

Another, related property of the ordering in $S_h(\tau)$ is exhibited in the following proposition (see Proposition 1.3 in [12]).

Proposition 4.7. *If $\{a_\beta\}$ is an increasing net in $S_h(\tau)$ such that $a_\beta \uparrow a \in S_h(\tau)$, then $x^*a_\beta x \uparrow x^*ax$ for all $x \in S(\tau)$.*

Next we discuss the Borel functional calculus (given by (5)) for operators $a \in S_h(\tau)$. For this purpose, we denote by $B_{bc}(\sigma(a))$ the $*$-algebra (with respect to complex conjugation) of all complex-valued Borel functions on $\sigma(a)$ which are bounded on all compact subsets of $\sigma(a)$. The proof of the first statement of the next proposition may be found in [31], Lemma 3.1; the second statement follows immediately from the properties of the functional calculus.

Proposition 4.8. *If* $a \in S_h(\tau)$, *then* $f(a) \in S(\tau)$ *for all* $f \in B_{bc}(\sigma(a))$. *Moreover, the map* $f \longmapsto f(a)$ *is a* $*$*-homomorphism from* $B_{bc}(\sigma(a))$ *into* $S(\tau)$ (*so, in particular, this map is positive*).

5. The measure topology in $S(\tau)$

The $*$-algebra $S(\tau)$ of all τ-measurable operators carries an important and useful vector space topology, the so-called (τ-) measure topology, which is Hausdorff, metrizable and complete (but, not locally convex in general).

As before, $(\mathcal{M}, \tau)$ is a fixed semi-finite von Neumann algebra on a Hilbert space $\mathcal{H}$. For convenience, we denote the set of all orthogonal projections in $\mathcal{M}$ by $P(\mathcal{M})$. Given $0 < \varepsilon, \delta \in \mathbb{R}$ we define $V(\varepsilon, \delta)$ to be the set of all $x \in S(\tau)$ for which there exists $p \in P(\mathcal{M})$ such that $\|xp\|_{B(\mathcal{H})} \leq \varepsilon$ and $\tau(\mathbf{1} - p) \leq \delta$. An alternative description of this set is given by

$$V(\varepsilon, \delta) = \left\{x \in S(\tau) : \tau\left(e^{|x|}(\varepsilon, \infty)\right) \leq \delta\right\}. \tag{6}$$

It can be shown that $V(\varepsilon, \delta)$ is balanced and absorbing. Furthermore, for $\varepsilon_j, \delta_j > 0$ $(j = 1, 2)$ we have $V(\varepsilon_1, \delta_1) + V(\varepsilon_2, \delta_2) \subseteq V(\varepsilon_1 + \varepsilon_2, \delta_1 + \delta_2)$ and $V(\varepsilon, \delta) \subseteq V(\varepsilon_1, \delta_1) \cap V(\varepsilon_2, \delta_2)$, where $\varepsilon = \min(\varepsilon_1, \varepsilon_2)$ and $\delta = \min(\delta_1, \delta_2)$. These properties imply that the collection $\{V(\varepsilon, \delta)\}_{\varepsilon,\delta>0}$ is a neighbourhood base at 0 for a vector space topology $\mathcal{T}_m$ on $S(\tau)$. Since $\bigcap_{\varepsilon,\delta>0} V(\varepsilon, \delta) = \{0\}$, this topology is Hausdorff. Moreover, $V(\varepsilon_1, \delta_1) V(\varepsilon_2, \delta_2) \subseteq V(\varepsilon_1\varepsilon_2, \delta_1\delta_2)$ for all $\varepsilon_j, \delta_j > 0$ and $V(\varepsilon, \delta)^* = V(\varepsilon, \delta)$, and so, $S(\tau)$ is also a topological $*$-algebra with respect to $\mathcal{T}_m$. The countable subcollection $\{V(1/n, 1/n)\}_{n=1}^{\infty}$ is also a base at 0 for $\mathcal{T}_m$ and hence, $\mathcal{T}_m$ is metrizable. Furthermore, it can be shown that $S(\tau)$ is complete with respect to $\mathcal{T}_m$. We collect these results (and some more) in the next theorem (for a proof, see, e.g., [36]).

Theorem 5.1. *The collection* $\{V(\varepsilon, \delta)\}_{\varepsilon,\delta>0}$ *is a neighbourhood base at* 0 *for a metrizable complete Hausdorff vector space topology* $\mathcal{T}_m$ *on* $S(\tau)$. *With respect to this topology,* $S(\tau)$ *is a topological* $*$*-algebra. Moreover,* $\mathcal{M}$ *is dense in* $S(\tau)$ *and the inclusion of* $\mathcal{M}$ (*with its norm topology*) *into* $S(\tau)$ *is continuous.*

The topology $\mathcal{T}_m$ is called the *measure topology* on $S(\tau)$ and convergence with respect to $\mathcal{T}_m$ is called *convergence in measure* (denoted by $x_n \overset{\mathcal{T}_m}{\to} x$). If $\{x_n\}_{n=1}^{\infty}$ is a sequence in $S(\tau)$, then it is immediately clear from (6) that

$$x_n \overset{\mathcal{T}_m}{\to} 0 \iff \lim_{n\to\infty} \tau\left(e^{|x_n|}(\varepsilon, \infty)\right) = 0 \ \forall \varepsilon > 0.$$

Furthermore, it is of some interest to note that the neighbourhoods $V(\varepsilon, \delta)$ are actually closed for the measure topology.

Example 5.2.

(a) *Let* (X, Σ, ν) *be a Maharam measure space. Let* $\mathcal{M} = L_\infty(\nu)$*, acting via multiplication on* $\mathcal{H} = L_2(\nu)$*, equipped with the trace given by* $\tau(f) = \int_X f d\nu$*,* $f \in L_\infty(\nu)^+$ *(see Example* 3.6 (ii)*). As we have mentioned in Example* 4.4*, the algebra* $S(\tau)$ *may be identified with the space* $S(\nu)$*. Via this identification, the measure topology in* $S(\tau)$ *corresponds to the usual topology of convergence in measure in* $S(\nu)$*, a neighbourhood base at* 0 *of which is given by the sets*

$$\{f \in S(\nu) : \nu(x \in X : |f(x)| > \varepsilon) \leq \delta\}, \quad \varepsilon, \delta > 0.$$

(b) *Let* $\mathcal{H}$ *be any Hilbert space and* $\mathcal{M} = B(\mathcal{H})$*, equipped with standard trace* τ *(see Example* 3.6 (ii)*). As observed in Example* 4.4 (ii)*,* $S(\tau) = B(\mathcal{H})$ *in this case. If* e *is an orthogonal projection with* $\tau(\mathbf{1} - e) < 1$*, then* $e = \mathbf{1}$ *and so,*

$$V(\varepsilon, \delta) = \left\{x \in B(\mathcal{H}) : \|x\|_{B(\mathcal{H})} \leq \varepsilon\right\}$$

for all $\varepsilon > 0$ *and* $0 < \delta < 1$*. Hence, the measure topology in* $S(\tau) = B(\mathcal{H})$ *coincides with the operator norm topology in* $B(\mathcal{H})$*.*

Next we discuss the relation between the partial ordering in $S_h(\tau)$ and the measure topology. First observe that the map $x \longmapsto \mathrm{Re}(x)$ is (uniformly) continuous (as $\mathrm{Re}(x) = 1/2(x + x^*)$) and so, $S_h(\tau)$ is a closed real subspace of $S(\tau)$. Another relevant observation in this respect is that the sets $V(\varepsilon, \delta)$ are *absolutely solid*: if $x \in V(\varepsilon, \delta)$ and $y \in S(\tau)$ with $|y| \leq |x|$, then $y \in V(\varepsilon, \delta)$. In the next proposition we collect some elementary properties (for the proof of (i) see [12], Proposition 1.4; the other statements follow immediately).

Proposition 5.3.

(i) *The positive cone* $S_h(\tau)^+$ *is closed in* $S_h(\tau)$*.*

(ii) *If* $\{a_n\}_{n=1}^\infty$ *is a sequence in* $S_h(\tau)$ *and* $a, b \in S_h(\tau)$ *are such that* $a_n \overset{\mathcal{T}_m}{\to} a$ *and* $a_n \leq b$ *for all* n*, then* $a \leq b$*.*

(iii) *If* $\{a_n\}_{n=1}^\infty$ *is an increasing sequence in* $S_h(\tau)$ *and* $a_n \overset{\mathcal{T}_m}{\to} a \in S_h(\tau)$*, then* $a = \sup_n a_n$ *in* $S_h(\tau)$*.*

(iv) *If* $\{x_n\}_{n=1}^\infty$ *and* $\{y_n\}_{n=1}^\infty$ *are two sequences in* $S(\tau)$ *such that* $y_n \overset{\mathcal{T}_m}{\to} 0$ *and* $|x_n| \leq |y_n|$ *for all* n*, then* $x_n \overset{\mathcal{T}_m}{\to} 0$*.*

In some sense, (iii) of the above proposition states that for increasing sequences, measure convergence implies order convergence. What about the converse: does $a_n \uparrow a$ in $S_h(\tau)$ imply that $a_n \overset{\mathcal{T}_m}{\to} a$? In general not (not even in the commutative situation). However, a restricted version is true. To formulate this result, we introduce the subspace $S_0(\tau)$ of $S(\tau)$ defined by

$$S_0(\tau) = \left\{x \in S(\tau) : \tau\left(e^{|x|}(\lambda, \infty)\right) < \infty \ \forall \lambda > 0\right\}. \tag{7}$$

In connection with definition (7), recall that for an operator $x \in S(\tau)$ we only know that $\tau\left(e^{|x|}(\lambda, \infty)\right) < \infty$ for *some* $\lambda > 0$ (see Definition 4.2). It can be shown that $S_0(\tau)$ is actually a two-sided closed ideal in $S(\tau)$. Moreover, $S_0(\tau)$ is absolutely solid in $S(\tau)$, that is, if $y \in S_0(\tau)$, $x \in S(\tau)$ and $|x| \leq |y|$, then $x \in S_0(\tau)$. The self-adjoint and positive elements in $S_0(\tau)$ are denoted by $S_{0,h}(\tau)$ and $S_0(\tau)^+$ respectively. This notation introduced, we can formulate the following "Lebesgue property" of the measure topology.

Proposition 5.4. *If $\{a_\beta\}$ is a decreasing net in $S_h^+(\tau)$ such that $a_\beta \downarrow 0$ and if there exists $a \in S_0(\tau)^+$ such that $a_\beta \leq a$ for all β, then $a_\beta \xrightarrow{\mathcal{T}_m} 0$.*

In the sense of the above proposition, one might say that the measure topology on $S_{0,h}(\tau)$ is a "Lebesgue topology" (that is, order convergence implies topological convergence). Next we would like to discuss in some detail some "Fatou type" properties of the measure topology.

Theorem 5.5. *Suppose that $\varepsilon, \delta > 0$, $a \in S_h(\tau)^+$ and that $\{a_\beta\}$ is a net in $S_h(\tau)^+$ such that $0 \leq a_\beta \uparrow a$ in $S_h(\tau)$. If $a_\beta \in V(\varepsilon, \delta)$ for all β, then $a \in V(\varepsilon, \delta)$.*

We shall indicate the proof of this result, which is based on the following two technical lemmas. For the notation used we refer to the end of Section 3.

Lemma 5.6. *If $a \in S_h(\tau)^+$, $0 < \varepsilon \in \mathbb{R}$ and $p \in P(\mathcal{M})$ such that $p \leq e^a(\varepsilon, \infty)$, then $p \precsim e^{pap}(\varepsilon, \infty)$.*

Proof. For notational convenience, put $b = pap$ and observe that

$$b = \left(a^{1/2}p\right)^* \left(a^{1/2}p\right) \quad \text{and so,} \quad b^{1/2} = \left|a^{1/2}p\right|.$$

Hence, $\mathcal{D}\left(b^{1/2}\right) = \mathcal{D}\left(\left|a^{1/2}p\right|\right) = \mathcal{D}\left(a^{1/2}p\right)$. We first show that $p \wedge e^b[0, \varepsilon] = 0$. To this end, let $q = p \wedge e^b[0, \varepsilon]$ and suppose that $q \neq 0$. Take $\xi \in \mathcal{H}$ such that $q\xi = \xi \neq 0$. This implies that $\xi = e^b[0, \varepsilon]\xi = e^{b^{1/2}}\left[0, \varepsilon^{1/2}\right]\xi$ and so, $\xi \in \mathcal{D}\left(b^{1/2}\right) = \mathcal{D}\left(a^{1/2}p\right)$. Since $\xi = p\xi$ and the algebraic product of $a^{1/2}$ and p is already closed, it follows that $\xi \in \mathcal{D}\left(a^{1/2}\right)$. Furthermore, $p \leq e^a(\varepsilon, \infty)$ and so, $\xi = e^a(\varepsilon, \infty)\xi = e^{a^{1/2}}\left(\varepsilon^{1/2}, \infty\right)\xi$. Using the properties of spectral measures, it is not difficult to show that this implies that $\left\|a^{1/2}\xi\right\|_{\mathcal{H}} > \varepsilon^{1/2}\|\xi\|_{\mathcal{H}}$. Hence,

$$\begin{aligned} \varepsilon^{1/2}\|\xi\|_{\mathcal{H}} &< \left\|a^{1/2}\xi\right\|_{\mathcal{H}} = \left\|a^{1/2}p\xi\right\|_{\mathcal{H}} = \left\|\left|a^{1/2}p\right|\xi\right\|_{\mathcal{H}} = \left\|b^{1/2}\xi\right\|_{\mathcal{H}} \\ &= \left\|b^{1/2}e^{b^{1/2}}\left[0, \varepsilon^{1/2}\right]\xi\right\|_{\mathcal{H}} \leq \varepsilon^{1/2}\|\xi\|_{\mathcal{H}}, \end{aligned}$$

which is a contradiction. Therefore, we may conclude that $p \wedge e^b[0, \varepsilon] = 0$ and this implies that $p \precsim e^b[0, \varepsilon]^\perp = e^b(\varepsilon, \infty)$ □

Using this observation we can show that the neighbourhoods $V(\varepsilon, \delta)$ are "locally determined" in the following sense.

Lemma 5.7. *Let $\varepsilon, \delta > 0$ be given. If $x \in S(\tau)$, then $x \in V(\varepsilon, \delta)$ if and only if $p|x|p \in V(\varepsilon, \delta)$ for all $p \in P(\mathcal{M})$ with $\tau(p) < \infty$.*

Proof. If $x \in V(\varepsilon,\delta)$, then it is easy to see that $p|x|p \in V(\varepsilon,\delta)$ for all $p \in P(\mathcal{M})$ (with $\tau(p) < \infty$). For the proof of the converse implication, suppose that $x \notin V(\varepsilon,\delta)$, that is, $\tau\left(e^{|x|}(\varepsilon,\infty)\right) > \delta$. Since the trace is semi-finite, there exists $p \in P(\mathcal{M})$ such that $p \leq e^{|x|}(\varepsilon,\infty)$ and $\delta < \tau(p) < \infty$. By Lemma 5.6, $p \precsim e^{p|x|p}(\varepsilon,\infty)$ and so, $\tau\left(e^{p|x|p}(\varepsilon,\infty)\right) \geq \tau(p) > \delta$, which shows that $p|x|p \notin V(\varepsilon,\delta)$. □

Now we can provide the proof of Theorem 5.5.

Proof. (of Theorem 5.5) Suppose that $p \in P(\mathcal{M})$ with $\tau(p) < \infty$. It follows from Proposition 4.7 that $0 \leq pa_\beta p \uparrow pap$ in $S_h(\tau)$. Since $\tau(p) < \infty$, we have $pap \in S_0(\tau)^+$, and so, it follows from Proposition 5.4 that $pa_\beta p \overset{\mathcal{T}_m}{\rightarrow} pap$. Since $pa_\beta p \in V(\varepsilon,\delta)$ for all β and $V(\varepsilon,\delta)$ is closed for the measure topology, we find that $pap \in V(\varepsilon,\delta)$. Via Lemma 5.7 we may conclude that $a \in V(\varepsilon,\delta)$. □

Recall that a subset W of a topological vector space $(V,\mathcal{T})$ is called bounded if for every neighbourhood U of 0 there exists $0 < \lambda \in \mathbb{R}$ such that $W \subseteq \lambda U$. Specializing this notion to the measure topology, we get the following definition.

Definition 5.8. *A subset W of $S(\tau)$ is called bounded in measure if for all $\varepsilon,\delta > 0$ there exists $\lambda > 0$ such that $W \subseteq \lambda V(\varepsilon,\delta)$.*

Using that $\lambda V(\varepsilon,\delta) = V(\lambda\varepsilon,\delta)$ for all $\lambda,\varepsilon,\delta > 0$ and the definition of the neighbourhoods $V(\varepsilon,\delta)$ we immediately obtain the following characterization of bounded sets in $S(\tau)$.

Proposition 5.9. *For a subset W of $S(\tau)$ the following statements are equivalent:*

(i) *W is bounded in measure;*
(ii) *for every $\delta > 0$ there exists $R > 0$ such that $W \subseteq V(R,\delta)$;*
(iii) *for every $\delta > 0$ there exists $R > 0$ such that $\tau\left(e^{|x|}(R,\infty)\right) \leq \delta$ for all $x \in W$.*

As an example, let us call a set $W \subseteq S(\tau)$ *order bounded* if there exists $a \in S_h(\tau)^+$ such that $|x| \leq a$ for all $x \in W$. We claim that W is bounded in measure. Indeed, let $\varepsilon,\delta > 0$ be given. Since $V(\varepsilon,\delta)$ is absorbing, there exists $\lambda > 0$ such that $a \in \lambda V(\varepsilon,\delta) = V(\lambda\varepsilon,\delta)$. Since the set $V(\lambda\varepsilon,\delta)$ is absolutely solid (that is, $y \in V(\lambda\varepsilon,\delta)$, $x \in S(\tau)$ and $|x| \leq |y|$ imply $x \in V(\lambda\varepsilon,\delta)$), it is clear that $W \subseteq \lambda V(\varepsilon,\delta)$. Hence, W is bounded in measure. As the next theorem shows, for increasing nets in $S_h(\tau)^+$, the converse also holds.

Theorem 5.10. *If $\{a_\beta\}$ is an increasing net in $S_h(\tau)^+$ which is bounded in measure, then $\sup_\beta a_\beta$ exists in $S_h(\tau)$.*

Proof. First we consider a special case. Suppose that $\{b_k\}_{k=1}^\infty$ is an increasing sequence of mutually commuting operators (that is, $b_k b_l = b_l b_k$ for all k and l) in $S_h(\tau)^+$ which is bounded in measure. We claim that $\sup_k b_k$ exists in $S_h(\tau)$. Indeed, let $\mathfrak{q}_k$ be the quadratic form corresponding to the operator b_k, that is,

$\mathcal{D}(\mathfrak{q}_k) = \mathcal{D}\left(b_k^{1/2}\right)$ and $\mathfrak{q}_k(\xi) = \left\|b_k^{1/2}\xi\right\|_{\mathcal{H}}^2$ for all $\xi \in \mathcal{D}(\mathfrak{q}_k)$. Defining $\mathfrak{q} : \mathcal{D}(\mathfrak{q}) \to [0,\infty)$ by

$$\begin{aligned} \mathcal{D}(\mathfrak{q}) &= \left\{\xi \in \bigcap_{k=1}^{\infty} \mathcal{D}(\mathfrak{q}_k) : \sup_k \mathfrak{q}_k(\xi) < \infty\right\}, \\ \mathfrak{q}(\xi) &= \sup_k \mathfrak{q}_k(\xi) = \lim_{k\to\infty} \mathfrak{q}_k(\xi), \quad \xi \in \mathcal{D}(\mathfrak{q}), \end{aligned}$$

it is easily verified that $\mathfrak{q}$ is a closed quadratic form (in the sense of [23], Section VI.2). The domain $\mathcal{D}(\mathfrak{q})$ is τ-dense (see Section 4) in $\mathcal{H}$. To prove this, we have to show that, given $\delta > 0$, there exists $p \in P(M)$ such that $p(\mathcal{H}) \subseteq \mathcal{D}(\mathfrak{q})$ and $\tau\left(p^{\perp}\right) \leq \delta$. Since $\{b_k\}_{k=1}^{\infty}$ is bounded in measure, there exists $R > 0$ such that $\tau\left(e^{b_k}(R,\infty)\right) \leq \delta$ for all k. Using that $b_k b_{k+1} = b_{k+1} b_k$ and $b_k \leq b_{k+1}$, it is easily verified that $e^{b_k}(R,\infty) \leq e^{b_{k+1}}(R,\infty)$. Therefore, the projection $q = \bigvee_{k=1}^{\infty} e^{b_k}(R,\infty)$ satisfies $\tau(q) \leq \delta$. Defining $p = \mathbf{1} - q = \bigwedge_{k=1}^{\infty} e^{b_k}[0,R]$ we have $\tau\left(p^{\perp}\right) \leq \delta$ and for $\xi \in p(\mathcal{H})$ we find that

$$\mathfrak{q}_k(\xi) = \left\|b_k^{1/2}\xi\right\|_{\mathcal{H}}^2 = \left\|b_k^{1/2} e^{b_k}[0,R]\,\xi\right\|_{\mathcal{H}}^2 = \left\|b_k^{1/2} e^{b_k^{1/2}}\left[0,R^{1/2}\right]\xi\right\|_{\mathcal{H}}^2 \leq R\,\|\xi\|_{\mathcal{H}}^2$$

and so, $\sup_k \mathfrak{q}_k(\xi) \leq R\,\|\xi\|_{\mathcal{H}}^2 < \infty$. Hence, $p(\mathcal{H}) \subseteq \mathcal{D}(\mathfrak{q})$, which shows that $\mathcal{D}(\mathfrak{q})$ is τ-dense (and so, norm dense in $\mathcal{H}$). Therefore, there exists a unique positive self-adjoint operator a in $\mathcal{H}$ such that $\mathcal{D}\left(a^{1/2}\right) = \mathcal{D}(\mathfrak{q})$ and $\left\|a^{1/2}\xi\right\|_{\mathcal{H}}^2 = \mathfrak{q}(\xi)$. Now it is readily verified that $a \in S_h(\tau)^+$ and that $b_k \uparrow a$ in $S_h(\tau)$.

Now we turn to the general case, where $\{a_\beta\}$ is an increasing net in $S_h(\tau)^+$ which is bounded in measure. For $k = 1, 2, \ldots$ we define

$$Y_k(a_\beta) = k a_\beta (a_\beta + k\mathbf{1})^{-1}.$$

The sequence $\{Y_k(a_\beta)\}_{k=1}^{\infty}$ is called the *Yosida approximation* of the operator a_β. Note that

$$\begin{aligned} Y_k(a_\beta) &= k\left(\mathbf{1} - k(a_\beta + k\mathbf{1})^{-1}\right) \\ &= a_\beta - a_\beta^2 (a_\beta + k\mathbf{1})^{-1}. \end{aligned}$$

It is not difficult to show that: (i) $Y_k(a_\beta) \in \mathcal{M}$ and $0 \leq Y_k(a_\beta) \leq k\mathbf{1}$ for all k; (ii) $0 \leq Y_k(a_\beta) \leq Y_{k+1}(a_\beta)$ for all k; (iii) $Y_k(a_\beta) \overset{\mathcal{T}_m}{\to} a_\beta$ as $k \to \infty$; (iv) $Y_k(a_\beta) \uparrow a_\beta$ in $S_h(\tau)$; (v) for fixed k we have $Y_k(a_\beta) \uparrow_\beta$ in $\mathcal{M}$.

Since $0 \leq Y_k(a_\beta) \uparrow \leq k\mathbf{1}$ in $\mathcal{M}$, there exists $b_k \in \mathcal{M}$ such that $Y_k(a_\beta) \uparrow_\beta b_k$ and $Y_k(a_\beta) \to_\beta b_k$ with respect to the strong operator topology (that is, $Y_k(a_\beta)\xi \to b_k\xi$ for all $\xi \in H$). It is clear that $b_k \leq b_{k+1}$ for all k. We claim that $b_k b_l = b_l b_k$ for all $k, l \geq 1$. Indeed, the nets $\{Y_k(a_\beta)\}_\beta$ and $\{Y_l(a_\beta)\}_\beta$ are uniformly bounded (by k and l, respectively) and converge strongly to b_k and b_l, respectively. This implies that $Y_k(a_\beta) Y_l(a_\beta) \to_\beta b_k b_l$ and $Y_l(a_\beta) Y_k(a_\beta) \to_\beta b_l b_k$ strongly. Since $Y_k(a_\beta) Y_l(a_\beta) = Y_l(a_\beta) Y_k(a_\beta)$ for all β, we may conclude that $b_k b_l = b_l b_k$. Next we show that $\{b_k\}_{k=1}^{\infty}$ is bounded in measure. Let $\delta > 0$ be given.

Since $\{a_\beta\}$ is bounded in measure, there exists $R > 0$ such that $a_\beta \in V(R,\delta)$ for all β. Using that $0 \le Y_k(a_\beta) \le a_\beta$, this implies that $Y_k(a_\beta) \in V(R,\delta)$ for all $k \ge 1$ and all β. Since $Y_k(a_\beta) \uparrow_\beta b_k$ in $S_h(\tau)$, it follows from Theorem 5.5 that $b_k \in V(R,\delta)$ for all $k \ge 1$. Hence, $\{b_k\}_{k=1}^\infty$ is bounded in measure.

From the first part of the proof it now follows that there exists $a \in S_h(\tau)^+$ such that $b_k \uparrow a$ in $S_h(\tau)$. It is easily verified that also $a_\beta \uparrow a$ in $S_h(\tau)$, which completes the proof of the theorem. □

We end this section mentioning some results concerning the continuity of the functional calculus. It follows from Proposition 4.8 that, for any $a \in S_h(\tau)$, the map $f \longmapsto f(a)$ is a $*$-homomorphism from $B_{bc}(\mathbb{R})$ into $S(\tau)$ (here $B_{bc}(\mathbb{R})$ denotes the $*$-algebra of all complex-valued Borel functions which are bounded on compact subsets of $\mathbb{R}$). The following result is relatively easy to prove (see [31], Proposition 3.2).

Theorem 5.11. *If $f \in B_{bc}(\mathbb{R})$ and $\{f_n\}_{n=1}^\infty$ is a sequence in $B_{bc}(\mathbb{R})$ such that $f_n \to f$ uniformly on compact subsets of $\mathbb{R}$, then $f_n(a) \overset{\mathcal{T}_m}{\to} f(a)$ for all $a \in S_h(\tau)$.*

The next theorem is less trivial. It is actually a special case of a more general result due to O.Ye. Tikhonov ([37]).

Theorem 5.12. *If $f \in C(\mathbb{R})$ and $a_n \overset{\mathcal{T}_m}{\to} a$ in $S_h(\tau)$, then $f(a_n) \overset{\mathcal{T}_m}{\to} f(a)$.*

Note that this theorem implies in particular that the absolute value map $x \longmapsto |x|$ is continuous on $S(\tau)$ with respect to the measure topology. Indeed, if $x_n \overset{\mathcal{T}_m}{\to} x$ in $S(\tau)$, then $x_n^* x_n \overset{\mathcal{T}_m}{\to} x^* x$ and now apply the above theorem with $f(\lambda) = \sqrt{|\lambda|}$.

6. Generalized singular value functions

In the setting of τ-measurable operators, the generalized singular value functions are the analogue (and actually, generalization) of the decreasing rearrangements of functions in the classical setting. As before, we assume that $(\mathcal{M},\tau)$ is a semi-finite von Neumann algebra on a Hilbert space $\mathcal{H}$. For $x \in S(\tau)$ the *distribution function* $d_x : [0,\infty) \to [0,\infty]$ is defined by

$$d_x(\lambda) = \tau\left(e^{|x|}(\lambda,\infty)\right), \quad \lambda \ge 0.$$

Note that it follows from the definition of τ-measurability that for each $x \in S(\tau)$ there exists $\lambda_0 \ge 0$ such that $d_x(\lambda) < \infty$ for all $\lambda > \lambda_0$. Furthermore, the function d_x is decreasing and right-continuous and $\lim_{\lambda\to\infty} d_x(\lambda) = 0$.

For $x \in S(\tau)$ the *generalized singular value function* $\mu(x) : [0,\infty) \to [0,\infty]$ is defined by

$$\mu(x;t) = \inf\{\lambda \ge 0 : d_x(\lambda) \le t\}, \quad t \ge 0.$$

Since $\lim_{\lambda\to\infty} d_x(\lambda) = 0$, it is clear that $\mu(x;t) < \infty$ for all $t > 0$ (and note that $\mu(x;0) < \infty$ if and only if $x \in \mathcal{M}$, in which case $\mu(x;0) = \|x\|_{B(\mathcal{H})}$.

The function $\mu(x)$ is decreasing and right-continuous. The notion of generalized singular value function for operators $x \in \mathcal{M}$ goes back to A. Grothendieck ([19]). A useful alternative description of the function $\mu(x)$ is the following.

Theorem 6.1. *(see [15]) If $x \in S(\tau)$, then*

$$\mu(x;t) = \inf\left\{\|xp\|_{B(\mathcal{H})} : p \in P(\mathcal{M}), p(\mathcal{H}) \subseteq \mathcal{D}(x), \tau(\mathbf{1}-p) \leq t\right\}$$

for all $t \geq 0$.

Let us consider two simple examples.

Example 6.2.

(i) *Let $\mathcal{H} = L_2(\nu)$, where (X, Σ, ν) is a Maharam measure space, and $\mathcal{M} = L_\infty(\nu)$, equipped with the trace τ given by $\tau(f) = \int_X f d\nu$ for $0 \leq f \in L_\infty(\nu)$ (see Example 3.6 (ii)). For any $f \in S(\tau) = S(\nu)$ the generalized singular value function coincides with the decreasing rearrangement as defined in Section 2.1.*

(ii) *Let $\mathcal{H}$ be any Hilbert space and $\mathcal{M} = B(\mathcal{H})$ equipped with the standard trace τ. If $x \in B(\mathcal{H})$ is a compact operator, then $|x| = (x^*x)^{1/2}$ is compact and self-adjoint. The eigenvalues of $|x|$ are called the singular values of x, denoted by $\{\mu_n(x)\}_{n=0}^\infty$. Here the numbers $\mu_n(x)$ are arranged in decreasing order and repeated according to multiplicity, so*

$$\|x\|_{B(\mathcal{H})} = \mu_0(x) \geq \mu_1(x) \geq \mu_2(x) \geq \cdots \downarrow 0.$$

It follows from the min-max formulas for the eigenvalues of self-adjoint compact operators (see, e.g., [32], Section 95) in combination with Theorem 6.1 that the generalized singular value function of x is given by $\mu(x;t) = \mu_n(x)$ whenever $n \leq t < n+1$ and $n = 0, 1, \ldots$. This example explains why in the general setting the function $\mu(x)$ is called the generalized singular value function.

There is a close connection between the measure topology and generalized singular value functions. Recall that the neighbourhood base $\{V(\varepsilon,\delta)\}_{\varepsilon,\delta>0}$ at zero for the measure topology is given by

$$V(\varepsilon,\delta) = \left\{x \in S(\tau) : \tau\left(e^{|x|}(\varepsilon,\infty)\right) \leq \delta\right\},$$

that is,

$$V(\varepsilon,\delta) = \{x \in S(\tau) : d_x(\varepsilon) \leq \delta\},$$

which implies that

$$\mu(x;t) = \inf\{\varepsilon > 0 : x \in V(\varepsilon,t)\} \tag{8}$$

for all $t > 0$. Conversely, for all $\varepsilon, \delta > 0$ we have

$$V(\varepsilon,\delta) = \{x \in S(\tau) : \mu(x;\delta) \leq \varepsilon\}. \tag{9}$$

Indeed, if $x \in V(\varepsilon, \delta)$, then it is clear from (8) that $\mu(x;\delta) \leq \varepsilon$. Conversely, if $\mu(x;\delta) \leq \varepsilon$, then it follows from the definition of $\mu(x;\delta)$ and the right-continuity of d_x that $d_x(\varepsilon) \leq \delta$ and so, $x \in V(\varepsilon, \delta)$. Note that this implies in particular that $x \in V(\mu', t)$ for all $x \in S(\tau)$, $t > 0$ and $\mu' > \mu(x;t)$ (and $x \in V(\mu(x;t), t)$ whenever $\mu(x;t) > 0$). These simple observations provide a way to transfer properties of the measure topology to properties of the generalized singular value function, and visa versa. For example, in the remarks preceding Proposition 5.3 it has been observed that the sets $V(\varepsilon, \delta)$ are absolutely solid, that is, $x \in V(\varepsilon, \delta)$, $y \in S(\tau)$ and $|y| \leq |x|$, imply that $y \in V(\varepsilon, \delta)$. Using (8), we see that, if $x, y \in S(\tau)$ and $|y| \leq |x|$, then $\mu(y) \leq \mu(x)$. As another example, the property $V(\varepsilon, \delta)^* = V(\varepsilon, \delta)$ immediately implies that $\mu(x^*) = \mu(x)$ for all $x \in S(\tau)$. It is not difficult to show that $yV(\varepsilon, \delta)z \subseteq V(\|y\| \|z\| \varepsilon, \delta)$ for all $y, z \in \mathcal{M}$. Consequently, $\mu(yxz) \leq \|y\| \|z\| \mu(x)$ for all $x \in S(\tau)$ and $y, z \in \mathcal{M}$. We present some other examples. In the next proposition we denote $\lim_{s \uparrow t} \mu(x;s) = \mu(x;t-0)$ for $t > 0$.

Proposition 6.3. (*cf.* [15], *Lemma* 3.4) *If $x \in S(\tau)$ and $\{x_n\}_{n=1}^{\infty}$ is a sequence in $S(\tau)$ such that $x_n \overset{\mathcal{T}_m}{\to} x$, then*

$$\mu(x;t) \leq \liminf_{n\to\infty} \mu(x_n;t) \leq \limsup_{n\to\infty} \mu(x_n;t) \leq \mu(x;t-0)$$

for all $t > 0$. In particular, $\mu(x;t) = \lim_{n\to\infty} \mu(x_n;t)$ for any $t > 0$ where $\mu(x;t)$ is continuous (and hence, $\mu(x_n) \to \mu(x)$ a.e. on $[0,\infty)$).

Proof. Since $x_n \overset{\mathcal{T}_m}{\to} x$, there exist $\varepsilon_n > 0$ and $\delta_n > 0$ such that $\varepsilon_n \downarrow 0$, $\delta_n \downarrow 0$ and $x - x_n \in V(\varepsilon_n, \delta_n)$ for all n. From the above observations it follows that

$$\begin{aligned} x &= x_n + (x - x_n) \in V(\mu(x_n,t) + \varepsilon_n, t) + V(\varepsilon_n, \delta_n) \\ &\subseteq V(\mu(x_n,t) + 2\varepsilon_n, t + \delta_n) \end{aligned}$$

and so, $\mu(x; t + \delta_n) \leq \mu(x_n, t) + 2\varepsilon_n$. Since $\mu(x)$ is right-continuous, this implies that $\mu(x;t) \leq \liminf_{n\to\infty} \mu(x_n;t)$. Take $0 < s < t$ and let $N \in \mathbb{N}$ be such that $s + \delta_n \leq t$ for all $n \geq N$. We find that

$$\begin{aligned} x_n &= x + (x_n - x) \in V(\mu(x;s) + \varepsilon_n, s) + V(\varepsilon_n, \delta_n) \\ &\subseteq V(\mu(x;s) + 2\varepsilon_n, s + \delta_n) \\ &\subseteq V(\mu(x;s) + 2\varepsilon_n, t) \end{aligned}$$

and hence, $\mu(x_n;t) \leq \mu(x;s) + 2\varepsilon_n$ for all $n \geq N$. This implies that

$$\limsup_{n\to\infty} \mu(x_n;t) \leq \mu(x;s).$$

Letting $s \uparrow t$, we get $\limsup_{n\to\infty} \mu(x_n;t) \leq \mu(x;t-0)$. □

Corollary 6.4. (*see, e.g.,* [12], *Lemma* 3.5) *If $\{a_\beta\}$ is a net in $S_h(\tau)^+$ such that $a_\beta \downarrow 0$ in $S_h(\tau)$ and there exists $a \in S_0(\tau)^+$ such that $0 \leq a_\beta \leq a$ for all β, then $\mu(a_\beta;t) \downarrow 0$ for all $t > 0$.*

Proof. From Proposition 5.4 we know that $a_\beta \xrightarrow{\mathcal{T}_m} 0$. Since the measure topology is metrizable, there exists a decreasing subsequence $\{a_{\beta_n}\}_{n=1}^{\infty}$ such that $a_{\beta_n} \xrightarrow{\mathcal{T}_m} 0$. It follows from Proposition 6.3 that $\mu(a_{\beta_n}; t) \downarrow 0$ as $n \to \infty$ for all $t > 0$, which implies that $\mu(a_\beta; t) \downarrow 0$ for all $t > 0$. □

In connection with the above result we mention that the elements in $S_0(\tau)$ may be characterized in terms of the generalized singular value function by

$$S_0(\tau) = \{x \in S(\tau) : \mu(x; t) \to 0 \quad \text{as} \quad t \to \infty\}, \tag{10}$$

as follows easily from the definition (see (7)).

Proposition 6.5. *(see, e.g., [12], Proposition 1.7) If* $0 \leq a_\beta \uparrow a$ *in* $S_h(\tau)$*, then* $\mu(a_\beta; t) \uparrow \mu(a; t)$ *for all* $t \geq 0$.

Proof. First we consider the case that $t > 0$. Since $\mu(a_\beta; t) \leq \mu(a; t)$ for all β, it is clear that $\alpha = \sup_\beta \mu(a_\beta; t) \leq \mu(a; t)$. Suppose that $\alpha < \mu(a; t)$ and take $\alpha_1 \in \mathbb{R}$ such that $\alpha < \alpha_1 < \mu(a; t)$. By (9), $\mu(a_\beta; t) \leq \alpha_1$ implies that $a_\beta \in V(\alpha_1, t)$ for all β. Hence, it follows from Theorem 5.5 that $a \in V(\alpha_1, t)$. Using (9) once again, we find that $\mu(a; t) \leq \alpha_1$, which is a contradiction.

Using that $\mu(a; 0) = \sup_{t>0} \mu(a; t)$ and $\mu(a_\beta; 0) = \sup_{t>0} \mu(a_\beta; t)$, the case $t = 0$ is now an immediate consequence of the above. □

Using the generalized singular value function we may also introduce the notion of submajorization (cf. Definition 2.4) for elements of $S(\tau)$. If $x, y \in S(\tau)$, then we say that x is submajorized by y, denoted by $x \prec\prec y$, whenever $\mu(x) \prec\prec \mu(y)$, that is,

$$\int_0^t \mu(x; s)\, ds \leq \int_0^t \mu(y; s)\, ds, \quad t \geq 0.$$

There are many useful submajorization inequalities involving the generalized singular value functions of element of $S(\tau)$, analogous to the classical inequalities for functions. We will not even try to list them all here but, we mention two of them for later reference.

Theorem 6.6. *If* $x, y \in S(\tau)$*, then:*

(i) $\mu(x + y) \prec\prec \mu(x) + \mu(y)$;
(ii) $\mu(x) - \mu(y) \prec\prec \mu(x - y)$.

Inequality (i) for the case of functions is classical and may probably be traced back to Hardy. Littlewood and Polya. For singular values of compact operators in Hilbert space, (i) was obtained by K. Fan ([16]). The general form for τ-measurable operators is due to Th. Fack and H. Kosaki ([15], Theorem 4.4). For the case of functions, inequality (ii) goes back to G.G. Lorentz and T. Shimogaki ([25]) and for singular values of compact operators in Hilbert space this inequality was obtained by A.S. Markus ([28]). The general case of (ii) was proved in [10].

7. Non-commutative Banach function spaces

As before, we assume that $(\mathcal{M},\tau)$ is a semi-finite von Neumann algebra on a Hilbert space $\mathcal{H}$. Let $E=E(0,\infty)$ be a symmetric Banach function space (see Definition 2.5) on $(0,\infty)$ (with respect to Lebesgue measure). With these ingredients we introduce

$$\begin{aligned} E(\tau) &= \{x\in S(\tau):\mu(x)\in E(0,\infty)\}, \\ \|x\|_{E(\tau)} &= \|\mu(x)\|_{E(0,\infty)}, \quad x\in E(\tau). \end{aligned}$$

The following result has been obtained in [10], [11]. We shall indicate the main steps of its proof.

Theorem 7.1. *With the above definitions we have:*

(i) *$E(\tau)$ is a linear subspace of $S(\tau)$ and $\|\cdot\|_{E(\tau)}$ is a norm on $E(\tau)$;*

(ii) *the embedding of $\left(E(\tau),\|\cdot\|_{E(\tau)}\right)$ into $(S(\tau),\mathcal{T}_m)$ is continuous;*

(iii) *$E(\tau)$ is complete with respect to $\|\cdot\|_{E(\tau)}$.*

Proof. (i) If $x,y\in E(\tau)$, then it follows from Theorem 6.6 (i), that $\mu(x+y)\prec\!\!\prec\mu(x)+\mu(y)$, which implies that $x+y\in E(\tau)$ and

$$\begin{aligned} \|x+y\|_{E(\tau)} &= \|\mu(x+y)\|_{E(0,\infty)}\leq\|\mu(x)+\mu(y)\|_{E(0,\infty)} \\ &\leq \|\mu(x)\|_{E(0,\infty)}+\|\mu(y)\|_{E(0,\infty)}=\|x\|_{E(\tau)}+\|y\|_{E(\tau)}. \end{aligned}$$

Now it is clear that $\|\cdot\|_{E(\tau)}$ is a norm on $E(\tau)$.

(ii) It is sufficient to show that the closed unit ball $B_{E(\tau)}$ of $E(\tau)$ is bounded in measure, that is, for every $\delta>0$ there exists $R>0$ such that $\mu(x;\delta)\leq R$ whenever $x\in B_{E(\tau)}$ (see Proposition 5.9 and (9)). Given $\delta>0$ and $x\in B_{E(\tau)}$, it follows from the inequality $0\leq\mu(x;\delta)\chi_{[0,\delta]}\leq\mu(x)$ (as $\mu(x)$ is decreasing on $[0,\infty)$) that $\mu(x;\delta)\leq\left\|\chi_{[0,\delta]}\right\|_{E(0,\infty)}^{-1}$. Hence, we may take $R=\left\|\chi_{[0,\delta]}\right\|_{E(0,\infty)}^{-1}$.

(iii) To show that $E(\tau)$ is complete with respect to $\|\cdot\|_{E(\tau)}$, suppose that $\{x_n\}_{n=1}^{\infty}$ is a Cauchy sequence in $E(\tau)$. It follows from (ii) that $\{x_n\}_{n=1}^{\infty}$ is Cauchy for the measure topology and so, there exists $x\in S(\tau)$ such that $x_n\overset{\mathcal{T}_m}{\to}x$ (see Theorem 5.1). Moreover, it follows from Theorem 6.6 (ii), that

$$\mu(x_m)-\mu(x_n)\prec\!\!\prec\mu(x_m-x_n)$$

and so, $\|\mu(x_m)-\mu(x_n)\|_{E(0,\infty)}\leq\|\mu(x_m-x_n)\|_{E(0,\infty)}=\|x_m-x_n\|_{E(\tau)}$ for all m and n. Hence, $\{\mu(x_n)\}_{n=1}^{\infty}$ is a Cauchy sequence in $E(0,\infty)$. Hence, there exists $f\in E(0,\infty)$ such that $\|\mu(x_n)-f\|_{E(0,\infty)}\to 0$. Furthermore, it follows from Proposition 6.3 that $x_n\overset{\mathcal{T}_m}{\to}x$ implies that $\mu(x_n)\to\mu(x)$ a.e. on $(0,\infty)$ and so, $\mu(x)=f\in E(0,\infty)$, that is, $x\in E(\tau)$. Applying the same argument to the Cauchy sequence $\{x-x_n\}_{n=1}^{\infty}$ (which converges to 0 in measure), we find that $\|\mu(x-x_n)\|_{E(0,\infty)}\to 0$, that is, $\|x-x_n\|_{E(\tau)}\to 0$. The proof is complete. □

The space $E(\tau)$ is called the non-commutative Banach function space corresponding to $E(0,\infty)$ and associated with $(\mathcal{M},\tau)$. From the definition it is clear that $x \in E(\tau)$ if and only if $|x| \in E(\tau)$, if and only if $x^* \in E(\tau)$ and that $\|x\|_{E(\tau)} = \||x|\|_{E(\tau)} = \|x^*\|_{E(\tau)}$. Furthermore we note that $E(\tau)$ is *symmetric*, that is, if $x \in S(\tau)$, $y \in E(\tau)$ and $x \prec\prec y$, then $x \in E(\tau)$ and $\|x\|_{E(\tau)} \leq \|y\|_{E(\tau)}$ (and so, in particular, $E(\tau)$ is an absolutely solid subspace of $S(\tau)$).

The real linear subspace of all self-adjoint elements in $E(\tau)$ is denoted by $E_h(\tau)$. The collection of all positive elements in $E_h(\tau)$ is denoted by $E_h(\tau)^+$, that is, $E_h(\tau)^+ = E(\tau) \cap S_h(\tau)^+$, which is a proper cone in $E_h(\tau)$. Hence, $E_h(\tau)$ has the structure of a partially ordered vector space with $E_h(\tau)^+$ as its positive cone. Since the embedding of $\left(E(\tau), \|\cdot\|_{E(\tau)}\right)$ into $(S(\tau), \mathcal{T}_m)$ is continuous and $S_h(\tau)^+$ is closed in $S_h(\tau)$ (see Proposition 5.3), it is clear that $E_h(\tau)^+$ is closed in $E_h(\tau)$. Therefore, $\left(E_h(\tau), \|\cdot\|_{E(\tau)}\right)$ is an *ordered Banach space* (for an exposition of the theory of ordered Banach spaces we refer the reader to, e.g., [2]; see also Chapter V in the book [33]). The positive cone $E_h(\tau)^+$ is generating (indeed, each $a \in E_h(\tau)$ can be decomposed as $a = a^+ - a^-$, where a^+ and a^- belong to $E_h(\tau)^+$ with $\|a^+\|_{E(\tau)}, \|a^-\|_{E(\tau)} \leq \|a\|_{E(\tau)}$). Furthermore, the norm in $E_h(\tau)$ is *monotone*, that is, $0 \leq a \leq b$ in $E_h(\tau)$ implies that $\|a\|_{E(\tau)} \leq \|b\|_{E(\tau)}$. This implies in particular that $E_h(\tau)^+$ is a *normal cone*. Consequently, any φ in the (real) Banach space dual $E_h(\tau)^*$ can be decomposed as $\varphi = \varphi_1 - \varphi_2$, with $\varphi_1, \varphi_2 \geq 0$. In other words, the dual cone of $E_h(\tau)^+$ is generating in $E_h(\tau)^*$. Moreover, a standard argument shows that any positive linear functional on $E_h(\tau)$ is automatically bounded.

Note that $E(\tau)$ is the complexification of $E_h(\tau)$, that is, $E(\tau) = E_h(\tau) \oplus iE_h(\tau)$. Indeed, any $x \in E(\tau)$ can be written as $x = \mathrm{Re}x + i\mathrm{Im}x$, with $\mathrm{Re}x$, $\mathrm{Im}x \in E_h(\tau)$ (and $\|\mathrm{Re}x\|_{E(\tau)}, \|\mathrm{Im}x\|_{E(\tau)} \leq \|x\|_{E(\tau)}$). This implies that $E_h(\tau)^*$ may be identified with a closed real subspace of $E(\tau)^*$. Indeed, let us call a functional $\varphi \in E(\tau)^*$ *self-adjoint* (or, hermitian) whenever $\varphi(x^*) = \overline{\varphi(x)}$ for all $x \in E(\tau)$ and denote by $E(\tau)^*_h$ the closed real subspace of $E(\tau)^*$ consisting of all self-adjoint functionals. It is easy to verify that the map $\varphi \longmapsto \varphi|_{E_h(\tau)}$ defines an isometric isomorphism form $E(\tau)^*_h$ onto $E_h(\tau)^*$. Via this isomorphism we may identify $E(\tau)^*_h$ with $E_h(\tau)^*$. Furthermore, with this identification, we have $E(\tau)^* = E_h(\tau)^* \oplus iE_h(\tau)^*$. Indeed, any $\varphi \in E(\tau)^*$ can be written as $\varphi = \varphi_1 + i\varphi_2$, where $\varphi_1, \varphi_2 \in E_h(\tau)^*$ are given by

$$\varphi_1(x) = \frac{1}{2}\left(\varphi(x) + \overline{\varphi(x^*)}\right), \ \varphi_1(x) = \frac{1}{2i}\left(\varphi(x) - \overline{\varphi(x^*)}\right), \ x \in E(\tau)^*.$$

This implies in particular that every $\varphi \in E(\tau)^*$ is a linear combination of four positive linear functionals.

As specific examples we mention the non-commutative L_p-spaces associated with $(\mathcal{M},\tau)$, that is, the spaces $L_p(\tau)$ corresponding to $L_p(0,\infty)$, for $1 \leq p \leq \infty$.

The norm in $L_p(\tau)$ is usually denoted simply by $\|\cdot\|_p$. In particular, $L_\infty(\tau) = \mathcal{M}$ and $\|x\|_\infty = \|x\|_{B(\mathcal{H})}$ for all $x \in L_\infty(\tau)$ (see the remarks preceding Theorem 6.1).

If we take $\mathcal{M} = B(\mathcal{H})$ with standard trace, then the spaces $E(\tau)$ correspond to the so-called *symmetrically normed ideals of compact operators*, the theory of which is developed in detail in the book [18]. In particular, in this case $L_p(\tau) = \mathfrak{S}_p$ for $1 \leq p < \infty$, the *p-Schatten ideals* of compact operators.

It follows from (3) that any non-commutative Banach function space satisfies

$$(L_1 \cap L_\infty)(\tau) \subseteq E(\tau) \subseteq (L_1 + L_\infty)(\tau)$$

with continuous embeddings. It is clear that $(L_1 \cap L_\infty)(\tau) = L_1(\tau) \cap L_\infty(\tau)$ and it can be shown that $(L_1 + L_\infty)(\tau) = L_1(\tau) + L_\infty(\tau)$. The restriction of the trace τ to $(L_1 \cap L_\infty)_h^+(\tau)$ is a positive linear functional and can be extended to a linear functional $\dot{\tau}$ on $(L_1 \cap L_\infty)(\tau)$, satisfying $\dot{\tau}(|x|) = \|x\|_1$ for all $x \in (L_1 \cap L_\infty)(\tau)$. Using that $(L_1 \cap L_\infty)(0,\infty)$ is dense in $L_1(0,\infty)$, it follows that $(L_1 \cap L_\infty)(\tau)$ is dense in $L_1(\tau)$ (see, e.g., Proposition 2.8 in [12]) and hence, $\dot{\tau}$ extends uniquely to a linear functional $\dot{\tau} : L_1(\tau) \to \mathbb{C}$. Moreover, $\dot{\tau}(|x|) = \|x\|_1$ for all $x \in L_1(\tau)$ and $\dot{\tau}$ is a positive functional on $L_1^+(\tau)$. For the details of this construction and further properties of this extended trace, which will be denoted again by τ, we refer the reader to Section 3 of [12].

Next we discuss some aspects of the duality theory of these non-commutative spaces. As above, we assume that $E(0,\infty)$ is a symmetric Banach function space on $(0,\infty)$.

Definition 7.2. *The Köthe dual space $E(\tau)^\times$ of $E(\tau)$ is defined by*

$$E(\tau)^\times = \left\{ y \in S(\tau) : xy \in L^1(\tau) \ \ \forall x \in E(\tau) \right\}.$$

It is clear that $E(\tau)^\times$ is a linear subspace of $S(\tau)$. It is not difficult to verify that $y \in E(\tau)^\times$ if and only if $yx \in L^1(\tau)$ for all $x \in E(\tau)$. Moreover, if $y \in E(\tau)^\times$ and $x \in E(\tau)$, then $\tau(xy) = \tau(yx)$. If, in addition, $x \geq 0$ and $y \geq 0$, then $\tau(xy) \geq 0$ (all these statements, and much more, can be found in Proposition 5.2 of [12]). For $y \in E(\tau)^\times$, we define the linear functional

$$\varphi_y : E(\tau) \to \mathbb{C}, \ \ \varphi_y(x) = \tau(xy), \ x \in E(\tau). \tag{11}$$

If $y \in E(\tau)^\times$ and $y \geq 0$, then φ_y is a positive functional, that is, $\varphi_y(x) \geq 0$ for all $x \in E_h^+(\tau)$. This observation can be used to show that the functional φ_y is bounded for every $y \in E(\tau)^\times$. The map $\Phi : E(\tau)^\times \to E(\tau)^*$ is linear and injective (which follows from $(L_1 \cap L_\infty)(\tau) \subseteq E(\tau)$). Now we define a norm $\|\cdot\|_{E(\tau)^\times}$ on $E(\tau)^\times$ by

$$\|y\|_{E(\tau)^\times} = \|\varphi_y\|_{E(\tau)^*}, \ \ y \in E(\tau)^\times.$$

We say that $E(\tau)^\times$ may be identified with a subspace of $E(\tau)^*$ *via trace duality* (which is given by (11). In the analysis of Köthe dual $E(\tau)^\times$ the following result plays a crucial role (see [12], Proposition 5.3).

Proposition 7.3. *If $y \in S(\tau)$, then $y \in E(\tau)^{\times}$ if and only if*

$$\sup\left\{\int_0^\infty \mu(x;t)\,\mu(y;t)\,dt : x \in E(\tau), \|x\|_{E(\tau)} \leq 1\right\} < \infty$$

and in this case we have

$$\|y\|_{E(\tau)^{\times}} = \sup\left\{\int_0^\infty \mu(x;t)\,\mu(y;t)\,dt : x \in E(\tau), \|x\|_{E(\tau)} \leq 1\right\}.$$

Using these observations, it can be shown that the normed linear space $\left(E(\tau)^{\times}, \|\cdot\|_{E(\tau)^{\times}}\right)$ has the following properties (see [12], Proposition 5.4):

(a) $(L_1 \cap L_\infty)(\tau) \subseteq E(\tau)^{\times} \subseteq (L_1 + L_\infty)(\tau)$, with continuous embeddings;
(b) the embedding of $\left(E(\tau)^{\times}, \|\cdot\|_{E(\tau)^{\times}}\right)$ into $(S(\tau), \mathcal{T}_m)$ is continuous;
(c) if $x \in S(\tau)$, $y \in E(\tau)^{\times}$ and $x \prec\prec y$, then $x \in E(\tau)^{\times}$ and $\|x\|_{E(\tau)^{\times}} \leq \|y\|_{E(\tau)^{\times}}$;
(d) if $\{y_\alpha\}$ is a net of positive elements in $E(\tau)^{\times}$ such that $0 \leq y_\alpha \uparrow$ and $\sup_\alpha \|y_\alpha\|_{E(\tau)^{\times}} < \infty$, then there exists a positive element $y \in E(\tau)^{\times}$ such that $y_\alpha \uparrow y$ and $\|y_\alpha\|_{E(\tau)^{\times}} \uparrow \|y\|_{E(\tau)^{\times}}$;
(e) $E(\tau)^{\times}$ is complete with respect to $\|\cdot\|_{E(\tau)^{\times}}$.

The important result for the identification of the Köthe dual $E(\tau)^{\times}$ is the following theorem (see [12], Theorem 5.6).

Theorem 7.4. *If $E = E(0,\infty)$ is a symmetric Banach function space on $(0,\infty)$ with Köthe dual space $E^{\times}$, then $E(\tau)^{\times} = E^{\times}(\tau)$ (with equality of norms).*

The linear functionals $\varphi \in E(\tau)^*$ which correspond to elements $y \in E(\tau)^{\times}$ via trace duality (11) have a characterization which is analogous to the commutative case (see Theorem 2.10).

Theorem 7.5. ([12], *Theorem* 5.11) *Suppose that $E = E(0,\infty)$ is a symmetric Banach function space on $(0,\infty)$. For $\varphi \in E(\tau)^*$ the following conditions are equivalent:*

(i) *φ is normal, that is, $x_\alpha \downarrow 0$ in $E_h(\tau)$ implies that $\varphi(x_\alpha) \to 0$;*
(ii) *φ is completely additive, that is, $e_\alpha \downarrow 0$ in $P(\mathcal{M})$ implies that $\varphi(xe_\alpha) \to 0$ and $\varphi(e_\alpha x) \to 0$ for all $x \in E(\tau)$;*
(iii) *there exists $y \in E(\tau)^{\times}$ such that $\varphi(x) = \tau(xy)$ for all $x \in E(\tau)$ (that is, $\varphi = \varphi_y$ in the notation of* (11)).

Via the same argument as used in the case of Banach lattices, it is easily see that every $\varphi \in E(\tau)^*$ is normal (briefly, $E(\tau)^* = E(\tau)^{\times}$) if and only if the norm in $E(\tau)$ is *order continuous*, that is, $x_\alpha \downarrow 0$ in $E_h(\tau)$ implies that $\|x_\alpha\|_{E(\tau)} \downarrow 0$. Another relevant observation in this connection is the following (see [12], Proposition 3.6, and [7]).

Proposition 7.6. *If the symmetric Banach function space $E = E(0,\infty)$ has order continuous norm, then the norm in $E(\tau)$ is also order continuous.*

Proof. The order continuity of the norm in $E(0,\infty)$ implies that $\mu(f;t) \to 0$ as $t \to \infty$ and so, $E(\tau) \subseteq S_0(\tau)$ (see (10)). Consequently, if $x_0 \geq x_\alpha \downarrow 0$ in $E_h(\tau)$, then it follows from Corollary 6.4 that $\mu(x_\alpha;t) \downarrow 0$ for all $t > 0$ and hence, $\|x_\alpha\|_{E(\tau)} = \|\mu(x_\alpha)\|_{E(0,\infty)} \downarrow 0$. □

We illustrate the above results with some explicit examples. If we take for example $E = L_p(0,\infty)$, with $1 \leq p < \infty$, the E has order continuous norm and so,

$$L_p(\tau)^* = L_p(\tau)^\times = L_p^\times(\tau) = L_q(\tau)$$

(identification via trace duality), where $p^{-1}+q^{-1} = 1$. Similarly, $\mathcal{M}^\times = L_\infty(\tau)^\times = L_1(\tau)$. Other examples are

$$\begin{aligned}(L_1(\tau)+L_\infty(\tau))^\times &= (L_1+L_\infty)^\times(\tau) = (L_1\cap L_\infty)(\tau) = L_1(\tau)\cap L_\infty(\tau),\\ (L_1(\tau)\cap L_\infty(\tau))^\times &= (L_1\cap L_\infty)^\times(\tau) = (L_1+L_\infty)(\tau) = L_1(\tau)+L_\infty(\tau).\end{aligned}$$

Of course, similar examples may be given using Orlicz spaces, Lorentz and Marcinkiewicz spaces.

We end this section with an interesting decomposition theorem for functionals in the Banach space dual $E(\tau)^*$, as was obtained in [14]. Let us first consider the situation for a Banach function space E on a (Maharam) measure space (X,Σ,ν). As before, we denote by E_n^* the collection of all normal (that is, order continuous) linear functionals on E, which is a band in the Banach space dual E^* (and may be identified with the Köthe dual $E^\times$). The disjoint complement of E_n^* in E^* will be denoted by E_s^* (sometimes this band is also denoted by E_{sn}^*) and the elements of E_s^* are termed *singular (normal) linear functionals*. Since $E^* = E_n^* \oplus E_s^*$, every $\varphi \in E^*$ has a unique decomposition $\varphi = \varphi_n + \varphi_s$, where $\varphi_n \in E_n^*$ and $\varphi_s \in E_s^*$ (and so, $\varphi_n \perp \varphi_s$). This decomposition can be viewed as an analogue of the so-called Yosida-Hewitt decomposition of measures. For the details we refer the reader to, e.g., Chapter 12 in [40] or Chapter 1 in [1]. From the definition it is clear that a functional $\varphi \in E^*$ is singular if and only if it follows from $|\psi| \leq |\varphi|$ and $\psi \in E_n^*$ that $\psi = 0$. Another useful characterization of singular functionals is that they vanish on large (order) ideals in E. To be more precise, an ideal (that is, absolutely solid linear subspace) $A \subseteq E$ is called *order dense* in E if for every $0 < u \in E$ there exists $v \in A$ such that $0 < v \leq u$. With this terminology, a linear functional $\varphi \in E^*$ is singular if and only $\varphi = 0$ on some order dense ideal in E (see [26], Theorem 50.4; this result also follows from Theorem 90.5 in [40]). We note that for this characterization it is essential that E_n^* separates the points of the Banach function space E (for example, on the Banach lattice $C[0,1]$ the functional of integration is singular and strictly positive).

Now we consider such a decomposition for functionals on a space $E(\tau)$, where $(\mathcal{M},\tau)$ is a semi-finite von Neumann algebra and $E = E(0,\infty)$ is a symmetric Banach function space on $(0,\infty)$. The concept of normal functional was already introduced in Theorem 7.5. A linear subspace $A \subseteq E(\tau)$ is called an order ideal if A is generated by its positive elements and if it follows from $0 \leq b \leq a$, $a \in A$ and $b \in E_h(\tau)$ that $b \in A$. Such an ideal A is called order dense in $E(\tau)$ if for every $0 < b \in E_h(\tau)$ there exists $a \in A$ such that $0 < a \leq b$. A linear functional $\varphi \in E(\tau)^*$ is said to be *singular* whenever φ vanishes on some order dense ideal in $E(\tau)$. Evidently, this notion of singularity agrees with the one for Banach function spaces, as follows from the above discussion. If $E = L_\infty(0,\infty)$, and so $E(\tau) = \mathcal{M}$, it also agrees with the usual definition of a singular functional on a von Neumann algebra (see, e.g., [34], Section III.2 or [22], Section 10.1), as follows from [14], Proposition 2.1, in combination with [34], Theorem III.3.8. Now we are in a position to formulate the decomposition theorem for elements of $E(\tau)^*$.

Theorem 7.7. ([14], *Corollary* 2.5) *If* $(\mathcal{M}, \tau)$ *is a semi-finite von Neumann algebra and* $E = E(0,\infty)$ *is a fully symmetric Banach function space on* $(0,\infty)$, *then every* $\varphi \in E(\tau)^*$ *has a unique decomposition* $\varphi = \varphi_n + \varphi_s$, *where* φ_n *is normal and* φ_s *is singular.*

For further details and interesting applications of this result, we refer the reader to [14].

8. Operator functions

As we have seen in the previous section, there are many results concerning non-commutative Banach function spaces which are analogous to the commutative theory (although most of the proofs are quite different!). However, there are some aspects of the non-commutative theory which are essentially different from the commutative situation. We shall illustrate this with some results concerning so-called *operator functions*. By an operator function we mean a map $a \longmapsto f(a)$, where $f : \mathbb{R} \to \mathbb{R}$ is an appropriate Borel function and the (non-commutative) variable a belongs to $E_h(\tau)$. If $f : \mathbb{R} \to \mathbb{R}$ is continuous, then we know by Theorem 5.12 that the map $a \longmapsto f(a)$ from $S_h(\tau)$ into itself, is continuous with respect to the measure topology. But, here we will be interested in Lipschitz-type norm estimates. To be more precise, we consider the following problem: under which assumptions, on the Banach function space $E = E(0,\infty)$ and on the function f, does there exists a constant $C > 0$ (depending on E and f) such that $\|f(a) - f(b)\|_{E(\tau)} \leq C \|a - b\|_{E(\tau)}$ for all $a, b \in E_h(\tau)$?

Let us say first a few words about the commutative situation. Suppose that E is any Banach function space on a (Maharam) measure space (X, Σ, ν) and let $a \in E$ be real-valued (we use here the symbol a for a function to keep the analogy with the above discussion). We may represent a by its spectral integral $a = \int_{\mathbb{R}} \lambda de^a(\lambda)$ as in (4). Note that we may consider a as a self-adjoint operator on the Hilbert space $L_2(\nu)$, acting via multiplication. The spectral measure of a is then given by $e^a(B) = \chi_{a^{-1}(B)}$ for all Borel sets $B \subseteq \mathbb{R}$. If $f : \mathbb{R} \to \mathbb{R}$ is a Borel function, then $f(a)$ is defined by

$$f(a) = \int_{\mathbb{R}} f(\lambda)\, de^{a}(\lambda)$$

(see (5)). Approximating f by simple functions, it is not difficult to see that $f(a) = f \circ a$ (the composition of f and a). Now suppose that the function f is Lipschitz continuous, that is, there exists a constant $C > 0$ such that $|f(\lambda) - f(\mu)| \leq C|\lambda - \mu|$ for all $\lambda, \mu \in \mathbb{R}$. If $a, b \in E$ are real-valued, then

$$|f(a)(x) - f(b)(x)| = |f(a(x)) - f(b(x))| \leq C|a(x) - b(x)|, \quad x \in X,$$

and hence,

$$|f(a) - f(b)| \leq C|a - b|. \tag{12}$$

Since E is an ideal in $L_0(\nu)$ and the norm on E is absolutely monotone, it follows that $f(a) - f(b) \in E$ and $\|f(a) - f(b)\|_E \leq C\|a - b\|_E$. This argument shows that in the commutative situation, it is more or less evident that Lipschitz continuity of f implies that the corresponding "operator function" is also Lipschitz continuous (with the same constant, independent of E). The crucial estimate is of course inequality (12). In the non-commutative situation, inequalities like (12) are not valid in general (if a and b do not commute) and, as it turns out, Lipschitz continuity of f is in general not enough to guarantee that the corresponding operator function satisfies a Lipschitz estimate.

As a special case, let us first consider the absolute value mapping corresponding to the function $f(\lambda) = |\lambda|$. In [13] the following result has been obtained.

Theorem 8.1. *Suppose that $1 < p < \infty$ and let $(\mathcal{M}, \tau)$ be a semi-finite von Neumann algebra. If $x, y \in S(\tau)$ such that $x - y \in L_p(\tau)$, then $|x| - |y| \in L_p(\tau)$ and*

$$\||x| - |y|\|_p \leq C_p \|x - y\|_p, \tag{13}$$

where $C_p > 0$ is a constant only depending on p.

In the case $\mathcal{M} = B(\mathcal{H})$, with standard trace (and so, $L_p(\tau) = \mathfrak{S}_p$, the p-Schatten ideal), the above result was obtained by E.B. Davies in [8] (see also [5]). Moreover, it was shown in [8] that an estimate like (13) fails for $p = 1, \infty$. We like to point out that it is sufficient to prove the above theorem for self-adjoint elements x and y only. Indeed, the general case is then obtained from this special case by considering the von Neumann algebra $M_2(\mathbb{C}) \otimes \mathcal{M}$, of all 2×2-matrices with entries in $\mathcal{M}$, and applying the result to the self-adjoint operators

$$\begin{bmatrix} 0 & x^* \\ x & 0 \end{bmatrix}, \begin{bmatrix} 0 & y^* \\ y & 0 \end{bmatrix}.$$

We leave the verification to the reader.

Furthermore, the result of Theorem 8.1 can be extended via interpolation techniques to a much larger class of spaces than the L_p-spaces. In fact, in [13], Theorem 3.4, it was shown that, if $E = E(0, \infty)$ is a symmetric Banach function space which is an (L_p, L_q)-interpolation space for some $1 < p \leq q < \infty$, then there exists a constant $C_E > 0$ (only depending on the space E) such that $\| |x| - |y| \|_{E(\tau)} \leq C_E \|x - y\|_{E(\tau)}$ for all $x, y \in E(\tau)$ with $x - y \in E(\tau)$, for all semi-finite von Neumann algebras $(\mathcal{M}, \tau)$ (and actually, this property characterizes the Banach function spaces which are (L_p, L_q)-interpolation space for some $1 < p \leq q < \infty$).

Finally we say a few words about more general operator functions $a \longmapsto f(a)$, $a \in E_h(\tau)$. For sake of simplicity we shall not state the results in full generality, but single out some important special cases (which follows from [29], Corollary 7.5 in combination with Proposition 8.5).

Theorem 8.2. *Suppose that $1 < p < \infty$, let $(\mathcal{M}, \tau)$ be a semi-finite von Neumann algebra and $f : \mathbb{R} \to \mathbb{R}$ be a function with weak derivative f' which is of bounded variation. There exists a constant $C_{p,f} > 0$ (only depending on p and the function f), such that*

$$\|f(a) - f(b)\|_p \leq C_{p,f} \|a - b\|_p$$

for all $a, b \in S_h(\tau)$ with $a - b \in L_p(\tau)$.

The function $f(\lambda) = |\lambda|$ satisfies the conditions of the above theorem and so, the result of Theorem 8.1 may be obtained via Theorem 8.2. Furthermore, also Theorem 8.2 actually holds for (L_p, L_q)-interpolation spaces with $1 < p, q < \infty$. In the paper [30] several results concerning the (Gâteaux) differentiability of operator functions have been obtained. All these results depend on the theory of so-called *double operator integrals*, originated by Birman and Solomyak in the setting of trace ideals and extended in [29] to the general setting of semi-finite von Neumann algebras. These double operator integrals and their relation to the UMD-property and R-bounded collections of operators, have been discussed also in detail in [38].

References

[1] C.D. Aliprantis, O. Burkinshaw, *Positive Operators*, Academic Press, Orlando, 1985.

[2] C.J.K. Batty, D.W. Robinson, Positive one-parameter semigroups on ordered Banach spaces, *Acta Appl. Math.* 1 (1984), 221–296.

[3] C. Bennett, R. Sharpley, *Interpolation of Operators,* Academic Press, Orlando, 1988.

[4] M.Š. Birman, M.Z. Solomyak, *Spectral theory of selfadjoint operators in Hilbert space*, D. Reidel Publishing Co., Dordrecht, 1987.

[5] M.Š. Birman, M.Z. Solomyak, Operator integration, perturbations and commutators, *Zap. Nauchn. Sem. Leningrad. Otdel. Mat. Inst. Steklov. (LOMI), Issled. Linein. Teorii Funktsii.* 17 (1989), 34–66.

[6] A.P. Calderón, Spaces between L^1 and L^∞ and the theorem of Marcinkiewicz, *Studia Math.* 26 (1966), 273–299.

[7] V.I. Chilin, F.A. Sukochev, Symmetric spaces on semi-finite von Neumann algebras, *Dokl. Akad. Nauk. SSSR* 13 (1990), 811–815 (Russian).

[8] E.B. Davies, Lipschitz continuity of functions of operators in the Schatten classes, *J. London Math. Soc.* 37 (1988), 148–157.

[9] J. Dixmier, *Von Neumann Algebras,* North-Holland Mathematical Library, Vol. 27, North-Holland, Amsterdam, 1981.

[10] P.G. Dodds, T.K. Dodds, B. de Pagter, Non-commutative Banach function spaces, *Math. Z.* 201 (1989), 583–597.

[11] P.G. Dodds, T.K. Dodds, B. de Pagter, A general Marcus inequality, *Proc. Centre Math. Anal. Austral. Nat. Univ.* 24 (1989), 47–57.

[12] Peter G. Dodds, Theresa K.-Y. Dodds, Ben de Pagter, Non-commutative Köthe duality, *Trans. Amer. Math. Soc.* 339 (1993), 717–750.

[13] P.G. Dodds, T.K. Dodds, B. de Pagter, F.A. Sukochev, Lipschitz Continuity of the Absolute Value and Riesz Projections in Symmetric Operator Spaces, *J. of Functional Analysis* 148 (1997), 28–69.

[14] P.G. Dodds, T.K. Dodds, F.A. Sukochev, O.Ye. Tikhonov, A Non-commutative Yosida-Hewitt Theorem and Convex Sets of Measurable Operators Closed Locally in Measure, *Positivity* 9 (2005), 457–484.

[15] Th. Fack, H. Kosaki, Generalized s-numbers of τ-measurable operators, *Pacific J. Math.* 123 (1986), 269–300.

[16] K. Fan, Maximum properties and inequalities for the eigenvalues of completely continuous operators, *Proc. Nat. Acad. Sci. U.S.A.* 37 (1951), 760–766.

[17] D.H. Fremlin, *Measure Theory, Volume 3: Measure Algebras*, Torres Fremlin, Colchester, 2002.

[18] I.C. Gohberg, M.G. Krein, *Introduction to the Theory of Linear Nonselfadjoint Operators*, Translations of Mathematical Monographs, Vol. 18, AMS, Providence, R.I., 1969.

[19] A. Grothendieck, Réarrangements de fonctions et inégalités de convexité dans les algèbres de von Neumann muni d'une trace, *Seminaire Bourbaki*, 1955, 113-01-113-13.

[20] P.R. Halmos, *A Hilbert Space Problem Book,* 2nd Ed., Graduate Texts in Math., Springer-Verlag, New York-Heidelberg-Berlin, 1982.

[21] R.V. Kadison, J.R. Ringrose, *Fundamentals of the theory of operator algebras, Volume I: Elementary Theory,* Academic Press, New York, 1983.

[22] R.V. Kadison, J.R. Ringrose, *Fundamentals of the theory of operator algebras, Volume II: Advanced Theory,* Academic Press, Orlando, 1986.

[23] T. Kato, *Perturbation Theory for Linear Operators*, Classics in Mathematics, Springer-Verlag, Berlin-Heidelberg-New York, 1995.

[24] S.G. Krein, Ju.I. Petunin, E.M. Semenov, *Interpolation of Linear Operators,* Translations of Math. Monographs, Vol. 54, Amer. Math. Soc., Providence, 1982.

[25] G.G. Lorentz and T. Shimogaki, Interpolation theorems for operators in function spaces, *J. Functional An.* 2 (1968), 31–51.

[26] W.A.J. Luxemburg, Notes on Banach function spaces XV, *Indag. Math.* 27 (1965), 415–446.

[27] W.A.J. Luxemburg, Rearrangement invariant Banach function spaces, *Proc. Sympos. in Analysis, Queen's Papers in Pure and Appl. Math.* 10 (1967), 83–144.

[28] A.S. Markus, The eigen- and singular values of the sum and product of linear operators, *Russian Math. Surveys* 19 (1964) 91–120.

[29] B. de Pagter, F.A. Sukochev, H. Witvliet, Double Operator Integrals, *J. of Functional Analysis* 192 (2002), 52–111.

[30] B. de Pagter, F.A. Sukochev, Differentiation of operator functions in non-commutative L_p-spaces, *J. of Functional Analysis* 212 (2004), 28–75.

[31] B. de Pagter, F.A. Sukochev, Commutator estimates and $\mathbb{R}$-flows in non-commutative operator spaces, *Proc. Edinburgh Math. Soc.*, to appear.

[32] F. Riesz, B. Sz.-Nagy, *Functional Analysis*, Frederick Ungar Publishing Co., New York, 1955.

[33] H.H. Schaefer (with M.P. Wolff), *Topological Vector Spaces* (2nd Edition), Springer-Verlag, New York, 1999.

[34] M. Takesaki, *Theory of Operator Algebras I*, Springer-Verlag, Berlin-Heidelberg-New York, 1979.

[35] M. Takesaki, *Theory of Operator Algebras II*, Springer-Verlag, Berlin-Heidelberg-New York, 2003.

[36] M. Terp, *L_p-spaces associated with von Neumann algebras*, Copenhagen University, 1981.

[37] O.Ye. Tikhonov, Continuity of operator functions in topologies connected with a trace on a von Neumann algebra, *Izv. Vyssh. Uchebn. Zaved. Mat.* (1987), 77–79 (in Russian; translated in *Sov. Math.* (*Iz. VUZ*) 31 (1987), 110–114.

[38] H. Witvliet, *Unconditional Schauder decompositions and multiplier theorems*, Ph.D. thesis, Delft University of Technology, 2000.

[39] A.C. Zaanen, *Integration,* North-Holland Publishing Company, Amsterdam, 1967.

[40] A.C. Zaanen, *Riesz Spaces II*, North-Holland Publishing Company, Amsterdam-New York-Oxford, 1983.

Ben de Pagter
Delft Institute of Applied Mathematics
Faculty EEMCS
Delft University of Technology
P.O. Box 5031
2600 GA Delft, The Netherlands
e-mail: `b.depagter@tudelft.nl`

Positivity
Trends in Mathematics, 229–254

Positive Operators on L^p-spaces

Anton R. Schep

1. Introduction

Throughout this paper we denote by L^p the Banach lattice of p-integrable functions on a σ-finite measure space $(X, \mathcal{B}, \mu)$, where $1 \leq p \leq \infty$. We will consider those aspects of the theory of positive linear operators, which are in some way special due to the fact the operators are acting on L^p-spaces. For general information about positive operators on Banach lattices we refer to the texts [1]. [20], and [36]. Our focus on L^p-spaces does not mean that in special cases some of the results can not be extended to a larger class of Banach lattices of measurable function such as Orlicz spaces or re-arrangement invariant Banach function spaces. However in many cases the results in these extensions are not as precise or as complete as in the case of L^p-spaces. We will discuss results related to the boundedness of positive linear operators on L^p-spaces. The most important result is the so-called Schur criterion for boundedness. This criterion is the most frequently used tool to show that a concrete positive linear operator is bounded from L^p to L^q. Then we will show how this result relates to the change of density result of Weis [33]. Next the equality case of Schur's criterion is shown to be closely related to the question whether a given positive linear operator attains its norm. We discuss in detail the properties of norm attaining operators on L^p-spaces and discuss as an example the weighted composition operators on L^p-spaces. Then we return to the Schur criterion and show how it can be applied to the factorization theorems of Maurey and Nikišin. Most results mentioned in this paper have appeared before in print, but sometimes only implicitly and scattered over several papers. Also a number of the proofs presented here are new.

2. Boundedness of positive linear operators

In this section we shall consider a positive operator T acting on a space of (equivalence classes of) measurable functions and give a necessary and sufficient condition for T to define a bounded linear operator from $L^p(Y, \nu)$ into $L^q(X, \mu)$ where $1 \leq q \leq p \leq \infty$ and obtain a bound for $\|T\|$. In what follows we follow initially closely [12], where the Schur criteria was proved as a factorization theorem, which allow us later on to derive the Maurey factorization theorem as an easy consequence. Note first that if we already know that a positive linear operator T maps L^p into L^q, then it is elementary that T is bounded, see, e.g., [1], Theorem 1.31 or [20], Proposition 1.3.5. For that reason we will need to consider initially more general domains and range spaces for the operators we consider. Let $L^0(X,\mu)$ denote the space of a.e. finite measurable functions on X and let $M(X,\mu)$ denote the space of extended real-valued measurable functions on X. Assume that T is defined on an *ideal* L of measurable functions, i.e., a linear subspace of $L^0(Y,\nu)$ such that if $f \in L$ and $|g| \leq |f|$ in L^0, then $g \in L$. By L_+ we denote the collection of nonnegative functions in L. A positive linear operator $T : L \to L^0(X,\mu)$ is called *order continuous* if $0 \leq f_n \uparrow f$ a.e. and $f_n, f \in L$ imply that $Tf_n \uparrow Tf$ a.e. We first prove that such operators have "adjoints".

Theorem 1. *Let L be an ideal of measurable functions on (Y,ν) and let T be a positive order continuous operator from L into $L^0(X,\mu)$. Then there exists an operator $T^t : L^0(X,\mu)_+ \to M(Y,\nu)_+$ such that for all $f \in L_+$ and all $g \in L^0(X,\mu)_+$ we have*

$$\int_X (Tf)g\,d\mu = \int_Y f(T^t g)\,d\nu.$$

Proof. Assume first that there exists a function $f_0 > 0$ a.e. in L. Let $g \in L^0(X,\mu)_+$. Then we define $\phi : L_+ \to [0,\infty]$ by $\phi(f) = \int (Tf)g\,d\mu$. Since $Tf_0 < \infty$ a.e. we can find $X_1 \subset X_2 \subset \cdots \uparrow X$ such that for all $n \geq 1$ we have

$$\int_{X_n} (Tf_0)g\,d\mu < \infty.$$

Let $L_{f_0} = \{h : |h| \leq cf_0 \text{ for some constant } c\}$ and define $\phi_n : L_{f_0} \to \mathbb{R}$ by

$$\phi_n(h) = \int_{X_n} (Th)g\,d\mu.$$

The order continuity of T now implies (through an application of the Radon–Nikodym theorem) that there exists a function $g_n \in L^1(Y, f_0 d\nu)$ such that for all $h \in L_{f_0}$ we have

$$\phi_n(h) = \int_Y hg_n\,d\nu,$$

see, e.g., [36], Theorem 86.3. Moreover we can assume that $g_1 \leq g_2 \leq \ldots$ a.e. Let $g_0 = \sup g_n$. An application of the monotone convergence theorem now gives

$$\int_X (Th)g\,d\mu = \int_Y hg_0\,d\nu$$

for all $0 \leq h \in L_{f_0}$. The order continuity of T and another application of the monotone convergence theorem now give

$$\int_X (Tf)g\,d\mu = \int_Y fg_0\,d\nu$$

for all $0 \leq f \in L$. If we put $T^t g = g_0$, then the theorem holds in case L contains a strictly positive f_0. In case no such f_0 exists in L, then we can find via Zorn's lemma a maximal disjoint system (f_n) in L^+ and apply the above argument to the restriction of T to the functions $f \in L$ with support in the support Y_n of f_n. We obtain that way functions g_n with support in Y_n so that for all such f we have

$$\int_X (Tf) g d\mu = \int_{Y_n} f g_n d\nu$$

Now define $T^t g = \sup g_n$ and one can easily verify that in this case the theorem again holds. This completes the proof of the theorem. $\square$

The above theorem allows us to define for any positive operator $T : L \to L^0(X,\mu)$ an adjoint operator T^*. Let $N = \{g \in L^0(X,\mu) : T^t(|g|) \ \in L^0(Y,\nu)\}$ and define $T^* g = T^t g^+ - T^t g^-$ for $g \in N$. It is easy to see that T^* is positive linear operator from N into $L^0(Y,\nu)$ such that

$$\int_X (Tf) g d\mu = \int_Y f(T^* g) d\nu$$

holds for all $0 \le f \in L$ and $0 \le g \in N$. Observe that in case $T : L^p \to L^q$ is a bounded linear operator and $1 \le p, q < \infty$ then T^* as defined as above is an extension of the Banach space adjoint. The above construction is motivated by the following example.

Example 2. Let $T(x,y) \ge 0$ be $\mu \times \nu$-measurable function on $X \times Y$. Let $L = \{f \in L^0(Y,\nu)$ such that $\int T(x,y)|f(y)| d\nu < \infty$ a.e.$\}$ and define T as the integral operator $Tf(x) = \int_Y T(x,y) f(y) d\nu(y)$ on L. Then one can check (using Tonelli's theorem) that $N = \{g \in L^0(X,\mu)$ such that $\int_Y T(x,y)|g(x)| d\mu < \infty$ a.e.$\}$ and that the operator T^* as defined above is the the integral operator $\int_X T(x,y) g(x) d\mu(x)$.

We now present a Hölder inequality for positive linear operators. The result is known in ergodic theory (see [16], Lemma 7.4). We include the short proof.

Theorem 3 (Abstract Hölder inequality). *Let L be an ideal of measurable functions on (Y,ν) and let T be a positive operator from L into $L^0(X,\mu)$. If $1 < p < \infty$ and $p' = \frac{p}{p-1}$, then we have*

$$T(fg) \le T(f^p)^{\frac{1}{p}} T(g^{p'})^{\frac{1}{p'}}$$

for all $0 \le f, g$ with $fg \in L$, $f^p \in L$ and $g^{p'} \in L$.

Proof. For any two positive real numbers x and y we have the inequality $x^{\frac{1}{p}} y^{\frac{1}{p'}} \le \frac{1}{p} x + \frac{1}{p'} y$, so that if $0 \le f, g$ with $fg \in L$, $f^p \in L$ and $g^{p'} \in L$, then for any $\alpha > 0$

$$\begin{aligned} T(fg) &= T\left((\alpha f)\left(\frac{1}{\alpha}\right) g\right) \le \frac{1}{p} T((\alpha f)^p) + \frac{1}{p'} T\left(\left(\frac{1}{\alpha} g\right)^{p'}\right) \\ &= \frac{1}{p} \alpha^p T(f^p) + \frac{1}{p'} \frac{1}{\alpha^{p'}} T(g^{p'}) \end{aligned}$$

Now for each $x \in X$ such that $T(f^p)(x) \ne 0$ choose the number α so that $\alpha^p T(f^p)(x) - \frac{1}{\alpha^{p'}} T(g^{p'})(x)$. $\square$

The following theorem gives a sufficient condition for a positive order continuous operator to be bounded between certain L^p-spaces.

Theorem 4. *Let L be an ideal of measurable functions on (Y,ν) and let T be a positive order continuous linear operator from L into $L^0(X,\mu)$. Let $1<q\le p<\infty$ and assume there exists $f_0\in L$ with $0<f_0$ a.e. and there exists $\lambda>0$ such that*

$$T^*(Tf_0)^{q-1}\le \lambda f_0^{p-1} \tag{1}$$

and in case $q<p$ also

$$Tf_0\in L^q(X,\mu). \tag{2}$$

Then T can be extended to a positive linear map from $L^p(Y,\nu)$ into $L^q(X,\mu)$ with

$$\|T\|_{p,q}\le \lambda^{\frac{1}{p}}\|Tf_0\|_q^{1-\frac{q}{p}} \tag{3}$$

in case $q<p$ and in case $p=q$

$$\|T\|_{p,p}\le \lambda^{\frac{1}{p}}. \tag{4}$$

If also $f_0\in L^p(Y,\nu)$, then

$$\|T\|_{p,q}\le \lambda^{\frac{1}{q}}\|f_0\|_p^{\frac{p-q}{q}}. \tag{5}$$

Proof. Define the positive linear operator $S: L^p(Y,\nu)\to L^0(X,\mu)$ by $Sf=(Tf_0)^{\frac{q-p}{p}}\cdot Tf$, note that $S=T$ in case $p=q$. Then it is straightforward to verify that $S^*(h)=T^*((Tf_0)^{\frac{q-p}{p}}\cdot h)$. This implies that

$$S^*(Sf_0)^{p-1}=S^*((Tf_0)^{\frac{q(p-1)}{p}})=T^*(Tf_0)^{q-1}\le\lambda f_0^{p-1},$$

i.e., S satisfies (1) with $p=q$. Let $Y_n=\{y\in Y:\frac{1}{n}\le f_0(y)\le n\}$. Then $L^\infty(Y_n,\nu)\subset L$. Let $0\le u\in L^\infty(Y_n,\nu)$. Then we have

$$\begin{aligned}\int (Su)^p d\mu &= \int S(uf_0^{-\frac{1}{p'}}f_0^{\frac{1}{p'}})^p d\mu\\ &\le \int S(u^p f_0^{-p+1})(Sf_0)^{\frac{p}{p'}}d\mu \text{ (Abstract Hölder inequality)}\\ &= \int u^p f_0^{-p+1}S^*(Sf_0)^{(p-1)}d\nu \le \int u^p f_0^{-p+1}\lambda f_0^{p-1}d\nu=\lambda\|u\|_p^p.\end{aligned}$$

Hence

$$\|Su\|_p\le\lambda^{\frac{1}{p}}\|u\|_p$$

for all $0\le u\in L^\infty(Y_n,d\nu)$. If $0\le u\in L$, let $u_n=\min(u,n)\chi_{Y_n}$. Then $u_n\uparrow u$ a.e. and $\|Su\|_p\le\lambda^{\frac{1}{p}}\|u\|_p$ holds for each u_n. The order continuity of T and the monotone convergence theorem imply that $\|S\|_{p,p}\le\lambda^{\frac{1}{p}}$. Note that in case $p=q$ this proves (4). In case $q<p$ define the multiplication operator M, by $Mh=(Tf_0)^{\frac{p-q}{p}}\cdot h$. Then (2) implies, by means of Hölder's inequality with $r=\frac{p}{q}, r'=\frac{p}{p-q}$, that $\|M\|_{p,q}\le\|Tf_0\|^{1-\frac{q}{p}}$. The inequality (3) follows now from the factorization $T=MS$. Inequality (5) follows from (3) by using the inequality $\|Tf_0\|_q\le\|T\|_{p,q}\|f_0\|_p$ and solving for $\|T\|_{p,q}$. This completes the proof of the theorem. □

In applications of the above theorem it is important to realize that f_0 does not have to be an element of $L^p(Y,\nu)$. We give some examples to illustrate this.

Example 5. Let $X = Y = [0,\infty)$ with $\mu = \nu$ equal to the Lebesgue measure and define the integral operator T by $Tf(x) = \frac{1}{x}\int_0^x f(t)dt$. An easy computation shows that for $1 < p < \infty$ the equality $T^*(Tf_0)^{p-1} \le \lambda f_0^{p-1}$ holds for some constant $\lambda = \lambda(\alpha)$, whenever $f_0(y) = y^\alpha$ for any $-1 < \alpha < 0$. One can verify that in this case $\alpha = -\frac{1}{p}$ gives the best upperbound for $\|T\|_p$, in which case $\lambda = (\frac{p}{p-1})^p$. The inequality $\|Tf\|_p \le \|T\|_p\|f\|$ is then the classical Hardy inequality.

Example 6. Let again $X = Y = [0,\infty)$ with $\mu = \nu$ equal to the Lebesgue measure and define the Laplace integral operator $\mathcal{L}$ by $\mathcal{L}(f)(x) = \int_0^\infty f(y)e^{-xy}\,dy$. Then obviously by the symmetry of the kernel e^{-xy} we have $\mathcal{L}^* = \mathcal{L}$. Let $1 < p < \infty$ and denote by p' the conjugate exponent of p. If we take $f_0(y) = y^{-\frac{1}{p'}}$, then it follows from $\mathcal{L}f_0(x) = \Gamma(\frac{1}{p})x^{-\frac{1}{p}}$ that

$$\mathcal{L}^*(\mathcal{L}f_0)^{p-1} = \Gamma\left(\frac{1}{p}\right)^{\frac{1}{p}} f_0^{p-1}. \tag{6}$$

Hence $\mathcal{L}$ defines a bounded operator on $L^p([0,\infty))$ with norm less or equal to $\Gamma(\frac{1}{p})$. Now $\mathcal{L}$ will have norm less or equal to $\Gamma(\frac{1}{p'})$ on $L^{p'}([0,\infty))$, so that the norm of $\mathcal{L}$ is less or equal to the minimum of $\Gamma(\frac{1}{p})$ and $\Gamma(\frac{1}{p'})$. As the Gamma function $\Gamma(x)$ is decreasing on $(0,1)$ we have that $\Gamma(\frac{1}{p}) < \Gamma(\frac{1}{p'})$ if $1 < p < 2$ and $\Gamma(\frac{1}{p'}) < \Gamma(\frac{1}{p})$ if $p > 2$. This shows that the equality in (6) does not imply that the norm of $\mathcal{L}$ is equal to $\Gamma(\frac{1}{p})$. Later on, when we discuss norm attainment of positive operators, we shall see that the situation is different when $f_0 \in L^p$.

The above theorem is an abstract version of what is called the *Schur test* for boundedness of integral operators (see [12] for the case $p = q = 2$ and see [9], Theorem 1.I for the case $1 < q \le p < \infty$). In these references the Schur test is formulated in a slightly different (but equivalent) form. We will only present this for the case $p = q$.

Theorem 7. *Let L be an ideal of measurable functions on (Y,ν) and let T be a positive order continuous linear operator from L into $L^0(X,\mu)$. Let $1 < p < \infty$ and assume there exists $f_0, g_0 \in L$ with $0 < f_0$, $0 \le g_0$ a.e. and there exists $C > 0$ such that*

$$T(f_0) \le Cg_0 \tag{7}$$

and

$$T^*(g_0^{p-1}) \le Cf_0^{p-1}. \tag{8}$$

Then T can be extended to a positive linear map from $L^p(Y,\nu)$ into $L^p(X,\mu)$ with

$$\|T\|_{p,p} \le C.$$

Proof. The proof is immediate from the previous theorem, since

$$T^*(Tf_0)^{p-1} \le C^{p-1}T^*(g_0^{p-1}) \le C^p f_0^{p-1}. \qquad \square$$

Remark. The above theorem can also be obtained as a consequence of the Riesz interpolation theorem, which should not be surprising as the Riesz interpolation theorem for positive operators can be proved via the Hölder inequality for positive operators. To see this observe first that if $Tf_0(x) \leq Cg_0(x)$ a.e., then $Tf_0(x) = 0$ a.e. on the set where $g_0(x) = 0$ a.e. Define now $\hat{T}$ as follows, $\hat{T}(f) = g_0(x)^{-1}T(ff_0)(x)$, where we define $\frac{0}{0} = 0$. Then $\hat{T}$ maps $L^p(Y, f_0^p d\nu)$ into $L^p(X, g_0^p d\mu)$ if and only if T maps $L^p(Y,\nu)$ into $L^p(X,\mu)$ and in that case $\|\hat{T}\| = \|T\|$. Now the adjoint of $\hat{T}$ with respect to the measure spaces $(Y, f_0^p d\nu)$ and $(X, g_0^p d\mu)$ is given by $\hat{T}^* g = \frac{1}{f_0^{p-1}} T^*(g g_0^{p-1})$. Now the hypotheses of the above theorem say that $\hat{T}(\mathbf{1}_Y) \leq C\mathbf{1}_X$ and $\hat{T}(\mathbf{1}_X) \leq C\mathbf{1}_Y$. It follows from the Riesz interpolation theorem that $\hat{T} : L^s(Y, f_0^p d\nu) \to L^s(X, g_0^p d\mu)$ for all $1 \leq s \leq \infty$ with the norm of $\hat{T}$ less or equal than C. We also observe that if f_0 satisfies $T^*(Tf_0)^{p-1} \leq \lambda f_0^{p-1}$, then f_0 and $g_0 = \lambda^{-\frac{1}{p}} Tf_0$ will satisfy the the conditions of the above theorem with $C = \lambda^{\frac{1}{p}}$, so that for $p = q$ the assumptions in the two above theorems are equivalent.

3. Necessity of the Schur boundedness test

We now discuss the converse to the above theorems, which is due to Gagliardo [9] (Theorem 1.II). We present a proof, which is a slight simplification of the proof given in [9].

Theorem 8. *Let $0 \leq T : L^p(Y,\nu) \to L^q(X,\mu)$ be a positive linear operator and assume $1 < p, q < \infty$. Then for all λ with $\lambda^{\frac{1}{q}} > \|T\|_{pq}$ there exists $0 < f_0$ a.e. in $L^p(Y,\nu)$ such that*

$$T^*(Tf_0)^{q-1} \leq \lambda f_0^{p-1}. \tag{9}$$

Proof. We can assume that $\|T\|_{p,q} = 1$. Then we assume that $\lambda > 1$. Now define $S : L^p(Y,\nu)_+ \to L^p(Y,\nu)_+$ by means of

$$Sf = (T^*(Tf)^{q-1})^{\frac{1}{p-1}}.$$

Then it is easy to verify that $\|f\|_p \leq 1$ implies that $\|Sf\|_p \leq 1$, also that $0 \leq f_1 \leq f_2$ implies that $Sf_1 \leq Sf_2$ and that $0 \leq f_n \uparrow f$ a.e. in L^p implies that $Sf_n \uparrow Sf$ a.e. Let now $0 < g$ a.e. in $L^p(Y,\nu)$ such that $\|g\|_p \leq 1$ and define $f_1 = \frac{\lambda-1}{\lambda} g$. For $n > 1$ we define $f_n = f_1 + \frac{1}{\lambda} Sf_{n-1}$. By induction we verify easily that $f_n \leq f_{n+1}$ and that $\|f_n\|_p \leq 1$ for all n. This implies that there exists f_0 in L^p such that $f_n \uparrow f_0$ a.e. and $\|f_0\|_p \leq 1$. Now $Sf_n \uparrow Sf_0$ implies that $f_0 = f_1 + \frac{1}{\lambda} Sf_0$. Hence $Sf_0 \leq \lambda f_0$, which is equivalent to the inequality (9) and $f_0 \geq f_1 > 0$ a.e., so that $f_0 > 0$ a.e. and the proof is complete. □

Corollary 9. *Let $0 \leq T : L^p(Y,\nu) \to L^p(X,\mu)$ be a positive linear operator and assume $1 < p < \infty$. Then for all $C > \|T\|_p$ there exist $f_0, g_0 \in L^p(Y,\nu)$ with*

$0 < f_0$, $0 \leq g_0$ *a.e. such that*

$$T(f_0) \leq Cg_0 \tag{10}$$

$$\text{and} \qquad T^*(g_0^{p-1}) \leq Cf_0^{p-1}. \tag{11}$$

In the next section we discuss when we can take $g_0 = f_0$ in these inequalities.

We conclude this section with a technical remark about the iteration used in the proof of Theorem 8.

Lemma 10. *Let S, λ, g and f_n be as in the proof of Theorem 8. Assume $Sg \leq \lambda g$. Then also $Sf_0 \leq \lambda f_0$.*

Proof. Let $\alpha = \frac{q-1}{p-1}$. Then $0 < \alpha$ and $S(cf) = c^\alpha Sf$ for all $f \geq 0$. By the proof of Theorem 8 we have

$$\begin{aligned} f_2 = f_1 + \frac{1}{\lambda}S(f_1) &= f_1 + \frac{1}{\lambda}\left(\frac{\lambda-1}{\lambda}\right)^\alpha Sg \\ &\leq f_1 + \left(\frac{\lambda-1}{\lambda}\right)^\alpha g = \frac{\lambda-1}{\lambda}g + \left(\frac{\lambda-1}{\lambda}\right)^\alpha g \\ &\leq \lambda g, \end{aligned}$$

since $(\frac{\lambda-1}{\lambda})^\alpha \leq 1 \leq \lambda + \frac{1}{\lambda} - 1$. By induction it now follows that $f_n \leq \lambda g$ and thus also $f_0 = \lim_{n\to\infty} f_n \leq \lambda g$. □

4. Change of measure and extrapolation

From the above corollary and the discussion following Theorem 7 it follows that if $T : L^p(Y,\nu) \to L^p(X,\mu)$ is a positive linear operator and if $1 < p < \infty$, then for all $C > \|T\|_p$ there exist $f_0, g_0 \in L^p(Y,\nu)$ with $0 < f_0$, $0 \leq g_0$ a.e. such that $\hat{T}(f) = g_0(x)^{-1}T(ff_0)(x)$ maps $L^s(Y, f_0^p d\nu)$ into $L^s(X, g_0^p d\mu)$ for all $1 \leq s \leq \infty$ with the norm of $\hat{T}$ less or equal than C. For applications to, e.g., spectral theory it is desirable that the extrapolated operator for T^n can be taken as $(\hat{T})^n$, which requires that we can take $f_0 = g_0$ in the above corollary. It was proved by Lutz Weis in [33] that we can always do this, except that his proof does not provide us with the same constants as above. In fact we can not always get the exact sharp results in this case, as we will see. The following theorem is a variation of Theorem 2.1 of [33].

Theorem 11. *Let $0 \leq T : L^p(Y,\nu) \to L^p(X,\mu)$ be a positive linear operator and assume $1 < p < \infty$. Then for all $C > \max\{\sqrt[p]{2}, \sqrt[p']{2}\}\|T\|_p$ there exist $f_0 \in L^p(Y,\nu)$ with $0 < f_0$ a.e. such that*

$$T(f_0) \leq Cf_0 \tag{12}$$

and

$$T^*(f_0^{p-1}) \leq Cf_0^{p-1}. \tag{13}$$

Proof. We can assume that $\|T\|_p = 1$. Now define $S : L^p(Y,\nu)_+ \to L^p(Y,\nu)_+$ by means of

$$Sf = \left(\frac{1}{2}T^*(f^{p-1})^{\frac{p}{p-1}} + \frac{1}{2}T(f)^p\right)^{\frac{1}{p}}.$$

Then it is easy to verify that $\|f\|_p \le 1$ implies that $\|Sf\|_p \le 1$, also that $0 \le f_1 \le f_2$ implies that $Sf_1 \le Sf_2$ and that $0 \le f_n \uparrow f$ a.e. in L^p implies that $Sf_n \uparrow Sf$ a.e. As in the proof of Theorem 8 it follows that for $\lambda > 1$ there exists $f_0 > 0$ a.e. such that $Sf_0 \le \lambda f_0$. This implies that $Tf_0 \le \lambda 2^{\frac{1}{p}} f_0$ a.e. and $T^* f_0^{p-1} \le \lambda^{p-1} 2^{\frac{1}{p'}} f_0^{p-1}$ a.e., which completes the proof of the theorem. □

The above theorem had been conjectured in the (infinite) matrix case by Vere-Jones [32], who also conjectured that one could take $C > \|T\|_p$. The first part of this conjecture was first proved by M. Koskela [17] in 1978 for the infinite matrix case, but with a worse constant than the one obtained by Weis. The method of proof of Koskela is to apply Theorem 8 to the operator $I + T$. Koskela also showed that already in the finite matrix case we can not take $C > \|T\|_p$, as conjectured by Vere-Jones. We will present a generalization of Koskela's result which will make this clear.

Theorem 12. *Let $0 \le T : L^p(X,\mu) \to L^p(X,\mu)$ be a positive linear operator with $1 < p < \infty$ and such that there exists a constant $a > 0$ with $T \ge aI$. Assume that for all $C > \|T\|_p$ there exists $f_0 \in L^p(X,\mu)$ with $0 < f_0$ a.e. such that*

$$T(f_0) \le Cf_0 \tag{14}$$

and

$$T^*(f_0^{p-1}) \le Cf_0^{p-1}. \tag{15}$$

Then $r(T) = \|T\|$.

Proof. Define $S = T - aI$. Then $S \ge 0$ and for $C > \|T\|$ there exists $f_0 > 0$ a.e. such that $Sf_0 \le (C-a)f_0$ and $S^* f_0^{p-1} \le (C-a)f_0^{p-1}$. It follows from Theorem 7 that $\|S\| \le C - a$. This implies that $\|S\| \le \|T\| - a = \|S + aI\| - a \le \|S\|$. Hence $\|S\| + a = \|S + aI\|$, i.e., the operator S satisfies the Daugavet equation. As L^p, for $1 < p < \infty$, is uniformly convex, it follows from the main result of [2] that $r(S) = \|S\|$. This implies that $r(T) = a + r(S) = a + \|S\| = a + \|T - aI\|$. By the same argument this implies that $r(T) = \epsilon + \|T - \epsilon I\|$ for all $0 < \epsilon < a$, so $r(T) = \|T\|$. □

Example 13. From the above theorem it is now clear how to get an example for which the second part of Vere-Jones conjecture fails. Let T be the positive operator on the two-dimensional $\ell^2(2)$ defined by the matrix

$$\begin{pmatrix} 1 & 0 \\ 1 & 1 \end{pmatrix}.$$

It is now not difficult to see that $r(T) = 1$ and $\|T\|_2 \ge \sqrt{2}$. Hence by the above theorem T must provide a counterexample for the conjecture of Vere-Jones. One easily checks that $C = 2$ is the best constant for which the inequalities (14) and (15) hold. As $\|T\| = \sqrt{\frac{3+\sqrt{5}}{2}} > \sqrt{2}$, it also shows that the factor $2^{\frac{1}{2}}$ in Weis' theorem is not sharp in this example. It seems an open question what is exactly the best constant in Theorem 11.

For $p = 2$ we have a class of operators, which includes the positive (in the sense of the lattice ordering) normal operators, for which Vere-Jones conjecture holds.

Theorem 14. *Let $0 \leq T : L^2(X, d\mu) \to L^2(X, d\mu)$ be a positive linear operator such that $T^*T \leq TT^*$, or $TT^* \leq T^*T$ in the lattice ordering. Then for all $C > r(T)$ there exists $f_0 \in L^2(X, \mu)$ with $0 < f_0$ a.e. such that*

$$T(f_0) \leq Cf_0 \tag{16}$$

and

$$T^*(f_0) \leq Cf_0. \tag{17}$$

In particular we have $r(T) = \|T\|$.

Proof. We will assume that $T^*T \leq TT^*$, the other case follows by duality. Let $C > r(T)$. Then the resolvent operator $R(C, T)$ satisfies $TR(C, T) = CR(C, T) - I \leq CR(C, T)$ and similarly $T^*R(C, T^*) \leq CR(C, T^*)$. Moreover the hypothesis $T^*T \leq TT^*$ implies that also $T^*R(C, T) \leq R(C, T)T^*$ by the Neumann series for $R(C, T)$. Now let $h_0 > 0$ a.e. in $L^2(X, d\mu)$ and put $f_0 = R(C, T)R(C, T^*)h_0$. Then by the above inequalities $Tf_0 \leq Cf_0$ and $T^*f_0 \leq Cf_0$. Moreover, since also $f_0 > 0$ a.e., this implies that $\|T\| \leq C$ for all $C > r(T)$, which shows that $r(T) = \|T\|$. □

Example 15. To have a non-normal operator T which satisfies the above conditions, we only need to take T to be the unilateral shift $T(\{\xi_n\}) = \{0, \xi_1, \xi_2, \dots\}$ on ℓ^2. In this case $T^*T = I$, while $TT^*(\{\xi_n\}) = \{0, \xi_2, \dots\}$, so that $TT^* \leq T^*T$.

We now indicate some applications of Theorem 11, which were obtained by the author in [27]. We start with an Egoroff type theorem.

Theorem 16. *Let $0 \leq T_\tau : L^p(X, \mu) \to L^p(X, \mu)$ with $1 \leq p \leq \infty$ and $\mu(X) < \infty$. Assume that the downward directed system $\{T_\tau\}$ satisfies $T_\tau \downarrow 0$. Then for all $\epsilon > 0$ there exists a measurable set $X_\epsilon \subset X$ with $\mu(X_\epsilon^c) < \epsilon$ such that if $1 < p \leq \infty$ we have $\|\chi_{X_\epsilon} T_\tau\| \downarrow 0$ and if $1 \leq p < \infty$ we have $\|T_\tau \chi_{X_\epsilon}\| \downarrow 0$.*

Proof. Let $0 \leq T_0 : L^p(X, \mu) \to L^p(X, \mu)$ such that $0 \leq T_\tau \leq T_0$ for all τ. Assume first that $1 < p < \infty$. From Theorem 11 it follows that there exists a strictly positive f_0 such that $\hat{T} = f_0^{-1} T_0 f_0$ is a positive operator from $L_s(X, f_0^{p-1} d\mu) \to L_s(X, f_0^{p-1} d\mu)$ for all $1 \leq s \leq \infty$. Define now $\hat{T}_\tau = f_0^{-1} T_\tau f_0$. It follows then from $0 \leq \hat{T}_\tau \leq \hat{T}_0$ that also $\hat{T}_\tau$ is a positive operator from $L_s(X, f_0^{p-1} d\mu) \to L_s(X, f_0^{p-1} d\mu)$ for all $1 \leq s \leq \infty$. In particular $\hat{T}_\tau$ is a positive operator from $L_\infty(X, f_0^{p-1} d\mu) \to L_\infty(X, f_0^{p-1} d\mu)$. From $T_\tau \downarrow 0$ it follows that $\hat{T}_\tau(\mathbf{1}) \downarrow 0$ a.e. on

X and thus there exist τ_n such that $\hat{T}_{\tau_n}(\mathbf{1}) \downarrow 0$ a.e. on X. From the classical Egoroff theorem we conclude that there exists for $\epsilon > 0$ a measurable set $X_\epsilon \subset X$ with $\mu(X_\epsilon^c) < \epsilon$ such that $\|\chi_{X_\epsilon}\hat{T}_{\tau_n}(\mathbf{1})\|_\infty \downarrow 0$. For any positive operator S on L^∞ we have that $\|S\| = \|S\mathbf{1}\|_\infty$. Hence $\|\chi_{X_\epsilon}\hat{T}_{\tau_n}\|_{L_\infty} \downarrow 0$. Applying now the Riesz interpolation theorem to the operator $\chi_{X_\epsilon}\hat{T}_{\tau_n}$ we obtain that $\|\chi_{X_\epsilon}\hat{T}_{\tau_n}\|_{L_p(X, f_0^{p-1}d\mu)} \downarrow 0$. From this it follows that $\|\chi_{X_\epsilon}T_{\tau_n}\|_{L_p(X,d\mu)} \downarrow 0$. In case $p = \infty$ we can just use the above argument without extrapolation and interpolation. The case $1 \le p < \infty$ follows by duality from the above result by applying it to the adjoints T_τ^*. □

We note that the above theorem gives a uniform way of proving some known results from the literature, i.e., we have the following corollary of results due to Semenov et al ([29, 30]).

Corollary 17. *Let $0 \le T : L^p(X,\mu) \to L^p(X,\mu)$ with $1 \le p \le \infty$. Then*

$$\inf\{\|\chi_P T \chi_Q\| : P, Q \subset X, \mu(P) > 0, \mu(Q) > 0\} = 0.$$

Proof. Without loss of generality we can assume that $\mu(X) < \infty$. Assume first that $1 \le p < \infty$. Let $P_n \downarrow \emptyset$ with $\mu(P_n) > 0$ and define $T_n = \chi_{P_n} T$. Then $T_n \downarrow 0$, so by the above theorem there exists for $\epsilon > 0$ a measurable set $X_\epsilon \subset X$ with $\mu(X_\epsilon^c) < \epsilon$ such that we have $\|T_n \chi_{X_\epsilon}\| \downarrow 0$, from which the desired results immediately follows. For $p = \infty$ the result follows by considering $T_n = T\chi_{P_n}$ and applying the corresponding result from the above theorem. □

Another application of Theorem 16 is an improvement of the above corollary in case $T \perp I$. As the result shows, we can restrict ourselves in the above corollary to diagonal operator blocks in this case.

Theorem 18. *Let $0 \le T : L^p(X,\mu) \to L^p(X,\mu)$ with $1 \le p \le \infty$. Then $T \perp I$ if and only if for all $Q \subset X$ with $\mu(Q) > 0$ we have*

$$\inf\{\|\chi_P T \chi_P\| : P \subset Q, \mu(P) > 0\} = 0.$$

Proof. Assume first $T \perp I$. Then we can assume that $Q = X$ and $\mu(X) < \infty$. Assume first that $1 \le p < \infty$. We denote by $\Pi = \{P_1, \ldots, P_n\}$ a partition of X in sets of positive measure and denote by $\mathcal{P}$ the set of all such partitions. The set $\mathcal{P}$ is downward directed, if we partially order it by refinement. For $\Pi \in \mathcal{P}$ we then define $S(\Pi) = \sum_{i=1}^n \chi_{P_i} T \chi_{P_i}$. It follows now from Theorem 1.1 of [25] that $S(\Pi) \downarrow_{\mathcal{P}} 0$ whenever $T \perp I$. Let $0 < \epsilon < \mu(X)$. Then by Theorem 16 there exists a measurable set $X_\epsilon \subset X$ with $\mu(X_\epsilon^c) < \epsilon$ such that we have $\|S(\Pi)\chi_{X_\epsilon}\| \downarrow_{\mathcal{P}} 0$. Hence there exists a partition $\Pi = \{P_1, \ldots, P_n\}$ of X such that $\|S(\Pi)\chi_{X_\epsilon}\| < \epsilon$. As $\mu(X_\epsilon) > 0$ there exists an $i \in \{1, \ldots, n\}$ such that $X_\epsilon \cap P_i$ has positive measure. If we put $P = X_\epsilon \cap P_i$ then $\|\chi_P T \chi_P\| < \epsilon$ and we conclude that

$$\inf\{\|\chi_P T \chi_P\| : P \subset X, \mu(P) > 0\} = 0.$$

The case $p = \infty$ is completely analogous and therefore left to the reader. For the converse, assume that T is not disjoint with I. Then there exists $Q \subset X$ and

$c > 0$ such that $T \geq c\chi_Q$. This implies that $\chi_P T \chi_P \geq c\chi_P$ for all $P \subset Q$, so $\|\chi_P T \chi_P\| \geq c > 0$ for all $P \subset Q$, which is a contradiction. □

We present another application of Theorem 16 to almost compact operators. For the purpose of this paper let us define almost compactness as follows. Let $1 < p < \infty$ and T a positive linear operator from $L^p(X, \mu)$ into itself. Then T is called *almost compact* if there exist measurable $X_n \subset X$ with $X_n \uparrow X$ such that $\chi_{X_n} T$ is a compact operator for all $n \geq 1$.

Theorem 19. *Let $1 < p < \infty$ and T a positive linear operator from $L^p(X, \mu)$ into itself. Then T is almost compact if and only if T is in the band generated by all the positive compact linear operators on $L^p(X, \mu)$. In particular, every positive integral operator on $L^p(X, \mu)$ is almost compact.*

Proof. If T is almost compact, then obviously T is in the band generated by the positive compact operators. Conversely, if $0 \leq T$ is in the band generated by the compact positive linear operators, then by the Dodds-Fremlin Theorem (see [1], or [36]) there exist $0 \leq T_n \uparrow T$ with each T_n compact. The result then follows immediately from Theorem 16. In case T is a positive integral operator, then T is in the band generated by the finite rank operators, so T is therefore almost compact by the above. □

One can extend the notion of almost compactness to any bounded linear operator on L^p. Weis [33] proved in that context that the collection of all norm bounded almost compact operators equals the norm closure of the integral operators on L^p. The above theorem can be viewed as a regular operator norm version of Weis' result. In the context of the above theorem it needs to observed that there exist positive linear compact operators on $L^p([0, 1])$ for $1 < p < \infty$, which are not integral. This was proved by Fremlin [7] for $p = 2$ and Wickstead [34, Theorem 3.4] extended Fremlin's construction to $1 < p < \infty$.

5. Norm attaining positive linear operators

In this section we discuss the relation between the attaining of its norm of a positive linear operator T and equality in the inequality of Theorem 4. Recall that a bounded linear operator $T : X \to Y$ between Banach spaces is called *norm attaining* if for some $0 \neq f \in X$ we have $\|Tf\|_Y = \|T\|\|f\|_X$.

Then for any bounded linear operator $T : L^p(Y, \nu) \to L^q(X, \mu)$ with $1 < p, q < \infty$ we call a function $0 \neq f \in L^p(X, \mu)$ a *critical point* of T if for some real number λ we have

$$T^*(\operatorname{sgn}(Tf)|Tf|^{q-1}) = \lambda \operatorname{sgn}(f)|f|^{p-1} \tag{18}$$

(such a function f is at least formally a solution to the Euler-Lagrange equation for the variational problem implicit in the definition of $\|T\|_{p,q}$). In the case that T is positive and $f \geq 0$ a.e. the equation 18 takes on the simpler form

$$T^*((Tf)^{q-1}) = \lambda f^{p-1} \tag{19}$$

The connection between critical points and the previous sections is contained in the next theorem.

Theorem 20. *Let L be an ideal of measurable functions on (Y,ν) and let T be a positive order continuous linear operator from L into $L^0(X,\mu)$. Let $1 < q \le p < \infty$ and assume there exists $f_0 \in L^p \cap L$ with $0 < f_0$ a.e. and there exists $\lambda > 0$ such that*

$$T^*(Tf_0)^{q-1} = \lambda f_0^{p-1}. \tag{20}$$

Then T can be extended to a positive linear map from $L^p(Y,\nu)$ into $L^q(X,\mu)$ with

$$\|T\|_{p,q} = \lambda^{\frac{1}{q}} \|f_0\|_p^{\frac{p}{q}-1}.$$

Proof. Integrating the equation (20) against f_0 gives that $\|Tf_0\|_q^q = \lambda \|f_0\|_p^p$ which shows $Tf_0 \in L^q$. From Theorem 4 it follows that T can be extended to a positive linear operator from L^p to L^q with

$$\|T\|_{p,q} \le \lambda^{\frac{1}{q}} \|f_0\|_p^{\frac{p}{q}-1}.$$

Now it follows from $\|Tf_0\|_q^q = \lambda \|f_0\|_p^p$ that

$$\|T\|_{p,q} \ge \lambda^{\frac{1}{q}} \|f_0\|_p^{\frac{p}{q}-1}$$

and the proof is complete. □

We now recall some general facts about smooth Banach spaces. Let E be a Banach space and let E^* denote its dual space. If $f^* \in E^*$ then we denote by $\langle f, f^* \rangle$ the value of f^* at $f \in E$. If $0 \ne f \in E$ then $f^* \in E^*$ *norms* f if $\|f^*\| = 1$ and $\langle f, f^* \rangle = \|f\|$. By the Hahn-Banach theorem there always exist such norming linear functionals. A Banach space E is called *smooth* if for every $0 \ne f \in E$ there exists a unique $f^* \in E^*$ which norms f. Geometrically this is equivalent with the statement that at each point f of the unit sphere of E there is a unique supporting hyperplane. It is well known that E is smooth if and only if the norm is Gâteaux differentiable at all points $0 \ne f \in E$ (see, e.g., [4]). If E is a smooth Banach space and $0 \ne f \in E$, then denote by $\Theta_E(f)$ the unique element of E^* that norms f. Note that $\|\Theta_E(f)\| = 1$. For basic properties of smooth Banach spaces and continuity properties of the map $f \mapsto \Theta_E(f)$ we refer to [4, part 3, Chapter 1].

The basic examples of smooth Banach spaces are the spaces $L^p(X,\mu)$ where $1 < p < \infty$. For $0 \ne f \in L^p(X,\mu)$ one can easily show that

$$\Theta_{L^p}(f) = \|f\|_p^{-(p-1)} \operatorname{sgn}(f)|f|^{p-1} \tag{21}$$

by considering when equality holds in Hölder's inequality. The following proposition shows the relation between critical points and norm attainment of operators on smooth Banach spaces.

Proposition 21. *Let $T : E \to F$ be a bounded linear operator between smooth Banach spaces. If T attains its norm at $0 \neq f \in E$ then there exists a real number α such that*

$$T^*(\Theta_F(Tf)) = \alpha\Theta_E(f) \tag{22}$$

and the norm of T is given by

$$\|T\| = \alpha.$$

Proof. Define $\Lambda_1, \Lambda_2 \in E^*$ by

$$\Lambda_1(h) = \langle h, \Theta_E(f)\rangle$$

$$\Lambda_2(h) = \frac{1}{\|T\|}\langle Th, \Theta_F(Tf)\rangle = \frac{1}{\|T\|}\langle h, T^*(\Theta_F(Tf))\rangle .$$

Then $\|\Lambda_1\| = 1$ (since $\|\Theta_E(f)\| = 1$) and $\Lambda_1(f) = \|f\|$, so Λ_1 norms f. Similarly $\|\Theta_F(Tf)\| = 1$ implies that $\|\Lambda_2\| \leq 1$, but using $\|Tf\| = \|T\|\|f\|$ we have $\Lambda_2(f) = \|f\|$. Therefore Λ_2 also norms f. The smoothness of E now implies that $\Lambda_1 = \Lambda_2$. Hence (22) holds with $\alpha = \|T\|$ as claimed. □

Applying the above proposition to positive operators on L^p-spaces we get the following result.

Theorem 22. *Let $1 < p, q < \infty$ and $0 \leq T : L^p \to L^q$ a positive linear operator, which attains its norm at $0 \leq f \in L^p$. Then f is a critical point of T, i.e.,*

$$T^*(Tf)^{q-1} = \lambda f^{p-1}$$

where

$$\lambda = \|T\|_{p,q}^q \|f\|_p^{q-p}.$$

It is well known that there exist positive linear operators between L^p spaces, who do not attain their norm, e.g., the operator $Tf(x) = xf(x)$ from $L^p([0,1])$ into itself does not attain its norm on L^p for $1 \leq p < \infty$. On the other hand every positive linear operator T from $L^\infty(Y,\nu)$ into $L^q(X,\mu)$ with $1 \leq q \leq \infty$ assumes its norm and $\|T\|_{\infty,q} = \|T(\mathbf{1})\|_q$. It was however proved by Lindenstraus ([18]) that the set of norm attaining operators from $L^p(Y,\nu)$ into $L^q(X,\mu)$ with $1 < p, q < \infty$ is norm dense in the collection of all bounded linear operators. Grząślewicz ([11]) observed that with essentially the same proof as in [18] the corresponding result for positive operators holds. In fact we have the following result.

Theorem 23. *Let $0 \leq T : L^p \to L^q$ be a positive linear operator and assume $1 < p, q < \infty$. Then for all $\epsilon > 0$ there exists a positive norm attaining operator $\hat{T} \geq T$ such that $\|\hat{T} - T\| < \epsilon$.*

Proof. We will only sketch the proof and for the verification of the details we refer the reader to [18]. Without loss of generality we can assume that $\|T\| = 1$ and that $0 < \epsilon < \frac{1}{3}$. Now choose $\epsilon_k > 0$ such that $2\sum_{i=1}^\infty \epsilon_i < \epsilon$, $2\sum_{i=k+1}^\infty \epsilon_i < \epsilon_k^2$, and $\epsilon_k < \frac{1}{10k}$. Then we construct inductively positive operators T_k, positive $f_k \in L^p$, and positive $g_k \in L^{q'}$ as follows. Define $T_1 = T$ and then $f_k \geq 0$ with $\|f_k\|_p = 1$ such that $\|T_k f_k\|_q \geq \|T_k\|_{p,q} - \epsilon_k^2$. Then find $0 \leq g_k \in L^{q'}$ with $\|g_k\|_{q'} = 1$ such that $\int g_k T_k(f_k)\,d\mu = \|T_k f_k\|_q$. Next define $T_{k+1} = T_k + \epsilon_k T_k^* g_k \otimes T_k f_k$. One can then check that $T_k \to \hat{T}$ in norm, $\|\hat{T} - T\| < \epsilon$ and that $\hat{T}$ attains its norm. Moreover $T_{k+1} \geq T_k \geq T$ implies that $\hat{T} \geq T$. □

The norm density of norm attaining operators for the remaining values of p and q have been proved too. The case $p = 1$ and $1 < q < \infty$ follows from a more general theorem due to Uhl ([31]), the case $p = q = 1$ was proved by Iwanik ([13]), while the case $p = 1$, $q = \infty$ was proved by Finet and Paya for the σ-finite case in [8] and in general by Paya and Saleh in [23]. The corresponding result for positive linear operators follows in these cases immediately as the operator norm $\|T\|_{p,q}$ equals the regular operator norm in all these cases. Now we discuss the structure of the set of norm attainers. Following [10] we define for a bounded operator $T : L^p \to L^q$

$$M(T) = \{f \in L^p : \|Tf\|_q = \|T\|_{p,q}\|f\|_p\}.$$

Hence T is norm attaining if and only if $M(T) \neq \{0\}$. We will be only considering $M(T)$ for positive linear operators. In this case we have the following proposition.

Proposition 24. *Let $1 < p, q < \infty$ and $0 \leq T : L^p(Y, \nu) \to L^q(X, \mu)$ a positive linear operator. Then the following hold.*

1. *If $f \in M(T)$, then $|f| \in M(T)$ and $|Tf| = T|f|$, $Tf^+ \wedge Tf^- = 0$.*
2. *If $0 \leq f \in M(T), 0 \leq g \in L^p$ and $f \wedge g = 0$ a.e, then $Tf \wedge Tg = 0$ a.e.*
3. *If $p = q$, $f \in M(T)$ and $f = f_1 + f_2$ with $f_1 \perp f_2$ and $Tf_1 \perp Tf_2$, then $f_1, f_2 \in M(T)$. In particular, if $p = q$, then $f \in M(T)$ implies that $f^+, f^- \in M(T)$.*

Proof. For the first part, note that $|Tf| \leq T|f|$, which implies immediately that $|f| \in M(T)$ and $T|f| = |Tf|$ for all $f \in M(T)$. Now $T|f| = |Tf|$ implies that

$$\|Tf^+ + Tf^-\|_q = \|T|f|\|_q = \|Tf\|_q = \|Tf^+ - Tf^-\|_q,$$

which implies that $Tf^+ \wedge Tf^- = 0$. Now let $0 \leq f \in M(T)$ and assume $f \neq 0$. We can then assume that $\|T\|_{p,q} = 1 = \|f\|_p$. Then from Theorem 22 it follows that $T^*(Tf)^{q-1} = f^{p-1}$. Multiplying both sides with g and integrating over Y we get

$$\int_X (Tf)^{q-1}(x) Tg(x)\,d\mu = \int_Y T^*(Tf)^{q-1}(y) g(y)\,d\nu = \int_Y f^{p-1}(y) g(y)\,d\nu = 0,$$

which implies that $Tf \wedge Tg = 0$ a.e. To prove part (3) we again assume $\|T\|_p = 1 = \|f\|_p$. Then we have that

$$1 = \|Tf\|_p^p = \|Tf_1\|_p^p + \|Tf_2\|_p^p \leq \|f_1\|_p^p + \|f_2\|_p^p = \|f\|_p^p = 1.$$

Hence $\|Tf_1\|_p^p = \|f_i\|_p^p$ for $i = 1, 2$. □

By $\mathrm{supp}(f)$ we denote the set $\{y \in Y : f(y) \neq 0\}$. Note this set should be considered as an element of the measure algebra of ν, as it is only defined up to a set of ν-measure zero. We denote by $\Lambda(T) = \{\mathrm{supp}(f) : f \in M(T)\}$.

Proposition 25. *Let $1 < p < \infty$ and $0 \leq T : L^p(Y, \nu) \to L^p(X, \mu)$ a positive linear operator. Then $f \in M(T)$ and $A \in \Lambda(T)$ implies that $f\chi_A, f\chi_{A^c} \in M(T)$.*

Proof. Let $0 \leq g \in M(T)$ with supp$g = A$. Then $g \wedge |f|\chi_{A^c} = 0$, so by (2) of the above proposition $Tg \wedge T(|f|\chi_{A^c}) = 0$. Now $|f|\chi_A$ is in the band generated by g, so $T(|f|\chi_A)$ is in the band generated by Tg, which implies that also $T(|f|\chi_A) \wedge T(|f|\chi_{A^c}) = 0$. Now the result follows from (3) of the above proposition. □

Theorem 26. *Let $1 < p < \infty$ and $0 \leq T : L^p(Y, \nu) \to L^p(X, \mu)$ a positive linear operator. Then $\Lambda(T)$ is a σ-ring with a largest element supp $f_0 = A_{\max}$.*

Proof. From the above proposition, part (3), it is immediate that if $A, B \in \Lambda(T)$, then $A \cap B \in \Lambda(T)$ and $A \setminus B \in \Lambda(T)$. It remains to show that $\Lambda(T)$ is closed under countable disjoint unions. Let $\{A_n\}$ be a disjoint collection in $\Lambda(T)$. Then we can find $0 \leq f_n \in M(T)$ with $\|f_n\|_p = 1$ such that $A_n =$ supp f_n. Now define $f = \sum_{n=1}^{\infty} \frac{1}{2^{\frac{n}{p}}} f_n$. Then $\|Tf\|_p = 1 = \|f\|_p$, so $f \in M(T)$ and thus supp $f = \cup_{n=1}^{\infty} A_n \in \Lambda(T)$. That $\Lambda(T)$ has now a largest element follows from the σ-finiteness of ν, as a maximal disjoint collection in $\Lambda(T)$ is countable. □

The main result about $M(T)$ is that $M(T)$ is a linear sublattice of L^p for a positive linear operator $T : L^p \to L^q$. This result is due to Grząślewicz ([10]) and Kan ([14]). Until now we followed [10], but now we switch to the approach of [14], which gives an description of $M(T)$ from which the result follows immediately. We introduce a larger σ-algebra than $\Lambda(T)$, which shares some of the properties of $\Lambda(T)$. For $T : L^p(Y, \nu) \to L^p(X, \mu)$ we define

$$\Lambda_0(T) = \{A : A \text{ measurable and } |T(f\chi_A)| \wedge |T(g\chi_{A^c})| = 0 \text{ for all } f, g \in L^p(Y, \nu)\}.$$

Theorem 27. *Let $0 \leq T : L^p(Y, \nu) \to L^p(X, \mu)$ with $1 < p < \infty$. Then $\Lambda_0(T)$ is a σ-algebra with $\Lambda(T) \subset \Lambda_0(T)$. Moreover for all $A \in \Lambda_0(T)$ and $f \in M(T)$ we have $\chi_A f \in M(T)$.*

Proof. Let $A \in \Lambda(T)$, where $A =$ supp(g_0) with $g_0 \in M(T)$. Let $f, g \in L^p$. Then $\chi_{A^c} g \perp g_0$ implies that $T(\chi_{A^c}) \perp T(g_0)$. Now $f\chi_A$ is in the band $\{g_0\}^{dd}$ generated by g_0, so $T(f\chi_A)$ is in the band generated by Tg_0, which implies that $T(f\chi_A) \perp T(g\chi_{A^c})$, i.e., $A \in \Lambda_0$. That $\lambda_0(T)$ is a σ-algebra is straightforward. If $A \in \Lambda_0(T)$ and $f \in M(T)$, then $\|Tf\|_p^p = \|T(f\chi_A)\|_p^p + \|T(f\chi_{A^c})\|_p^p = \|T\|_p^p \|f\|_p^p = \|T\|_p^p(\|f\chi_A\|_p^p + \|f\chi_{A^c}\|_p^p)$. This implies $\|T(f\chi_A)\|_p = \|T\|_p \|f\chi_A\|_p$, so $\chi_A f \in M(T)$. □

Corollary 28. *Let $0 \leq T : L^p(Y, \nu) \to L^p(X, \mu)$ with $1 < p < \infty$ and $f \in M(T)$. Then $mf \in M(T)$ for all $\Lambda_0(T)$ measurable functions m such that $mf \in L^p(Y, \nu)$.*

Proof. The proof is immediate from the above theorem and the p-additivity of the norms in case m is a $\Lambda_0(T)$ measurable simple function. The general case follows now from the Dominated Convergence Theorem. □

For the remaining part of this section we shall assume that T is positive linear operator from $L^p(Y,\nu) \to L^p(X,\mu)$ such that $Tf \neq 0$ a.e. in case $f \geq 0$ and $f \neq 0$ a.e. We can always achieve this by removing a measurable set from Y corresponding to the absolute null ideal of T.

Lemma 29. *Let* $0 \leq T : L^p(Y,\nu) \to L^p(X,\mu)$ *with* $1 < p < \infty$ *and* $f \in M(T)$.

1. *sgn* f *is* $\Lambda_0(T)$ *measurable.*
2. *If* $0 \leq g \in M(T)$ *and* $g \in \{f\}^{dd}$, *then* $\frac{g}{f}|_{suppf}$ *is* $\Lambda_0(T)$ *measurable.*

Proof. In case f is real-valued, then the $\Lambda_0(T)$ measurability of sgn$f$ follows immediate from the fact that also $f^+, f^- \in M(T)$. For the complex case we refer to [14]. Likewise we refer for the second part to [14]

□

Now we can state and proof the main theorem about $M(T)$.

Theorem 30. *Let* $0 \leq T : L^p(Y,\nu) \to L^p(X,\mu)$ *with* $1 < p < \infty$. *Then* $M(T)$ *is a closed linear sublattice of* $L^p(Y,\nu)$. *Moreover, the restriction of* T *to* $M(T)$ *is a disjointness preserving linear operator.*

Proof. Let $0 < f_0 \in M(T)$ such that $A_{\max} = \text{supp} f_0$ is the maximal element of $\Lambda(T)$. Let now $g \in M(T)$. Then by the above lemma $m_1 = \text{sgn} g$ is $\Lambda_0(T)$ measurable and also $m_2 = \frac{g}{f_0}\chi_{A_{\max}}$ is $\Lambda_0(T)$ measurable and $g = m_1 m_2 f_0$. Combining this with Corollary 28 we see that

$$M(T) = \{m f_0 \in L^p(Y,\nu) : m \text{ is } \Lambda_0(T)_{|A_{\max}} \text{ measurable}\}.$$

This description shows immediately that $M(T)$ is a closed linear sublattice of $L^p(Y,\nu)$. □

We now discuss the case $0 \leq T : L^p(Y,\nu) \to L^q(X,\mu)$, where $1 \leq q < p < \infty$. In the case $1 < q < p < \infty$ it was recently proved by G. Sinnamon [24] that every positive linear operator attains its norm. In his proof Sinnamon introduced a more complicated version of the iteration of the proof of Theorem 8. Our proof shows that one can use the original iteration as well. A consequence of the following theorem is that the structure of the $\Lambda(T)$ is very simple in case $1 < q < p < \infty$.

Theorem 31. *Let* $1 \leq q < p < \infty$ *and* $0 \leq T : L^p(Y,\nu) \to L^q(X,\mu)$ *a positive linear operator. Then* T *attains its norm.*

Proof. We can assume that $\|T\|_{p,q} = 1$. Assume first that $1 < q < p < \infty$. Let $Sf = (T^*(Tf)^{q-1})^{\frac{1}{p-1}}$ as in the proof of Theorem 8. Let $\lambda_0 > 1$ and let $\phi_0 = f_0$ with $\|\phi_0\|_p \leq 1$ such that $S(\phi_0) \leq \lambda_0 \phi_0$ as in the proof of Theorem 8. Define now $\lambda_n = \lambda_{n-1}^{\frac{1}{2}}$ for $n \geq 1$ and ϕ_n the positive solution of $\phi_n = \frac{\lambda_n - 1}{\lambda_n}\phi_{n-1} + \frac{1}{\lambda_n} S\phi_n$ given by the proof of Theorem 8. By Lemma 10 we have $\phi_{n+1} \leq \lambda_n \phi_n$ for all $n \geq 0$. Moreover by Theorem 8 we have that $1 \leq \lambda_n \|\phi_n\|_p^{p-q}$, so that $\lim_{n\to\infty} \|\phi_n\|_p = 1$. Now

$$\phi_0 \geq \frac{1}{\lambda_0}\phi_1 \geq \cdots \geq \frac{1}{\lambda_0 \cdots \lambda_{n-1}}\phi_n$$

for all $n \geq 1$ and $\lambda_0 \ldots \lambda_{n-1} = \lambda_0 \lambda_0^{\frac{1}{2}} \ldots \lambda_0^{\frac{1}{2^n}} \uparrow \lambda_0^2$. Hence $\tilde{\phi} = \lim_{n\to\infty} \frac{1}{\lambda_0 \ldots \lambda_{n-1}} \phi_n$ exists in norm and pointwise a.e. and $\|\tilde{\phi}\|_p = \frac{1}{\lambda_0^2}$. This implies that $\phi = \lambda_0^2 \tilde{\phi} = \lim_{n\to\infty} \phi_n$ exists in norm and pointwise a.e. and $\|\phi\|_p = 1$. By taking limits in $\phi_n = \frac{\lambda_n - 1}{\lambda_n} \phi_{n-1} + \frac{1}{\lambda_n} S\phi_n$ we see that $S\phi = \phi$, i.e., T attains its norm at ϕ. Now assume $q = 1$. Then let $f_0 = (T^*\mathbf{1})^{p'-1}$. One easily verifies then that $\|Tf_0\|_1 = \|T^*\mathbf{1}\|_{p'}^{p'} = \|f_0\|_p = \|T^*\|^{p'} = 1$. Hence T attains its norm at f_0. □

By supp T we denote the largest measurable set A (up to a set of measure zero) such that $0 \leq f$ and $Tf = 0$ implies that supp $f \cap A = \emptyset$.

Theorem 32. *Let $1 < q < p < \infty$ and $0 \leq T : L^p(Y, \nu) \to L^q(X, \mu)$ a positive linear operator. Then $\Lambda(T) = \{\emptyset, \text{supp } T\}$. Moreover $M(T)$ contains a unique $f \geq 0$ of norm one and if $g \in M(T)$, then $|g| = \|g\|_p f$.*

Proof. We assume again that $\|T\|_{p,q} = 1$. Let $A \in \Lambda(T)$ with $\nu(A) > 0$. Then there exists $0 \leq f \in M(T)$ of norm one with supp $f = A$. Let $0 \leq h \in L^p(Y, \nu)$ of norm one such that $f \wedge h = 0$. Then by Proposition 24 we have that $Tf \wedge Th = 0$. This implies that for all $t > 0$ we have that

$$1 + t^q \|Th\|_q^q = \|T(f + th)\|_q^q \leq \|f + th\|_p^q = (1 + t^p)^{\frac{q}{p}}.$$

This implies that

$$\|Th\|_q^q \leq \lim_{t\downarrow 0} \frac{(1 + t^p)^{\frac{q}{p}} - 1}{t^q} = 0.$$

Hence $Th = 0$. This implies that supp $T \subset A$. On the other hand $\|Tf\|_q = 1 = \|f\|_p$ implies that $A \subset$ supp T. Hence $\Lambda(T) = \{\emptyset, \text{supp } T\}$. Let $0 \leq f \in M(T)$ with $\|f\|_p = 1$ and assume again $\|T\|_{p,q} = 1$. Then by Theorem 22 we have

$$T^*(Tf)^{q-1} = f^{p-1}.$$

Now

$$\begin{aligned}
\|g\|_p^q = \int_X |Tg|^q \, d\mu &\leq \int_X (T(|g| f^{-\frac{1}{q'}} f^{\frac{1}{q'}})) \, d\mu \\
&\leq \int_X T(|g|^q f^{-\frac{q}{q'}}) T(f)^{\frac{q}{q'}} \, d\mu \\
&= \int_Y |g|^q f^{1-q} T^*(Tf)^{q-1} \, d\nu \\
&= \int_Y |g|^q f^{1-q} f^{p-1} \, d\nu = \int_Y |g|^q f^{p-q} \, d\nu \\
&\leq \left(\int_Y |g|^p \, d\nu \right)^{\frac{q}{p}} \left(\int_Y f^p \, d\nu \right)^{1-\frac{q}{p}} = \|g\|_p^q.
\end{aligned}$$

From the equality case of Hölder's inequality it now follows that $|g| = cf$, which implies that $c = \|g\|_q$. □

Remark. Note that we did not discuss in this section the structure of $M(T)$ in case $q > p$. This is for a good reason. The above results do not hold in that case as can be seen from the identity map $I : \ell_p \to \ell_q$, which has norm one for $1 \leq p \leq q \leq \infty$. One can verify easily that if $p < q$, then $M(I) = \{\alpha e_n : \alpha \in \mathbb{R}\}$, where $\{e_n\}$ denotes the standard unit basis of ℓ_p.

6. Weighted composition operators

Let as before (X, μ) and (Y, ν) be σ-finite measure spaces and denote by Σ the σ-algebra of μ-measurable sets and by Λ the σ-algebra of ν-measurable sets. Let ϕ denote a measurable mapping from $X \to Y$, i.e., $\phi^{-1}(A) \in \Lambda$ for all $A \in \Sigma$. Such a measurable mapping is called null-preserving if $\nu(\phi^{-1}(A)) = 0$ for all $A \in \Sigma$ with $\mu(A) = 0$. For a null-preserving ϕ and $h \in L^0(X, \mu)$ we define the weighted composition operator $Tf(x) = h(x)f(\phi(x))$ a.e. Note that T is an order continuous linear mapping from $L^0(Y, \nu)$ into $L^0(X, \mu)$. Moreover T has a modulus $|T|$ given by $|T|f(x) = |h(x)|f(\phi(x))$ a.e. It is clear that T is bounded from $L^p(Y, \mu)$ into $L^q(X, \mu)$ and $\|T\|_{p.q} = \|\,|T|\,\|_{p,q}$. Moreover T attains its norm at f if and only if T attains its norm at $|f|$ if and only if $|T|$ attains its norm at $|f|$ (this is immediate from $|Tf| = |T|f|| = |T|(|f|))$. Therefore we restrict ourselves to positive weighted composition operators. We note that weighted composition operators are disjointness preserving, i.e., if $f \perp g$, then $Tf \perp Tg$. In fact, on standard Borel spaces every disjointness preserving operator is a weighted composition operator and in general one can always represent a disjointness preserving operator as a weighted composition operator on function spaces on the Stone spaces of the measure algebras. To keep things more concrete we will restrict ourselves to operators which are weighted composition operators on the given spaces. For the boundedness we will restrict ourselves to the case $1 \leq p \leq q \leq \infty$ as we will see that in case $p > q$ there exist no non-zero weighted composition (or disjointness preserving) operators in the non-atomic case. We first state a simple lemma.

Lemma 33. *Let a, b, c, d, e, f be positive real numbers, such that $a = b + c$ and $d = e + f$. Then*

$$\frac{a}{d} \leq \max\left\{\frac{b}{e}, \frac{c}{f}\right\}.$$

Theorem 34. *Let $1 \leq p < q < \infty$ and $T : L^p(Y, \nu) \to L^q(X, \mu)$ a weighted composition operator. If ν is non-atomic, then $T = 0$.*

Proof. Let $f \neq 0$ in $L^p(Y, \nu)$. Let $A = \{y \in Y : f(y) \neq 0\}$. Put $A_1 = A$. If we write $A_1 = B \cup C$ a disjoint union of measurable sets, then $\|f\chi_{A_1}\|_p^p = \|f\chi_B\|_p^p + \|f\chi_C\|_p^p$. If ν is non-atomic, then can find B (and thus C) such that $\|f\chi_B\|_p^p = \|f\chi_C\|_p^p = \frac{1}{2}\|f\chi_{A_1}\|_p^p$. As T preserves disjointness we have also $\|T(f\chi_{A_1})\|_q^q = \|T(f\chi_B)\|_q^q + \|T(f\chi_C)\|_q^q$. By the above lemma we can take A_2 equal to either B or C so that

$$\frac{\|T(f\chi_{A_2})\|_q^q}{\|f\chi_{A_2}\|_p^p} \geq \frac{\|T(f\chi_{A_1})\|_q^q}{\|f\chi_{A_1}\|_p^p}.$$

By induction we construct a sequence A_n with

$$A_{n+1} \subset A_n, \quad \|f\chi_{A_{n+1}}\|_p^p = \frac{1}{2}\|f\chi_{A_n}\|_p^p,$$

and

$$\frac{\|T(f\chi_{A_{n+1}})\|_q^q}{\|f\chi_{A_{n+1}}\|_p^p} \geq \frac{\|T(f\chi_{A_n})\|_q^q}{\|f\chi_{A_n}\|_p^p}.$$

Assume now that $T(f) \neq 0$ a.e. Then

$$\frac{\|T(f\chi_{A_1})\|_q^q}{\|f\chi_{A_1}\|_p^p} > 0$$

for all $n \geq 1$. This implies that

$$\|T\|_{p,q}^q \geq \frac{\|T(f\chi_{A_n})\|_q^q}{\|f\chi_{A_n}\|_q^q} \geq \frac{\|T(f\chi_{A_n})\|_q^q}{\|f\chi_{A_n}\|_p^p}\frac{1}{\|f\chi_{A_n}\|_p^{q-p}} \uparrow \infty,$$

which is a contradiction. Hence $T(f) = 0$ a.e. for all $f \neq 0$ a.e. and thus $T = 0$. □

The idea of the above proof was taken from [14], where a more general result was proved. Let $0 \leq h \in L^0(X, \Sigma, \mu)$ and $\phi : X \to Y$ a null-preserving measurable mapping as above. To discuss the boundedness of the associated weighted composition operator T we introduce a measure. Define $\nu_{T,p}(A) = \int_{\phi^{-1}(A)} h^p \, d\mu$ for all $A \in \Lambda$. Then it is straight forward that $\nu_{T,p}$ is a measure on Λ, and $\nu_{T,p} \ll \nu$. For $0 \leq f \in L^0(Y, \nu)$ we have now the following formula for the Radon-Nikodym derivative of $\nu_{T,p}$:

$$\int_Y f(y) \frac{d\nu_{T,p}}{d\nu} \, d\nu = \int_X h(x)^p f(\phi(x)) \, d\mu.$$

Theorem 35. *Let $1 \leq p < \infty$, $0 \leq h \in L^0(X, \mu)$ and $\phi : X \to Y$ a null-preserving measurable mapping. Then the following are equivalent for the weighted composition operator $Tf(x) = h(x)f(\phi(x))$.*

1. *T is bounded from $L^p(Y, \nu)$ into $L^p(X, \mu)$.*
2. *There exists a constant $C > 0$ such that*

$$\int_{\phi^{-1}(A)} h^p(x) \, d\mu \leq C\mu(A)$$

for all $A \in \Lambda$.

3. *The Radon-Nikodym derivative $\frac{d\nu_{T,p}}{d\nu} \in L^\infty(Y, \nu)$.*

Moreover $\|T\|_p = \left\|\frac{d\nu_{T,p}}{d\nu}\right\|_\infty^{\frac{1}{p}}$.

Proof. The equivalence of (1) and (2) is immediate from the p-additivity of the norm and that (2) is the same as $\|T(\chi_A)\|_p \leq C\|\chi_A\|_p$. The equivalence of (2) and (3) is an easy exercise about Radon-Nikodym derivatives. □

Next we deal with the case $1 \leq q < p < \infty$.

Theorem 36. *Let $1 \leq q < p < \infty$, $0 \leq h \in L^0(X,\mu)$ and $\phi : X \to Y$ a null-preserving measurable mapping. Then the following are equivalent for the weighted composition operator $Tf(x) = h(x)f(\phi(x))$.*

1. *T is bounded from $L^p(Y,\nu)$ into $L^q(X,\mu)$.*
2. *The Radon-Nikodym derivative $\frac{d\nu_{T,q}}{d\nu} \in L^r(Y,\nu)$, where $\frac{1}{r} = \frac{1}{q} - \frac{1}{p}$.*

Moreover $\|T\|_{p,q} = \left\| \left(\frac{d\nu_{T,q}}{d\nu}\right)^{\frac{1}{q}} \right\|_r$.

Proof. Assume (2) holds. Let $f \in L^p(Y,\nu)$. Then using Hölder's inequality we have

$$\begin{aligned}\int_X |Tf(x)|^q \, d\mu &= \int_X h(x)^q |f(\phi(x))|^q \, d\mu = \int_Y |f(y)|^q \frac{d\nu_{T,q}}{d\nu} \, d\nu \\ &\leq \|f\|_p^q \left\| \frac{d\nu_{T,q}}{d\nu} \right\|_{\frac{p}{p-q}} = \|f\|_p^q \left\| \left(\frac{d\nu_{T,q}}{d\nu}\right)^{\frac{1}{q}} \right\|_r^q,\end{aligned}$$

where $\frac{1}{r} = \frac{1}{q} - \frac{1}{p}$. This shows that (1) holds and that $\|T\|_{p,q} \leq \left\| \left(\frac{d\nu_{T,q}}{d\nu}\right)^{\frac{1}{q}} \right\|_r$. Now assume (1) holds. Let $0 \leq g \in L^{\frac{p}{q}}(Y,\nu)$ with $\|g\|_{\frac{p}{q}} \leq 1$. Then we have

$$\begin{aligned}\int_Y g \left(\frac{d\nu_{T,q}}{d\nu}\right) d\nu &= \int_X h(x)^q g(\phi(x)) \, d\mu \\ &= \int_X T(g^{\frac{1}{q}})^q \, d\mu \\ &\leq \|T\|_{p,q}^q \|g^{\frac{1}{q}}\|_p^q \leq \|T\|_{p,q}^q.\end{aligned}$$

Hence by the converse of Hölder's inequality we have $\left\| \frac{d\nu_{T,q}}{d\nu} \right\|_{\frac{p}{p-q}} \leq \|T\|_{p,q}^q$, which is the same as $\left\| \left(\frac{d\nu_{T,q}}{d\nu}\right)^{\frac{1}{q}} \right\|_r \leq \|T\|_{p,q}$. □

Corollary 37. *Let $1 \leq q < p < \infty$. Then every weighted composition operator $Tf(x) = h(x)f(\phi(x))$ from $L^p(Y,\nu)$ into $L^q(X,\mu)$ attains its norm at*

$$f = \left(\frac{d\nu_{T,q}}{d\nu}\right)^{\frac{1}{p-q}}.$$

Proof. We can assume $T \geq 0$ and that $\|T\|_{p,q} = 1$. Then taking $f = \left(\frac{d\nu_{T,q}}{d\nu}\right)^{\frac{1}{p-q}}$ we can check as in the above proof that $\|Tf\|_q = 1 = \|f\|_p$, which shows that T attains its norm at f. □

7. Factorization theorems of positive linear operators

We begin with a theorem due to Maurey ([19]).

Theorem 38. *Let $0 \le T : L^p(Y,\nu) \to L^q(X,\mu)$ a positive linear operator and assume $1 \le q < p < \infty$. Then there exists $0 < g$ a.e. in $L^r(X,\mu)$ with $\frac{1}{r} = \frac{1}{q} - \frac{1}{p}$ such that $\frac{1}{g} \cdot T : L^p(Y,\nu) \to L^p(X,\mu)$.*

Proof. From Theorem 8 it follows that there exists $0 < f_0 \in L^p(Y,\nu)$ such that

$$T^*(Tf_0)^{q-1} \le \lambda f_0^{p-1}$$

holds. Define now, as in the proof of Theorem 4, $Sf = (Tf_0)^{\frac{q-p}{p}} \cdot Tf$. Then

$$S^*(Sf_0)^{p-1} = S^*((Tf_0)^{\frac{q(p-1)}{p}}) = T^*(Tf_0)^{q-1} \le \lambda f_0^{p-1},$$

implies that S is a bounded linear map from $L^p(Y,\nu)$ into $L^p(X,\mu)$ and $g = (Tf_0)^{\frac{p-q}{p}} \in L^r(X,\mu)$. □

Next we will discuss Nikišin's theorems for positive linear operators with domain $L^p(Y,\nu)$. Although the factorization theorems will hold for σ-finite measures μ, the terminology will be simpler if we restrict ourselves to finite measures. Therefore we will assume for the rest of the section that $\mu(X) < \infty$. Recall then that a collection H of measurable functions is bounded in measure in $L^0(X,\mu)$, if for all $\epsilon > 0$ there exists a constant M such that $\mu\{x \in X : |h(x)| \ge M\} < \epsilon$ for all $h \in H$. We start with a simple lemma, which holds in fact for positive linear operator defined on any Banach lattice.

Lemma 39. *Let $1 \le p \le \infty$ and $0 \le T : L^p(Y,\nu) \to L^0(X,\mu)$ a positive linear operator. Then the image under T of the unit ball of $L^p(Y,\nu)$ is bounded in measure.*

Proof. Assume the lemma is false. Then there exists $\epsilon > 0$ such that for all $n \ge 1$ there exists $f_n \in L^p$ with $\|f_n\|_p \le 1$ such that $\mu\{x \in X : |Tf_n(x)| \ge \epsilon 2^n\} \ge \epsilon$. Let $f_0 = \sum_{n=1}^{\infty} \frac{|f_n|}{2^n}$. Then $Tf_0 \ge \sum_{n=1}^{\infty} \frac{|Tf_n|}{2^n}$ implies that $\frac{|Tf_n|}{2^n} \to 0$ a.e., and thus in measure. This contradicts however that $\mu\{x \in X : \frac{|Tf_n(x)|}{2^n}| \ge \epsilon\} \ge \epsilon$ for all n. □

Theorem 40. *Let $H \subset L^0(X,\mu)$ be a convex, solid and bounded in measure. Then there exists a strictly positive $\phi \in L^0(X,\mu)$ such that*

$$\int_X |h|\phi \, d\mu \le 1$$

for all $h \in H$.

Proof. Let M be the closure in $L^2(X,\mu)$ of $H \cap L^2(X,\mu)$. Let $A \subset X$ with $\mu(A) > 0$. Then we claim there exists $m \in \mathbb{N}$ such that $m\chi_A \notin M$. If this would not be true, then for all $m \ge 1$ we can find $f_n \in H \cap L^2$ such that $\|f_m - m\chi_A\|_2 \le 1$. Then $\frac{f_m}{m} \to 0$ in measure, since H is bounded in measure, but on the other hand $\frac{f_m}{m} \to \chi_A$ in L^2, which implies that $\mu(A) = 0$. Hence there exist $m \ge 1$ such that

$m\chi_A \notin M$. By the Hahn-Banach theorem there exists $g \in L^2(X,\mu)$ such that $\int_X gh\,d\mu \le 1$ for all $h \in M$ and $m\int_A g\,d\mu > 1$. Now sgn $gh \in M$ for $h \in M$ implies that $\int_X |g|\,|f|\,d\mu \le 1$. Now $\int_A |g|\,d\mu > \frac{1}{m}$ implies that we can a constant $c > 0$ such that $B = \{x \in A; |g(x)| \ge \frac{1}{c}\}$ has positive measure. This implies that $\int_B |h|\,d\mu \le c$ for all $h \in M$. Now for $h \in H$ there exist $h_n \in M$ such that $|h_n| \uparrow |h|$. It follows from the Monotone Convergence Theorem that $\int_B |h|\,d\mu \le c$ for all $h \in H$. We have therefore shown now that for all $A \subset X$ with $\mu(A) > 0$ there exist $B \subset A$ with $\mu(B) > 0$ and $c \in \mathbb{R}$ such that $\int_B |h|\,d\mu \le c$ for $h \in H$. Now by using Zorn's lemma we can find disjoint $X_n \subset X$ with $\cup_{n=1}^{\infty} X_n = X$ and $c_n \in \mathbb{R}$ such that $\int_{X_n} |h|\,d\mu \le c_n$ for all $h \in H$. Now $\phi = \sum_{n=1}^{\infty} \frac{1}{2^n c_n}\chi_{X_n}$ satisfies the conclusion of the theorem. □

The above proof is a slight modification of the proof in [20] that the associate space of a Banach function space separates the points of the Banach function space. That the above theorem is essentially the same as that theorem is clear from the results in [28], where it was shown that convex solid sets bounded in measure are unit balls of normed Köthe function spaces.

Theorem 41. *Let $0 \le T : L^p(Y,\nu) \to L^0(X,\mu)$ a positive linear operator and assume $1 \le p < \infty$. Then there exists $0 < g$ a.e. in $L^0(X,\mu)$ such that $\frac{1}{g}\cdot T : L^p(Y,\nu) \to L^p(X,\mu)$.*

Proof. Combining the above lemma and theorem we see that there exists a strictly positive $\phi \in L^0(X,\mu)$ such that $\phi\cdot T : L^p(Y,\nu) \to L^1(X,\mu)$. The result now follows from the $q = 1$ case of Maurey's theorem 38. □

In Harmonic Analysis various variations of the above theorems have been proved as weighted norm inequalities. We will just indicate two of those instances, where these results are included in the above results. Let $1 < p < \infty$ and weights v, w (i.e., non-negative measurable functions), then a positive operator T is said to satisfy a weighted L^p-inequality if

$$\int_X (Tf(x))^p\, w(x)d\mu \le C\int_Y f(y)^p\, v(y)d\nu$$

for all $0 \le f$. Note that we can either absorb the weight in the measures or in the operator. Either way, the following theorem, which is called the Rubio de Francia Algorithm in [5], becomes a special case of Theorems 7 and 11.

Theorem 42. *Let $Q(x,y) \ge 0$ and let T be the operator $Tf(x) = \int_X Q(x,y)f(y)\,dy$. Let $p > 1$, with $\frac{1}{p} + \frac{1}{q} = 1$. Then $T : L^p(X, vd\nu) \to L^p(X, wd\mu)$ is a bounded operator if and only if there exists a positive $\alpha \in L^{pq}(X, d\nu)$, with*

1. $w^{\frac{1}{p}}T(\alpha^q v^{-\frac{1}{p}}) \le C\alpha^q$ *a.e. and*
2. $v^{-\frac{1}{p}}T^*(\alpha^p \lambda^{\frac{1}{p}}) \le C\alpha^p$ *a.e.,*

where T^* is the adjoint of T.

The following theorem was proved in [15]. Let T be as in the above theorem a positive integral operator.

Theorem 43. *Let $1 < p < \infty$ and w a weight on X. Then there exists a weight $v \in L^0(X, \nu)$ such that T is bounded from $L^p(X, v d\nu)$ to $L^p(X, w d\mu)$ if and only if there is a positive $\phi \in L^0$ with*

$$\int_X (T\phi)^p \, w d\mu < \infty.$$

Proof. The condition $T(\phi) \in L^p(X, w d\mu)$ is equivalent with the statement that T^* is bounded from $L^{p'}(X, w d\mu) \to L^1(X, \phi d\nu)$. The weighted norm inequality follows now from the $q = 1$ case of Maurey's theorem 38. □

8. Some open problems

Connected to the topics covered in this paper we have the following open problems:

1. When is a positive operator T bounded from $L^p(Y, \nu)$ to $L^q(X, \mu)$ for $1 \leq p < q \leq \infty$? There seems to be no analogue of the Schur criterion in this case.
2. Does every positive operator T from $L^p(Y, \nu)$ to $L^q(X, \mu)$ with $1 < p < q < \infty$ attain its norm?
3. If $T : L^p(Y, \nu) \to L^p(X, \mu)$ is positive linear operator with norm $\|T\|_p = 1$ and $1 < p < \infty$, does there exist $0 < f_0 \in L^0(Y, \nu)$ such that

$$T^*(Tf_0)^{p-1} \leq f_0^{p-1}$$

   a.e.? The example of Hardy's operator show that we can not find in general such f_0 with $f_0 \in L^p(Y, \nu)$, but it is an open problem whether we can find such $f_0 \in L^0(Y, \nu)$.
4. What is the best constant C for which Theorem 11 holds for all positive operators T? I.e., let $0 \leq T : L^p(Y, \nu) \to L^p(X, \mu)$ be a positive linear contraction, what is the smallest C such that there exist $f_0 \in L^p(Y, \nu)$ with $0 < f_0$ a.e. such that

$$T(f_0) \leq C f_0$$

   and

$$T^*(f_0^{p-1}) \leq C f_0^{p-1}?$$

5. Let $T : L^p(Y, \nu) \to L^p(X, \mu)$ be a positive linear operator with norm $\|T\|_p = 1$ and $1 < p < \infty$. Does there exist for all $\epsilon > 0$ a function $0 \leq f_0 \in L^0(Y, \nu)$ with $\|f_0\|_p = 1$ such that

$$T^*(Tf_0)^{p-1} \geq (1 - \epsilon) f_0^{p-1}$$

   a.e.? In general we don't know the answer to this question, but for positive integral operators we have the following result.

Theorem 44. *Let $T : L^p(Y, \nu) \to L^p(X, \mu)$ be a positive linear integral operator with norm $\|T\|_p = 1$ and $1 < p < \infty$. Then there exists $0 \le f_0 \in L^0(Y, \nu)$ with $\|f_0\|_p = 1$ such that*

$$T^*(Tf_0)^{p-1} \ge (1 - \epsilon) f_0^{p-1}$$

a.e.

Proof. Let $X_n \uparrow X$ and $Y_n \uparrow Y$ with $\mu(X_n) < \infty$ and $\nu(Y_n) < \infty$. Let $T(x, y)$ denote the kernel of T. Define the integral operators T_n by the kernels $T_n(x, y) = \min\{T(x, y), n\}\chi_{X_n \times Y_n}(x, y)$. Then $T_n \uparrow T$, so $\|T_n\|_p \uparrow \|T\|_p = 1$ and each T_n is a compact operator. Let N be such that $\|T_N\|_p > (1 - \epsilon)^{\frac{1}{p}}$. Then the fact that T_N attains its norm, implies that there exist $0 \le f_0 \in L^p(Y, \nu)$ such that

$$T_N^*(T_N f_0)^{p-1} \ge (1 - \epsilon) f_0^{p-1}.$$

The result follows now as $T^* \ge T_N^*$ and $T \ge T_N$. □

9. Concluding remarks

In this paper we primarily dealt with consequences of the Schur criterion

$$T^*(Tf_0)^{q-1} \le \lambda f_0^{p-1}$$

for positive operators T and with consequences of the equality case of the above inequality. There are several other important results about positive linear operators we did not discuss. Let us indicate two such topics. First there are the results about extending positive or regular operators defined on a subspace of $L^p(Y, \nu)$ into $L^p(X, \mu)$. For positive operators K. Donner [6] obtained the necessary and sufficient conditions, while Pisier [22] obtained the result for regular operators. Secondly there is the result that each closed sublattice of L^p is the range of a positive contractive projection and Ando's Theorem [3], which says that this property characterizes L^p or $c_0(\Gamma)$ among all Banach lattices. For a proof of these results we refer to [20] (pg. 134 137).

References

[1] Y.A. Abramovitch, C.D. Aliprantis, *An Invitation to Operator Theory*, Graduate Studies in Mathematics, **50**, AMS, 2002.

[2] Y. Abramovitch, C.D. Aliprantis, O. Burkinshaw, The Daugavet equation in uniform convex Banach spaces, *J. Funct. Anal.* **97** (1991), 215–230.

[3] T. Ando, Banachverbände und positive Projektionen, *Math. Z.* **109**, 121–103.

[4] B. Beauzamy, *Introduction to Banach spaces and their geometry*, North-Holland, 1982.

[5] S. Bloom, Solving Weighted Norm Inequalities using the Rubio de Francia Algorithm, *Proc. AMS* **101** (1987), 306–312.

[6] K. Donner, Extension of Positive Operators and Korovkin Theorems, *Lect. Notes in Math.* **904**, (1982), 1–182.

[7] D.H. Fremlin, A positive compact operator, *Manuscripta Math.* **15** (1975), 323–327.

[8] C. Finet and R. Paya, Norm attaining operators from L_1 into L_∞, *Israel J. Math.* **108** (1998), 139–143.

[9] E. Gagliardo, On integral transformations with positive kernel, *Proc. AMS*, **16** (1965), 429–434.

[10] R. Grząślewicz, On isometric domains of positive operators on L^p–spaces, *Colloq. Math.* **52**, (1987), 251–261.

[11] R. Grząślewicz, Approximation theorems for positive operators on L^p-spaces, *J. Approx. Theory*, **63** (1990), 123–136.

[12] R. Howard and A.R. Schep, Norms of positive operators on L^p–spaces, *Proc. AMS* **109** (1990), 135–146.

[13] A. Iwanik, Norm attaining operators on Lebesgue spaces, *P*ac. J. Math. **83**, (1979), 381–386.

[14] C-H. Kan, Norming vectors of linear operators between L_p spaces, *Pac. J. of Math.* **150**, (1990), 309–327.

[15] R. Kerman and E. Sawyer, On weighted norm inequalities for positive linear operators, *Proc. AMS* **105** (1989), 589–593.

[16] U. Krengel, *Ergodic Theorems*, De Gruyter, 1985

[17] M. Koskela, A characterization of non-negative matrix operators on ℓ_p to ℓ_q with $\infty > p \geq q > 1$, *Pac. J. of Math.*, **75**, (1978), 165–169.

[18] J. Lindenstraus, On operators which attain their norm, *Isr. J. of Math.*, **1** (1963), 139–148.

[19] B. Maurey, Théorèmes de factorisation pour les opérateurs linéaires à valeurs dans les espaces L^p, *Astérisque* **11**, 1974.

[20] Peter Meyer-Nieberg, *Banach lattices*, Springer-Verlag, 1991.

[21] E.M. Nikišin, Resonance theorems and superlinear operators, *Russ. Math. Surv.* **25** (1970), 124–187.

[22] G. Pisier, Complex interpolation and regular operators between Banach lattices, *Arch. Math.* **62** (1994), 261–269.

[23] R. Paya and Y. Saleh, Norm attaining operators from $L_1(\mu)$ into $L_\infty(\nu)$, *Arch. Math.* **75**, (2000), 380–388.

[24] G. Sinnamon, Schur's lemma and the best constants in weighted norm inequalities, *Le Matematiche* **57** (2002), 185–204.

[25] A.R. Schep, Positive diagonal and triangular operators, *Journ. of Operator Theory*, **3**(1980), 165–178.

[26] A.R. Schep, Factorization of positive multilinear maps, *Illin J. of Math.* **28** (1984), 579–591.

[27] A.R. Schep, Daugavet type inequalities for operators on L^p-spaces, *Positivity* **7** (2003), 103–111.

[28] A.R. Schep, Convex Solid subsets of $L_0(X, \mu)$, *Positivity* **9** (2005), 491–499.

[29] E.M. Semenov and B.S. Tsirel'son, The problem of smallness of operator blocks in L_p spaces, *Z. Anal. Anwendungen*, **2** (1983), 367–373. (Russian)

[30] E.M. Semenov and A.M. Shteĭnberg, Norm estimates of operator blocks in Banach lattices *Math. USSR Sbornik*, **54** (1986), 317–333.

[31] J.J. Uhl, Norm attaining operators on $L_1[0,1]$ and the Radon-Nikodym property, *Pac. J. Math.* **63**, (1976), 293–300.

[32] D. Vere-Jones, Ergodic properties of non-negative matrices-II, *Pac. J. of Math.*, **26**, (1968), 601–620.

[33] L. Weis, Integral operators and changes of density, *Indiana Univ. Math. J.* **31**(1982), 83–96.

[34] A.W. Wickstead, Positive compact operators on Banach lattices: some loose ends. *Positivity*, **4** (2000), 313–325.

[35] A.C. Zaanen, *Integration*, North-Holland, 1967.

[36] A.C. Zaanen, *Riesz Spaces II*, North-Holland, 1983.

Anton R. Schep
Department of Mathematics
University of South Carolina
Columbia, SC 29208, USA
e-mail: schep@math.sc.edu

Positivity
Trends in Mathematics, 255–279

Regular Operators between Banach Lattices

A.W. Wickstead

Introduction

If X and Y are Banach lattices then there are several spaces of linear operators between them that may be studied. $\mathfrak{L}(X,Y)$ is the space of all *norm bounded* operators from X into Y. There is no reason to expect there to be any connection between the order structure of X and Y and that of $\mathfrak{L}(X,Y)$. $\mathfrak{L}^r(X,Y)$ is the space of *regular* operators, i.e., the linear span of the positive operators. This at least has the merit that when it is ordered by the cone of positive operators then that cone is *generating*. $\mathfrak{L}^b(X,Y)$ is the space of *order bounded* operators, which are those that map order bounded sets in X to order bounded sets in Y. We always have $\mathfrak{L}^r(X,Y) \subseteq \mathfrak{L}^b(X,Y) \subseteq \mathfrak{L}(X,Y)$ and both inclusions may be proper.

I will look at the following problems, which are a small selection of those that we might have considered, and attempt to give a snapshot survey of our state of knowledge of these problems at the time of writing.

(1) When are all bounded operators regular?
(2) What can we say about the order structure of $\mathfrak{L}^r(X,Y)$?
(3) When are all order bounded operators regular?
(4) What kind of duality theory is there for regular operators?
(5) What happens for the classical Banach lattices?

I will take it for granted that the reader is familiar with the basic theory of Banach lattices and elementary properties of linear operators between them, as presented in [6], [17] or [21]. I will not attempt to give complete proofs of the results presented here, except where they are not easily otherwise accessible.

1. Banach lattice terminology

In this section we summarise the basics of Banach lattice terminology, both to make this paper more self-contained and because this terminology is not completely standard.

A *vector lattice* or *Riesz space* E is a real vector lattice on which there is a lattice ordering such that

(1) For all $x, y, z \in E$, $x \le y \Rightarrow x + z \le y + z$
(2) For all $x, y \in E$ and $0 \le \lambda \in \mathbb{R}$, $x \le y \Rightarrow \lambda x \le \lambda y$.

The set of *positive elements* in E is $E_+ = \{x \in E : x \ge 0\}$. Every element x of a vector lattice may be written in a minimal way as the difference of two positive elements $x = x^+ - x^-$, where $x^+ = x \vee 0$ is the *positive part* of x and $x^- = (-x) \vee 0$ is the *negative part* of x. The *modulus* of x is $|x| = x \vee (-x)$. x and y are said to be *disjoint* or *orthogonal*, written $x \perp y$, if $|x| \wedge |y| = 0$.

A *vector sublattice* H of E is a vector subspace such that $x, y \in H \Rightarrow x \vee y, x \wedge y \in H$, where these lattice operations are computed in E. Beware that is is possible for a vector subspace of a vector lattice to be a lattice for the inherited ordering without being a vector sublattice.

A *strong order unit* in a vector lattice E is an element $e \in E_+$ such that for every $x \in E$ there is $\lambda \in \mathbb{R}$ such that $-\lambda e \le x \le \lambda e$. A *weak order unit* is an element $u \in E_+$ such that if $x \in E$ and $x \perp u$ then $x = 0$. If E is the space of all bounded continuous real-valued functions on $\mathbb{R}$, with the pointwise order, then a strong order unit is any e such that there is $\epsilon > 0$ with $e(t) \ge \epsilon$ for all $t \in \mathbb{R}$ whilst a weak order unit is any w such that $w(t) > 0$ on a dense subset of $\mathbb{R}$.

A vector lattice is *Archimedean* if $x, y \in E$ and $nx \le y$ for all $n \in \mathbb{N}$ implies that $x \le 0$. A vector lattice is *Dedekind complete* (resp. *Dedekind σ-complete*) if every non-empty (countable) subset which is bounded above has a least upper bound. It follows automatically that non-empty (countable) subsets which are bounded below have greatest lower bounds.

An *atom* in an Archimedean vector lattice E is $a \in E_+$ such that $0 \le b \le a$ implies that b is a real multiple of a, whilst E is said to be *atomic* if the only element of E that is disjoint from every atom is the zero element. In the real space $L_p(\mu)$ the atoms correspond to atoms of the measure and the space is atomic if and only if the measure is purely atomic.

A *Banach lattice* is a real Banach space X which is a vector lattice such that $x, y \in X$ with $|x| \le |y| \Rightarrow \|x\| \le \|y\|$. Banach lattices must be Archimedean but need not even be Dedekind σ-complete. There is a theory of complex Banach lattices but we will not need to go into that in this paper. The classical Banach spaces, in the real case, are all Banach lattices under a natural (pointwise or pointwise almost everywhere) order and their usual norm, including the spaces $C(K)$, for K a compact Hausdorff space, and $L_p(\mu)$.

A Banach lattice X is an *AM-space* if $\|x \vee y\| = \|x\| \vee \|y\|$ for all $x, y \in X_+$. A classical result of Kakutani in [13] tells us that these are precisely those Banach lattices which are isometrically order isomorphic to a closed sublattice of some $C(K)$. Similarly, *AL-spaces* are Banach lattices X such that whenever $x, y \in X_+$, $\|x+y\| = \|x\|+\|y\|$. In [12] Kakutani proved that these are isometrically isomorphic to spaces $L_1(\mu)$, for suitable measures μ.

Banach lattices X, in which every subset $A \subset X_+$ which is downward directed to 0 has $\inf\{\|a\| : a \in A\} = 0$, are said to have an *order continuous norm*. If $1 \le p \le \infty$ then an infinite-dimensional Banach lattice $L_p(\mu)$ has an order continuous norm if and only if $p < \infty$. If Σ is a locally compact Hausdorff space then $C_0(\Sigma)$, the continuous real-valued functions on Σ which vanish at infinity, has an order continuous norm if and only if Σ is discrete. A *KB-space* is a Banach lattice in which every monotone norm bounded sequence is convergent. KB-spaces must have an order continuous norm, but even though c_0 has an order continuous norm it fails to be a KB-space.

The norm in a Banach lattice X is said to be *Fatou* if whenever $A \subset X_+$ is upward directed with supremum b then $\|b\| = \sup\{\|a\| : a \in A\}$. The norm is *weakly Fatou* if there is $K \in \mathbb{R}$ such that whenever $A \subset X_+$ is upward directed with supremum b then $\|b\| \le K \sup\{\|a\| : a \in A\}$. A Banach lattice has a *Levi* norm if every norm bounded upward directed set of positive elements has a supremum. Every Banach lattice with a Levi norm must have a weakly Fatou norm, but it need not be Fatou.

2. When are all bounded operators regular?

With most of the problems we shall consider concerning pairs of Banach lattices we would like to characterize the pairs (X, Y) for which a certain property holds. That is often not possible. What often *is* possible is to characterize those range spaces Y such that the desired property holds for all pairs (X, Y) and those domains X such that the property holds for all pairs (X, Y). We start by considering such properties in this setting. What about the nicest possible range spaces first of all?

Theorem 2.1. *If Y is a Dedekind complete Banach lattice then the following are equivalent:*

(1) *Y has a strong order unit.*
(2) *For all Banach lattices X and Y, $\mathfrak{L}(X,Y) = \mathfrak{L}^r(X,Y)$.*

The fact that (1)⇒(2) dates back to Kantorovich's work in [14], whilst that (2)⇒(1) is proved in [5]. Without the hypothesis of Dedekind completeness the implication (1)⇒(2) fails. For example, Lemma 2.10 of [29] shows that if K is a compact Hausdorff space in which there is a sequence with no convergent subsequence then $\mathfrak{L}(C(K), c) \neq \mathfrak{L}^r(C(K), c)$. Similarly, condition (2) alone is not enough to force Dedekind completeness of Y, as is shown by an example in [3]. However we do have:

Theorem 2.2. *If Y is any Banach lattice then the following are equivalent:*

(1) *Y is Dedekind complete with a strong order unit.*
(2) *For all Banach lattices X, $\mathfrak{L}(X,Y)$ is a vector lattice.*

Proof. The proof that (2) implies Dedekind completeness of Y is based on the proof of a similar result in a vector lattice setting which is contained in [3]. As Y

is certainly uniformly complete, in order to prove that it is Dedekind complete it suffices, by a well-known result of Veksler and Gejler [24] to prove that any disjoint family of positive elements $(y_i)_{i\in I} \in Y$, which has an upper bound $y \in Y$, must have a supremum.

Let $c(I)$ denote the space of all real-valued functions f on I with the property that there is a real α such that for all $\epsilon > 0$ the set $\{i \in I : |f(i) - \alpha| > \epsilon\}$ is finite. We write ℓ_f for this (unique) real α. Under the supremum norm and pointwise partial order $c(I)$ is a Banach lattice. Define an operator $T : c(I) \to Y$ by

$$T(f) = \sum_{i\in I} (f(i) - \ell_f) y_i.$$

This series has only countably many non-zero terms and is Cauchy because if $\epsilon > 0$ and $F = \{i \in I : |f(i) - \ell_f| > \epsilon\}$ then for any finite set $G \subset \mathbb{N} \setminus F$ we have

$$\left|\sum_{i\in G} (f(i) - \ell_f) y_i\right| \le \sum_{i\in G} |f(i) - \ell_f| y_i \le \epsilon \sum_{i\in G} y_i = \epsilon \bigvee_{i\in G} y_i \le \epsilon y$$

so that $\left\|\sum_{i\in G}(f(i) - \ell_f)y_i\right\| \le \epsilon\|y\|$. As Y is norm complete, the series converges. Let us also define $U : c(I) \to Y$ by

$$U(f) = T(f) + \ell_f y.$$

If $f \ge 0$ then $(U - T)(f) = \ell_f y \ge 0$ so that $U \ge T$. Also, for any $f \in c(I)_+$ and $i \in I$ we have $\ell_f \ge \ell_f - f(i), 0$ so that

$$\ell_f y \ge \ell_f y_i \ge (\ell_f - f(i)) y_i.$$

As the (y_i) are disjoint, we must have $\ell_f y \ge \sum_{i\in I}(\ell_f - f(i))y_i$. I.e. $(U - T)(f) \ge -T(f)$ so that $U - T \ge -T$ and hence $U \ge 0$. It follows T is regular and therefore has a positive part, T^+.

Writing $\mathbf{1}_I$ for the constantly one function in $c(I)$, we claim that $T^+(\mathbf{1}_I)$ is the supremum in Y of the family (y_i), which will complete the proof. Firstly note that $T(\mathbf{1}_I)$ is an upper bound for the (y_i) as if we let $\boldsymbol{e}_i$ denote the function that takes the value 1 at i and 0 on the rest of I, then $\mathbf{1}_I \ge \boldsymbol{e}_i \ge 0$ so that

$$T^+(\mathbf{1}_I) \ge T^+(\boldsymbol{e}_i) \ge T(\boldsymbol{e}_i) = y_i.$$

On the other hand the previous paragraph constructed a positive majorant U for T with $U(\mathbf{1}_I) = y$. Since $U \ge T^+$, we must have $y \ge U(\mathbf{1}_I) \ge T^+(\mathbf{1}_I)$. As y was any upper bound for the family (y_i), we see that $T^+(\mathbf{1}_I)$ is indeed the supremum of the family (y_i). □

Before looking at the dual result, let us record a result with a slightly technical statement. If $1 \le p < \infty$ then ℓ_p is *finitely lattice representable* in a Banach lattice X if for each $n \in \mathbb{N}$ there is a vector sublattice X_n of X and an order isomorphism T_n of X_n onto ℓ_p^n such that

$$\sup_{n\in\mathbb{N}} \|T_n\|\|T_n^{-1}\| < \infty.$$

Theorem 2.3. *Let X and Y be Banach lattices such that every operator in the operator norm closure of the finite rank operators from X into Y is regular.*

(1) *If, for some $p \in [1,\infty)$, ℓ_p is finitely lattice representable in Y then X is isomorphic to an AL-space.*
(2) *If, for some $p \in [1,\infty)$, ℓ_p is finitely lattice representable in X^* then Y is isomorphic to an AM-space.*

This (in a slightly strengthened form) was proved by Abramovich and Janovskii in [4], whilst a slightly weaker result was proved by Cartwright and Lotz in [7].

Theorem 2.4. *If X is a Banach lattice then the following are equivalent:*

(1) *X is isomorphic to an atomic AL-space.*
(2) *For all Banach lattices Y, $\mathfrak{L}(X,Y) = \mathfrak{L}^r(X,Y)$.*
(3) *For all Banach lattices Y, $\mathfrak{L}(X,Y)$ is a vector lattice.*

Proof. The only non-obvious part of the proof is that (3)⇒(1). By Theorem 2.3 X is certainly isomorphic to an AL-space. If X is not purely atomic then it contains an closed sublattice H which is isomorphic to $L_1([0,1])$. By Theorem 2.7.3 of [17] there is a positive projection $P : X \to H$.

Let (r_n) denote the sequence of Rademacher functions on $[0,1]$. Define an operator $T : L_1[0,1] \to c_0$ by $Tx = \sum_{n=1}^{\infty} r_n(x)\boldsymbol{e}_n$ where $\boldsymbol{e}_n$ denotes the nth standard basis vector in c_0. The Rademacher functions converge weak* to 0 when considered, as we do here, as elements of $L_1[0,1]^*$ so that $Tf \in c_0$. T is bounded as $\|Tx\| = \sup_{n=1}^{\infty} \|r_n(x)\| \le \sup_{n=1}^{\infty} \|r_n\|_\infty \|x\|_1 = \|x\|_1$. Note that $T(r_n) = \boldsymbol{e}_n$ for all $n \in \mathbb{N}$ so that $T(r_1 + r_n) = \boldsymbol{e}_1 + \boldsymbol{e}_n$. As r_1 is constantly one and $|r_n| = r_1$ for all $n \in \mathbb{N}$, we see that $r_1 + r_n$ is positive. It follows that if we had $U \ge T, 0$ then

$$U(2r_1) \ge U(r_1 + r_n) \ge T(r_1 + r_n) = \boldsymbol{e}_1 + \boldsymbol{e}_n$$

so that $U(2r_1) \ge \boldsymbol{e}_n$ for all $n \in \mathbb{N}$, which is inconsistent with $U(2r_1)$ lying in c_0. It follows that T is not regular after all. It follows easily that $T \circ P : X \to c_0$ is bounded but not regular. □

In many cases that are known when $\mathfrak{L}(X,Y) = \mathfrak{L}^r(X,Y)$ then either X is an AL-space or Y is an AM-space, at least up to isomorphism. Krivine in [16] has shown that for any infinite-dimensional Banach lattice either some ℓ_p, for some $p \in [1,\infty)$, or c_0 is finitely lattice representable in it. This shows that there is only a small gap between Theorem 2.3 and what is needed to obtain a theorem from this observation. It is, however, a gap that cannot be filled. This was established in an example due to Abramovich in [1]. The following example is a slightly simplified version of that. The author would like to thank Yuri Abramovich for providing him with a translation of the original example – typical of his generosity of spirit!

The example depends on an estimate for the relationship between the regular and operator norms of operators from finite-dimensional Banach lattices into an ℓ_p space. The following proof is attributed by Abramovich to B.S. Tsirel'son.

Lemma 2.5. *Let X be an n-dimensional Banach lattice and $p \in [1,\infty)$. For any $T \in \mathcal{L}(X,\ell_p)$ we have $\|T\|_r \le 2^{n/p}\|T\|$.*

Proof. Let $e_1, e_2, \dots, e_n$ be a Hamel basis for X consisting of disjoint atoms. If $T \in \mathcal{L}(X,\ell_p)$ let $z_k = Te_k$. It is clear that $|T|e_k = |z_k|$, so that

$$\|T\|_r = \big\||T|\big\| = \sup\{\big\||T|x\big\|_p : \|x\| \le 1\}$$
$$= \sup\{\big\|\sum_{k=1}^n \lambda_k |z_k|\big\|_p : \|x\| \le 1, x = \sum_{k=1}^n \lambda_k e_k\},$$

where $\|\cdot\|_p$ denotes the ℓ_p-norm. Given that X is finite-dimensional and that $|T|$ is positive, this supremum is attained and at a positive element of X. Thus there is $y \in X_+$ with $\|y\| = 1$ at which this supremum is attained. We can write $y = \sum_{k=1}^n \mu_k e_k$ where each $\mu_k \ge 0$.

Let Γ be the collection of all n-tuples $\gamma = (\varepsilon_1, \varepsilon_2, \dots, \varepsilon_n)$, with each $\varepsilon_k = \pm 1$, so that the cardinality of Γ is 2^n. For each $\gamma \in \Gamma$ let

$$\mathbb{N}_\gamma = \{m \in \mathbb{N} : \operatorname{sign}\big(z_k(m)\big) = \varepsilon_k, k = 1, 2, \dots, n\}.$$

Clearly if $\gamma_1 \ne \gamma_2$ then $\mathbb{N}_{\gamma_1} \cap \mathbb{N}_{\gamma_2} = \emptyset$. Note also that we have $\bigcup_{\gamma\in\Gamma} \mathbb{N}_\gamma = \mathbb{N}$. We now see that

$$\|T\|_r = \big\|\sum_{k=1}^n \mu_k |z_k|\big\|_p = \left\|\sum_{\gamma\in\Gamma}\sum_{k=1}^n \mu_k |z_k \chi_{\mathbb{N}_\gamma}|\right\|_p$$
$$= \left(\sum_{\gamma\in\Gamma} \big\|\mu_k |z_k \chi_{\mathbb{N}_\gamma}|\big\|_p^p\right)^{1/p} \le 2^{n/p} \max_{\gamma\in\Gamma} \left\|\sum_{k=1}^n \mu_k z_k \chi_{\mathbb{N}_\gamma}|\right\|_p .$$

If this maximum is attained at $\gamma_0 = (\varepsilon_1^0, \varepsilon_2^0, \dots, \varepsilon_n^0)$ then we see that

$$\|T\|_r \le 2^{n/p} \left\|\sum_{k=1}^n \mu_k z_k \chi_{\mathbb{N}_{\gamma_0}}|\right\|_p = 2^{n/p} \left\|\sum_{k=1}^n \mu_k \varepsilon_k^0 z_k \chi_{\mathbb{N}_{\gamma_0}}\right\|_p$$
$$\le 2^{n/p} \left\|\sum_{k=1}^n \mu_k \varepsilon_k^0 z_k\right\|_p \le 2^{n/p} \sup\{\left\|\sum_{k=1}^n \lambda_k z_k\right\|_p : \|\sum \lambda_k e_k\| \le 1\}$$
$$= 2^{n/p}\|T\|.$$
□

Corollary 2.6. *Let X be an n-dimensional Banach lattice and $p \in (1,\infty)$. For any $T \in \mathcal{L}(\ell_p, X)$ we have $\|T\|_r \le 2^{n(p-1)/p}\|T\|$.*

Proof. Apply the preceding lemma to $T^* \in \mathcal{L}(\ell_{p'}, X^*)$, where $\frac{1}{p} + \frac{1}{p'} = 1$. □

Example 2.7. *For any $\epsilon > 0$ there are Banach lattices X and Y such that*

(1) *X is not isomorphic to an AL-space.*
(2) *Y is not isomorphic to an AM-space.*
(3) *$\mathcal{L}(X,Y) = \mathcal{L}^r(X,Y)$ with $\|T\|_r \le (1+\epsilon)\|T\|$ for all $\in \mathcal{L}(X,Y)$.*

Proof. Recall that for any Banach lattice X, the M-characteristic of X is defined to the supremum of the norms of finite disjoint suprema of positive elements of X each of norm at most one. I.e.

$$p_M(X) = \sup\{\|x\| : x = x_1 \vee x_2 \vee \cdots \vee x_n, x_i \in X_+,\ x_i \perp x_j\ (i \neq j), \\ \|x_i\| \leq 1, n \in \mathbb{N}\}.$$

Clearly $1 \leq p_M(X) \leq \infty$ and it is well known that $p_M(X) < \infty$ if and only if X is isomorphic to an AM-space. It is routine to verify that $p_M(\ell_p^n) = n^{1/p}$.

This example is based on two sequences of finite-dimensional Banach lattices (X_n) and (Y_n) such that $\sup_n p_M(X_n^*) = \infty$, $\sup_n p_M(Y_n) = \infty$ and for all $T \in \mathcal{L}(X_m, Y_n)$ we have $\|T\|_r \leq (1+\epsilon)\|T\|$ for any $m, n \in \mathbb{N}$. Once these sequences are given, we set X to be the ℓ_1-sum of the spaces X_n and Y to be the ℓ_∞-product of the spaces Y_n. We have $p_M(Y) \geq p_M(Y_n)$ for all $n \in \mathbb{N}$ so that $p_M(Y) = \infty$ and therefore Y is not isomorphic to an AM-space. Similarly we see that, as X^* may be identified with the ℓ_∞-product of the spaces X_n^*, $p_M(X^*) = \infty$ so that X^* is not isomorphic to an AM-space and therefore X is not isomorphic to an AL-space.

We may represent a bounded operator $T : X \to Y$ as an infinite matrix of operators (T_{ij}) where $T_{ij} : X_i \to Y_j$ with $T(x_j)_{j=1}^\infty = (\sum_{j=1}^\infty T_{ij}x_j)_{i=1}^\infty$. It is routine to verify that the norm of T is precisely $\sup_{i=1}^\infty \sum_{j=1}^\infty \|T_{ij}\|$. The matrix $(|T_{ij}|)$ may now be verified easily to be a bounded operator from X into Y, to be the modulus of T and, given that each $\||T_{ij}|\| \leq (1+\epsilon)\|T_{ij}\|$, to have norm at most $(1+\epsilon)\|T\|$. Once we prove the existence of these two sequences of Banach lattices, the proof will be complete.

We start the definition of the sequences $X_n = \ell_{p_n}^{k_n}$ and $Y_n = \ell_{q_n}^{k_n}$ by choosing $k_1 = 1$, $p_1 = \infty$ and $q_1 = 1$, so that $X_1 = Y_1 = \mathbb{R}$. Clearly if $T \in \mathcal{L}(X, Y_1)$ or $T \in \mathcal{L}(X_1, Y)$ then $\||T|\| = \|T\|$, no matter what X and Y we take. The sequence of integers (k_n) and of reals (q_n) will increase to ∞, whilst the sequence (p_n) will decrease to 1.

If X_k and Y_k have been defined for $1 \leq k < n$ then we proceed to define Y_n by choosing $q_n > q_{n-1}$ such that $2^{k_{n-1}/q_n} \leq 1+\epsilon$ and hence $2^{k_j/q_n} \leq 1+\epsilon$ for $k = 1, 2, \ldots, n-1$. It follows from Lemma 2.5 that for all $T \in \mathcal{L}(X_k, \ell_q)$ we will have $\||T|\| \leq (1+\epsilon)\|T\|$, for all of $k = 1, 2, \ldots, n-1$ and for any $q \geq q_n$. In particular this inequality will hold with $q = q_m$ for any $m \geq n$ and also if we replace ℓ_{q_m} by $\ell_{q_n}^k$ for any integer k. I.e. it will hold for all operators in $\mathcal{L}(X_k, Y_m)$ for $1 \leq k < n$ and $m \geq n$. Now that we have chosen q_n we choose $k_n > n$ so that $k_n^{1/q_n} \geq n$ and hence $p_M(y_n) \geq n$.

We next choose $1 < p_n < p_{n-1}$ so that $2^{k_n(p_n-1)/p_n} \leq 1+\epsilon$ and hence $2^{k_j(p_n-1)/p_n} \leq 1+\epsilon$ for $1 \leq j \leq n$. By 2.6 we have $\||T|\| \leq (1+\epsilon)\|T\|$ for all $T \in \mathcal{L}(\ell_{p_n}, Y_k)$ where $1 \leq k \leq n$. Finally, if necessary, we decrease p_n, whilst keeping $p_n > 1$, so that $k_n^{(p_n-1)/p_n} \geq n$. I.e. $p_M(X_n^*) = p_M(\ell_{p_n}^{k_n *}) = p_M(\ell_{(p_n-1)/p_n}^{k_n}) \geq n$. These sequences now have all the desired properties. □

Given that there is no complete characterization of exactly when all norm bounded operators are regular, and as the "universal domains" and "universal range spaces" are so specialized, it is worth seeking extra cases where we have equality. We will look at what happens when we try increasing these universal domains/ranges in a natural manner.

Theorem 2.8. *The following conditions on a Banach lattice Y are equivalent:*

(1) *Y has a Levi norm.*
(2) *Whenever the Banach lattice X is isomorphic to an AL-space, $\mathfrak{L}(X,Y) = \mathfrak{L}^r(X,Y)$.*
(3) *Whenever the Banach lattice X is isomorphic to an AL-space, $\mathfrak{L}(X,Y)$ is a vector lattice.*

In [15], the fact that (1)⇒(3) was established under the slightly stronger hypothesis that Y was a KB-space. Synnatzschke established the full force of that implication in [23]. Clearly (3)⇒(2) and the fact that (2)⇒(1) is Theorem 3.5 of [5].

At the other extreme we have:

Theorem 2.9. *The following conditions on a Banach lattice X are equivalent:*

(1) *X is atomic with an order continuous norm.*
(2) *If Y is isomorphic to an AM-space then $\mathfrak{L}(X,Y) = \mathfrak{L}^r(X,Y)$.*
(3) *If Y is isomorphic to an AM-space then $\mathfrak{L}(X,Y)$ is a lattice.*

Proof. The proofs that (1)⇒(3)⇒(2) are routine, whilst the fact that (2)⇒(1) follows directly from Theorem 2.3. □

Let us record here that there are versions of Theorems 2.1 and 2.8 in which the condition on the range space is weakened at the expense of imposing a restriction on the size of order intervals in the domain. Details of these results will be published elsewhere.

3. The order structure of $\mathfrak{L}^r(X,Y)$

The space of all regular operators does at least have a *generating cone*, i.e., every regular operator is the difference of two positive operators, but in general there is little more that can be said about its order structure. It certainly need not be a lattice.

Definition 3.1. A Banach lattice has *property* $(*)$ if, for every sequence (f_n) in X_+^* which converges $\sigma(X^*,X)$ to $f \in X_+^*$ as $n \to \infty$, we have $|f_n - f| \to 0$ for $\sigma(X^*,X)$ as $n \to \infty$.

The connection between this property and the order structure of spaces of regular operators is given by the following result of van Rooij, Theorem 8.2 of [20].

Theorem 3.2. *A Banach lattice X has property $(*)$ if and only if $\mathfrak{L}^r(X,c)$ is a lattice.*

The space $X = L_1([0,1])$ does not have property $(*)$ (consider the sequence of Rademacher functions in $X^* = L_\infty([0,1])$, (r_n), for which $r_n \to 0$ weak* but $|r_n| = \mathbf{1}$ for all n) so that $\mathcal{L}^r(L_1([0,1]), c)$ is not a lattice.

As in the first section, we can seek the range spaces (or domains) which always give us a vector lattice of regular operators or we can attempt to tackle the much more difficult problem of describing exactly which pairs of Banach lattices make $\mathcal{L}^r(X,Y)$ be a vector lattice. We look at the easier problems first. We established the following result in Theorem 2.2.

Theorem 3.3. *A Banach lattice Y is Dedekind complete if and only if, for every Banach lattice X, $\mathcal{L}^r(X,Y)$ is a vector lattice.*

In fact in this case we can say more, namely that the lattice operations are given by the so-called *Riesz-Kantorovich* formulae, typical of which is:

$$|T|(x) = \sup\{Tx' : -x \le x' \le x\}$$

for $x \in X_+$. Dually, in Theorem 10.2 of [20], van Rooij established:

Theorem 3.4. *A Banach lattice X is atomic with an order continuous norm if and only if, for every Banach lattice Y, $\mathcal{L}^r(X,Y)$ is a vector lattice.*

Again the lattice operations are described by the Riesz-Kantorovich formulae.

There is still no complete description of when $\mathcal{L}^r(X,Y)$ is a vector lattice, but there has been some progress on this in recent years. The importance of condition $(*)$ was established by van Rooij when he proved in Theorem 8.12 of [20] that:

Theorem 3.5. *If X and Y are Banach lattices and $\mathcal{L}^r(X,Y)$ is a vector lattice then* either *X has property $(*)$* or *Y is Dedekind σ-complete.*

A major stumbling block to further progress here is that we don't know exactly what property $(*)$ amounts to in a Banach lattice setting. We do have partial results. Theorem 3.1 of [8] actually establishes a result (apparently) slightly stronger than this:

Theorem 3.6. *If X is a Banach lattice which is atomic with an order continuous norm then it has an property $(*)$. If X has property $(*)$ and* (a) *is Dedekind σ-complete or* (b) *is separable or* (c) *$X = C_0(\Sigma)$, where Σ is a locally compact Hausdorff space, then it is atomic with an order continuous norm.*

Putting together the preceding two results plus a careful analysis of the proof that $\mathcal{L}^r(X,Y)$ is a vector lattice when Y is Dedekind complete gives us:

Theorem 3.7. *If X and Y are Banach lattices and X is separable then $\mathcal{L}^r(X,Y)$ is a vector lattice if and only if* either *X is atomic with an order continuous norm* or *Y is Dedekind σ-complete.*

The separability allows us to go from property $(*)$, which is not obviously enough to guarantee that we have a lattice of operators, to being atomic with an order continuous norm, which is. In order even to state a generalisation of the last result, we need to introduce two new definitions.

Definition 3.8. The *density character* of a subset S of a topological space is the smallest cardinal $\mathfrak{a}$ such that S contains a dense subset of cardinality $\mathfrak{a}$.

Definition 3.9. A vector lattice Y is Dedekind $\mathfrak{a}$-complete if every non-empty subset of cardinality at most $\mathfrak{a}$, which is bounded above, has a supremum, whilst we say that Y is Dedekind $<\mathfrak{a}$-complete if every non-empty subset of cardinality strictly less that $\mathfrak{a}$, and which is bounded above, has a supremum.

Definition 3.10. If $\mathfrak{a}$ is an infinite cardinal we say that a Banach lattice X is $\mathfrak{a}$-homogeneous if X is non-atomic and whenever $x, y \in X$ with $x \leq y$ and $x \neq y$ then the density character of the order interval $[x, y]$ is $\mathfrak{a}$.

Recall that a measure algebra (Σ, μ) is naturally metrisable by defining $d(A, B) = \mu(A \setminus B) + \mu(B \setminus A)$. The measure algebra (Σ, μ) is said to be $\mathfrak{a}$-homogeneous if every subset with non-zero measure has density character $\mathfrak{a}$. If $\mathbf{2} = \{0, 1\}$ and γ is the measure on the subsets of $\mathbf{2}$ with $\gamma(\{0\}) = \gamma(\{1\}) = 1/2$ and $\mathfrak{a}$ is any infinite cardinal then the measure algebra consisting of $\gamma^{\mathfrak{a}}$ on the measurable subsets of $\mathbf{2}^{\mathfrak{a}}$ is, up to isomorphism, the only $\mathfrak{a}$-homogeneous measure algebra. It is a routine exercise to show that a measure algebra (Σ, μ) is $\mathfrak{a}$-homogeneous if and only if $L_1(\mu)$ is $\mathfrak{a}$-homogeneous.

Much is known about Banach lattices X with an order continuous norm, both explicitly and implicitly because, if X had a weak order unit then we could find a probability measure μ such that $L_\infty(\mu) \subseteq X \subseteq L_1(\mu)$, where the embedding of $L_\infty(\mu)$ is onto a dense ideal in X and that of X^* is onto a dense ideal in $L_1(\mu)$, [17], Theorem 2.7.8. It follows from the theorem of Amemiya, [17] Theorem 2.4.8, that the norms in $L_1(\mu)$ and in X generate the same topology on order intervals in X so that X is $\mathfrak{a}$-homogeneous if and only if the measure algebra of μ is $\mathfrak{a}$-homogeneous. In general, by producing a maximal disjoint family in X_+ we obtain an embedding of X onto a dense ideal in an ℓ_1-sum of spaces $L_1(\mu)$.

Recall that Maharam's representation of measure algebras in terms of homogeneous measure algebras allows the following very concrete description of AL-spaces. See, for example, [22] §26 for details. In the statement of the following result, $\mathfrak{a}X$ denotes the ℓ_1-direct sum of $\mathfrak{a}$ many copies of X.

Theorem 3.11. *Let Y be an AL-space. There exists a unique well-ordered family $(\mathfrak{a}_\sigma)_{-1 \leq \sigma < \tau}$ such that:*

(1) *for each $\sigma \geq 0$, each $\mathfrak{a}_\sigma$ is equal to 0, or to 1, or is uncountable.*
(2) *$\{\sigma : \mathfrak{a}_\sigma \neq 0\}$ is cofinal in τ, and*
(3) *Y is isometrically order isomorphic to $\ell_1^{\mathfrak{a}_{-1}} \oplus_1 \ell_1\big(\mathfrak{a}_\sigma L_1(\mathbf{2}^{\aleph_\sigma}, \gamma^{\aleph_\sigma}); 0 \leq \sigma < \tau\big)$.*

Corollary 3.12. *Let X be a Banach lattice with an order continuous norm. There is a unique ordinal τ, a cofinal subset Σ of τ and a pairwise disjoint collection $(X_\sigma)_{\sigma \in \Sigma}$ of bands in X such that X_σ is $\aleph_\sigma$-homogeneous and $X = \mathrm{at}(X) \oplus \sum_{\sigma \in \Sigma} X_\sigma$, where $\mathrm{at}(X)$ is the band generated by the atoms in X.*

Definition 3.13. If e is a weak order unit for a Banach lattice X with an order continuous norm then an *e-integral* is $\phi \in X_+$ with $\phi(e) = 1$ and $\phi(x) > 0$ if $0 \neq x \in X_+$.

Definition 3.14. Let X be an Banach lattice with an order continuous norm, with a weak order unit e and an e-integral ϕ. A non-empty family of pairs $\{(e_\alpha, \phi_\alpha) : \alpha \in \mathfrak{a}\} \subset X \times X^*$, where $\mathfrak{a}$ is an ordinal, is said to be a (ϕ, e)-*Rademacher system* if

(1) For all $\alpha \in \mathfrak{a}$, $|\phi_\alpha| = \phi$ and $|e_\alpha| = e$.

(2) For all $\alpha, \beta \in \mathfrak{a}$, $\phi_\alpha(e) = \phi(e_\alpha)$ and $\phi_\alpha(e_\beta) = \begin{cases} 1 & \text{if } \alpha = \beta \\ 0 & \text{if } \alpha \neq \beta. \end{cases}$

(3) If $\alpha \neq \beta$ then $|\phi_\alpha - \phi_\beta|(e) = \phi(|e_\alpha - e_\beta|) = \frac{1}{2}$.

(4) For all $x \in X$, all $f \in X^*$ and all $\epsilon > 0$ the sets $\{\alpha \in \mathfrak{a} : |\phi_\alpha(x)| \geq \epsilon\}$ and $\{\alpha \in \mathfrak{a} : |f(e_\alpha)| \geq \epsilon\}$ are finite.

(5) If $m \in \mathbb{N}$, $\sigma : \{1, 2, \dots, m\} \to \{+, -\}$ and we set $\psi_\alpha = \bigwedge_{j=1}^m \phi_{\alpha+j}^{\sigma(j)}$ then we have
   (a) $\psi_\alpha(e) = 2^{-m}$ for all $\alpha \in \mathfrak{a}$, and
   (b) for all $x \in X$ and all $\epsilon > 0$ the set $\{\alpha \in \mathfrak{a} : |\psi_\alpha(x) - 2^{-m}\phi(x)| \geq \epsilon\}$ is finite.

At present, we will limit ourselves to proving the existence of sufficiently large such systems and some very basic properties.

Theorem 3.15. *If $\mathfrak{a}$ is an infinite cardinal and X is an $\mathfrak{a}$-homogeneous Banach lattice with an order continuous norm, a weak order unit e and an e-integral ϕ then there is a (ϕ, e)-Rademacher system of cardinality $\mathfrak{a}$.*

Proof. By the $\mathfrak{a}$-homogeneity we can find injective lattice homomorphisms $\Pi : X \to L_1(\mathbf{2}^{\mathfrak{a}}, \gamma^{\mathfrak{a}})$ and $\Theta : X^* \to L_1(2^{\mathfrak{a}}, \gamma^{\mathfrak{a}})$, such that $\Pi(e) = \mathbf{1}$ and $\Theta(\phi) = \mathbf{1}$, where $\mathbf{1}$ is the constantly one function on $2^{\mathfrak{a}}$. For each $\alpha \in \mathfrak{a}$, let r_α denote the function on $\mathbf{2}^{\mathfrak{a}}$ which is 1 if the αth component is 0 and is -1 if that component is 1. This certainly gives us a family of cardinality $\mathfrak{a}$ and with $|r_\alpha| = \mathbf{1}$ for each $\alpha \in \mathfrak{a}$. For each $\alpha \in \mathfrak{a}$ let $\phi_\alpha = \Theta^{-1}(r_\alpha)$, so that $|\phi_\alpha| = \phi$, and $e_\alpha = \Pi^{-1}(r_\alpha)$, so that $|e_\alpha| = e$. The statements in (2) and (3) are clear from the representation.

It is clear that $(r_\alpha)_{\alpha\in\mathfrak{a}}$ is an orthonormal system in $L_2(\mathbf{2}^{\mathfrak{a}}, \gamma^{\mathfrak{a}})$, so that for all $y \in L_2(\mathbf{2}^{\mathfrak{a}}, \gamma^{\mathfrak{a}})$ we have

$$\sum_{\alpha\in\mathfrak{a}} \left| \int r_\alpha y \, d\gamma^{\mathfrak{a}} \right|^2 < \infty.$$

In particular, for each $\epsilon > 0$, the set $F_y = \{\alpha \in \mathfrak{a} : |\int r_\alpha y d\gamma^{\mathfrak{a}}| \geq \epsilon/2\}$ is finite. If we take any $x \in X$ we can find $y \in L_\infty(\mathbf{2}^{\mathfrak{a}}, \gamma^{\mathfrak{a}}) \subset L_2(\mathbf{2}^{\mathfrak{a}}, \gamma^{\mathfrak{a}})$ with $\|\mathcal{P}(x) - y\|_1 < \epsilon/2$. For all α outside the finite set F_α we have

$$\begin{aligned} |\phi_\alpha(x)| \left| \int r_\alpha \Pi(x) d\gamma^{\mathfrak{a}} \right| &\leq \left| \int r_\alpha y d\gamma^{\mathfrak{a}} \right| + \left| \int r_\alpha \big(\Pi(x) - y\big) d\gamma^{\mathfrak{a}} \right| \\ &< \epsilon/2 + \int |r_\alpha| |\Pi(x) - y| d\gamma^{\mathfrak{a}} \\ &< \epsilon/2 + \|\Pi(x) - y\|_1 < \epsilon \end{aligned}$$

which completes half the proof of (4). The proof of the other half is similar.

It is clear that $\bigwedge_{j=1}^m r_{\alpha+j}^{s(j)} = \prod_{j=1}^m t_{\alpha+j}^{s(j)}$ is the characteristic function of a set of $\gamma^{\mathfrak{a}}$-measure 2^{-m}. It follows that

$$\psi_\alpha(e) = \bigwedge_{j=1}^m \phi^{s(j)})\alpha + j(e) = \bigwedge_{j=1}^m \Theta^{-1}(r_{\alpha+j}^{s(j)}(\Pi^{-1}(\mathbf{1}))$$

$$= \Theta^{-1}\left(\bigwedge_{j=1}^m r_{\alpha+j}^{s(j)}\right)(\Pi^{-1}(\mathbf{1})) = \int \bigwedge_{j=1}^m r_{\alpha_j}^{s(j)} \times \mathbf{1}\, d\gamma^{\mathfrak{a}} = 2^{-m},$$

which establishes (5) (a).

If we set $t_\alpha = r_{\alpha+q_1} r_{\alpha+q_2} \cdots r_{\alpha+q_k}$ (where $q_1 < q_2 < \cdots < q_k$) and $\alpha < \beta$ then t_α is orthogonal to t_β in $L_2(\mathbf{2}^{\mathfrak{a}}, \gamma^{\mathfrak{a}})$, as the product $t_\alpha t_\beta$ is a product of powers of r_δ, not all of which are $\mathbf{1}$ as $r_{\alpha+q_1}$ only occurs once. It follows that (t_α) is an orthonormal sequence in $L_2(\mathbf{2}^{\mathfrak{a}}, \gamma^{\mathfrak{a}})$. Again it follows that for any $\epsilon > 0$ and $y \in L_2(\mathbf{2}^{\mathfrak{a}}, \gamma^{\mathfrak{a}})$ the set $\{\alpha \in \mathfrak{a} : |\int t_\alpha y\, d\gamma^{\mathfrak{a}}| \geq \epsilon\}$ is finite. Note that

$$\bigwedge_{j=1}^m r_{\alpha+j}^{s(j)} = \prod_{j=1}^m r_{\alpha+j}^{s(j)} = 2^{-m} \prod_{j=1}^m (\mathbf{1} s(j) r_{\alpha+j})$$

(remembering that each $s(j)$ is either $+$ or $-$.) This expands to a sum $\mathbf{1} + \sum_{n=1}^N t_\alpha^n$ where each t_α^n is a family of the form considered above. It follows that the family $\{\alpha \in \mathfrak{a} : \left|\int \left(\bigwedge_{j=1}^m r_{\alpha+j}^{s(j)} - \mathbf{1}\right)(y)\, d\gamma^{\mathfrak{a}}\right| \geq \epsilon\}$ is finite. The transition to the statement (5) (b) proceeds as in the proof of part (4). □

Full details of the preceding result and of the following theorem may be found in [10].

Theorem 3.16. *Let X be a Banach lattice with an order continuous norm and Y be any Banach lattice. The following are equivalent:*

(1) *Either X is atomic with an order continuous norm or Y is Dedekind $<\mathfrak{a}$-complete where $\mathfrak{a}$ is the smallest cardinal that is greater than the density character of every order interval in X.*
(2) *$\mathcal{L}^b(X,Y)$ is a Banach lattice under the regular norm in which all lattice operations satisfy the Riesz-Kantorovich formulae.*
(3) *$\mathcal{L}^r(X,Y)$ is a vector lattice.*

Proof. The only new part of this result is that (3)⇒(1). If X is not atomic choose $\mathfrak{b} < \mathfrak{a}$ such that X contains a homogenous band, B, with order intervals all having density character $\mathfrak{b}$. If $0 \neq e \in B_+$ and $\phi \in B_+$ is an e-integral there is a (ϕ, e)-Rademacher system $\{(e_\beta, \phi_\beta) : \beta \in \mathfrak{b}\}$. We show that Y must be $\mathfrak{b}$-complete, by proving that every order bounded disjoint family $(y_\beta)_{\beta\in\mathfrak{b}}$ in Y_+ has a supremum, using a well-known result due to Veksler and Gejler, [24]. Let y be an upper bound for this family.

Define operators $S, T : X \to Y$ by $S(x) = \sum_{\beta\in\mathfrak{b}} \phi_\beta(x) y_\beta$ and $T(x) = \phi(x) y$. The series for S converges because for each $x \in X$ and each $\epsilon > 0$ the set $\{\beta \in \mathfrak{b} :$

$|\phi_\beta(x)| \geq \epsilon\}$ is finite and because of the order boundedness of the family $(y_\beta)_{\beta\in\mathfrak{b}}$. If $x \in X_+$ and F is a finite subset of $\mathfrak{b}$ then we have

$$\begin{aligned} T(x) - \sum_{\beta\in F} \phi_\beta(x)y_\beta &= \phi(x)y - \sum_{\beta\in F} \phi_\beta(x)y_\beta \geq \phi(x)y - \sum_{\beta\in F} \phi(x)y_\beta \\ &= \phi(x)y - \phi(x)\sum_{\beta\in F} y_\beta \geq \phi(x)y - \phi(x)y = 0 \end{aligned}$$

and taking the limit we see that $T(x) - S(x) \geq 0$. We can see similarly that $T(x) + S(x) \geq 0$ if $x \in X_+$ so that $S \in \mathfrak{L}^r(X,Y)$. As we are assuming that $\mathfrak{L}^r(X,Y)$ is a lattice, we can deduce that $T \geq |S|$. For any $\beta \in \mathfrak{b}$ we have

$$|S|(e) = |S|(|e_\beta|) = |S|(e_\beta^+ + e_\beta^-) \geq S(e_\beta^+) + (-S)(e_\beta^-) = S(e\beta^+ - e_\beta^-) = S(e_\beta)$$

But $S(e_\beta) = \sum_{\alpha\in\mathfrak{b}} \phi_\alpha(e_\beta)y_\beta = y_\beta$, so that $|S|(e)$ is an upper bound for the family of all $(y_\beta)_{\beta\in\mathfrak{b}}$. On the other hand, we have seen that $|S|(e) \leq T(e) = \phi(e)y = y$, where y was any upper bound for $(y_\beta)_{\beta\in\mathfrak{b}}$, so we see that $|S|(e)$ is the supremum of that family, as claimed. Thus Y is Dedekind $\mathfrak{b}$-complete and hence is Dedekind $<\mathfrak{a}$-complete. □

The formulation of the preceding result involving $<\mathfrak{a}$-completeness seems rather artificial. We could hope to replace it by $\mathfrak{b}$-completeness where $\mathfrak{b}$ is the supremum of density characters of order intervals in X. There is no prospect of proving such a result unconditionally, but there is with a set-theoretic assumption. Recall that an uncountable cardinal $\mathfrak{a}$ is *weakly inaccessible* if it is a regular limit cardinal. Their existence cannot be proved inside Zermelo-Fraenkel set theory together with the Axiom of Choice. In [10], Elliott established:

Theorem 3.17. *Assume that there is no weakly inaccessible cardinal. Let X be a Banach lattice with an order continuous norm and Y be any Banach lattice. The following are equivalent:*

(1) *Either X is atomic with an order continuous norm or Y is Dedekind $\mathfrak{a}$-complete where $\mathfrak{a}$ is the smallest cardinal that is greater than or equal to the density character of every order interval in X.*
(2) *$\mathfrak{L}^b(X,Y)$ is a Banach lattice under the regular norm in which all lattice operations satisfy the Riesz-Kantorovich formulae.*
(3) *$\mathfrak{L}^r(X,Y)$ is a vector lattice.*

He also showed that the equivalence of the three conditions in the theorem is actually equivalent to the non-existence of weakly inaccessible cardinals. We omit the details of this proof.

Again, in these last cases, if $\mathfrak{L}^r(X,Y)$ is a vector lattice then the lattice operations are all given by the Riesz-Kantorovich formulae. There is now enough evidence that I would lean towards the conjecture that if $\mathfrak{L}^r(X,Y)$ is a vector lattice then the lattice operations in $\mathfrak{L}^r(X,Y)$ are all given by the Riesz-Kantorovich formulae. I still have an open mind about whether or not it is true that if a single operator $T \in L^r(X,Y)$ has a modulus then it is given by the Riesz-Kantorovich formula.

What about weaker conditions on the order structure? The only obvious one that seems worth considering is:

Definition 3.18. An ordered vector space X has the *Riesz Separation Property* if whenever $x_1, x_2, z_1, z_2 \in X$ with $x_1, x_2 \leq z_1, z_2$ there is $y \in X$ with $x_1, x_2 \leq y \leq z_1, z_2$. It is equivalent that X has the *Riesz Decomposition Property* which states that if $0 \leq x \leq y_1 + y_2$ and $y_1, y_2 \geq 0$ then there are $x_1, x_2 \geq 0$ such that $x = x_1 + x_2$ and $0 \leq x_k \leq y_k$ $(k = 1, 2)$.

This is of interest because if $\mathfrak{L}^r(X,Y)$ is normed by the *regular* norm, $\|T\|_r = \inf\{\|U\| : U \geq \pm T\}$, then $\mathfrak{L}^r(X,Y)^*$ is a lattice if and only if $\mathfrak{L}^r(X,Y)$ has the Riesz Separation Property.

Definition 3.19. We say that a Banach lattice X has the *countable interpolation property* if, given countable subsets A and C of X such that if $a \in A$ and $c \in C$ then $a \leq c$, then there is $b \in X$ such that $a \leq b \leq c$ for all $a \in A$ and $c \in C$.

The countable interpolation property is also known as the *Cantor property.* It is well known that $C(K)$ has the Cantor property if and only if K is an F-space, i.e., the closures of two disjoint open F_σ sets are disjoint. This kind of property is quite strong but falls short of Dedekind σ-completeness. The following result was established by Danet [9] and Wickstead [26].

Theorem 3.20. *The following conditions on a Banach lattice Y are equivalent:*

(1) *Y has the Cantor property.*
(2) *For every separable Banach lattice X, $\mathfrak{L}^r(X,Y)$ has the Riesz Separation Property.*
(3) *$\mathfrak{L}^r(c,Y)$ has the Riesz Separation Property.*

4. When are all order bounded operators regular?

This is certainly true if Y is Dedekind complete, but we can do slightly better. There are no essentially new ideas needed to prove this result.

Theorem 4.1. *Let X and Y be Banach lattices such that whenever $\mathfrak{a}$ is the density character of an order interval in X then Y is Dedekind $\mathfrak{a}$-complete, then $\mathfrak{L}^r(X,Y) = \mathfrak{L}^b(X,Y)$.*

In fact, in such a case $\mathfrak{L}^b(X,Y)$ will actually be a lattice. There are very few general results in this area. The few that are known concern the special case when $Y = C(K)$, for K a compact Hausdorff space. The following result is Theorem 4.1 of [25]

Theorem 4.2. *If K is a compact Hausdorff space such that, for every Banach lattice X, $\mathfrak{L}^r\big(X, C(K)\big) = \mathfrak{L}^b\big(X, C(K)\big)$ then there is $n \in \mathbb{N}$ such that, for any disjoint open sets $U_1, U_2 \ldots, U_{n+1}$ in K, $\bigcap_{k=1}^{n+1} \overline{U_k} = \emptyset$.*

If we take two disjoint infinite Stonean spaces S_1 and S_2 and identify a non-isolated point in each of the spaces then we obtain a space K that is not Stonean (so that $C(K)$ is not Dedekind complete) yet has the property described in the theorem, for $n = 2$, so that every order bounded operator from any Banach lattice into $C(K)$ is actually regular (see [3]). By suitable choices of K we can ensure that $C(K)$ either is or is not Dedekind σ-complete. See [11] and [25] for details of these examples. In particular, let us note that it is possible for the order bounded and regular operators to coincide without being a vector lattice.

Just as there is a natural norm on the space of regular operators, under which it is norm complete, so also there is a natural norm on the space of order bounded operators, namely the *order bound norm* defined by

$$\|T\|_b = \inf\{M \in \mathbb{R} : \forall x \in X_+ \exists y \in Y_+ \text{ with } T([-x,x]) \subseteq [-y,y] \text{ and } \|y\| \leq M\|x\|\},$$

under which $\mathfrak{L}^b(X,Y)$ is norm complete, see [27] for details.

If $\mathfrak{L}^r(X,Y) = \mathfrak{L}^b(X,Y)$ then the regular and order bound norms must be equivalent. Their non-equivalence can sometimes be used to prove that not all order bounded operators are regular.

Example 4.3. For all $n \in \mathbb{N}$, $\mathfrak{L}^r(c^n, c) = \mathfrak{L}^b(c^n, c)$ but $\mathfrak{L}^r(\ell_\infty(c^n), c) \neq \mathfrak{L}^b(\ell_\infty(c^n), c)$.

The proof given in [27] uses the fact that we can find an operator T in $\mathfrak{L}^r(c^n, c)$ with $\|T\|_b = 2$ and $\|T\|_r \geq n$. This operator T already shows that the regular and order bound norms need not be equal even when $\mathfrak{L}^r(X,Y) = \mathfrak{L}^b(X,Y)$.

5. What kind of duality theory is there for regular operators?

If $T \in \mathfrak{L}^r(X,Y)$ then we certainly have $T^* \in \mathfrak{L}^r(Y^*, X^*)$ and $\mathfrak{J} \circ T \in \mathfrak{L}^r(X, Y^{**})$, where $\mathfrak{J} : Y \to T^{**}$ is the natural embedding of Y into its bidual.

Example 5.1. If r_n denotes the nth Rademacher function on $[0,1]$, define $T : L_1([0,1]) \to c_0$ by $Tx = \sum_{n=1}^\infty \left(\int_0^1 r_n(t)x(t)\,dt\right) e_n$, where e_n is the nth standard basis vector in c_0, which it is routine to see is not regular. However both $T^* : c_0^* = \ell_1 \to L_1([0,1])^* = \mathfrak{L}_\infty([0,1])$ and $\mathfrak{J} \circ T : L_1([0,1]) \to c_0^{**} = \ell_\infty$ are regular.

This already shows that some care needs to be taken with duality for regular operators. It turns out though that T^* and $\mathfrak{J} \circ T$ both behave in pretty much the same way. We will prove here rather more than is contained in [8] so will give complete proofs. We start by giving a result about Banach spaces which is surely well known, but for which we know no reference.

Lemma 5.2. *Let X and Y be Banach spaces.*

(1) *If $U \in \mathfrak{L}(Y^*, X^*)$ then $U^*{}_{|X} \in \mathfrak{L}(X, Y^{**})$ and $(U^*{}_{|X})^*{}_{|Y^*} = U$.*
(2) *If $V \in \mathfrak{L}(X, Y^{**})$ then $V^*{}_{|Y^*} \in \mathfrak{L}(Y^*, X^*)$ and $(V^*{}_{|Y^*})^*{}_{|X} = V$.*
(3) *If $T \in \mathfrak{L}(X.Y)$ then $(J_Y \circ T)^*{}_{|Y^*} = T^*$.*

Proof. To start with it is clear that if $U \in \mathfrak{L}(Y^*, X^*)$ then ${U^*}_{|X} \in \mathfrak{L}(X, Y^{**})$ and that if $V \in \mathfrak{L}(X, Y^{**})$ then ${V^*}_{|Y^*} \in \mathfrak{L}(Y^*, X^*)$. Let $f \in Y^*$, $\hat{f}$ denote its canonical image in Y^{***}, and let $x \in X$ then

$$({U^*}_{|X})^*(\hat{f})(x) = \hat{f}(U^* x) = (Uf)(x)$$

whilst

$$({V^*}_{|Y^*})^*_X(x)(f) = (V^*)(f)(x) = f(Vx),$$

establishing (1) and (2). For (3), note that

$$\begin{aligned}(\mathfrak{J}_Y) \circ T)^*(\hat{f}) &= \hat{f}\big((\mathfrak{J}_Y \circ T)(x)\big) = \big((\mathfrak{J}_Y \circ T)(x)\big)(f) = \big(\mathfrak{J}(Tx)\big)(f) \\ &= f(Tx) = (T^* f)(x)\end{aligned}$$

so that we do indeed have $(J_Y \circ T)_{|Y^*} = T^*$. □

Corollary 5.3. *If X and Y are Banach spaces then the linear operators $\mathfrak{L}(Y^*, X^*) \ni U \mapsto {U^*}_{|X} \in \mathfrak{L}(X, Y^{**})$ and $L(X, Y^{**}) \ni V \mapsto (V^*)_{|Y^*} \in \mathfrak{L}(Y^*, X^*)$ are mutually inverse isometric bijections.*

Proof. Statements (1) and (2) in Lemma 5.2 tells us that these maps are mutually inverse bijections. As, for example, $\|(U^*)_{|X}\| \le \|U^*\| = \|U\|$ both maps are contractions and, being mutually inverse, must be isometries. □

Corollary 5.4. *If X and Y are Banach lattices then the linear operators*

$$\mathfrak{L}^r(Y^*, X^*) \ni U \mapsto {U^*}_{|X} \in \mathfrak{L}^r(X, Y^{**}) \quad \text{and}$$
$$L^r(X, Y^{**}) \ni V \mapsto (V^*)_{|Y^*} \in \mathfrak{L}^r(Y^*, X^*)$$

are mutually inverse bijective order isomorphisms which are isometries for both the operator and regular norms.

Proof. It is clear that both maps are positive and, being bijective and inverses of each other, both are order isomorphisms. We know that they are isometries for the operator norm from the preceding corollary. That the maps are isometries for the regular norm follows immediately. □

Corollary 5.5. *If X and Y are Banach lattices and $T \in \mathfrak{L}(X, Y)$ then $T^* \in \mathfrak{L}^r(Y^*, X^*)$ if and only if $\mathfrak{J}_Y \circ T \in \mathfrak{L}^r(X, Y^{**})$.*

If T is regular then both of the operators T^* and $\mathfrak{J} \circ T$ are into a Dedekind complete range space so that both operators have a modulus. If T itself has a modulus, then it would be nice if we knew that $|T^*| = |T|^*$ or that $|\mathfrak{J} \circ T| = \mathfrak{J} \circ |T|$. Unfortunately neither of these equalities holds in general. The same holds, of course, for the positive and negative parts of these operators.

Example 5.6. Let K be an infinite compact Hausdorff space and let k_0 be any non-isolated point in K. Define $T : C(K) \to C(K)$ be defined by $Tf = f - f(k_0)\mathbf{1}_K$. As k_0 is not isolated, for any $f \in C(K)_+$, we have $f = \sup\{g \in C(K) : 0 \le g \le f \text{ and } g(k_0) = 0\}$. Clearly $I \ge T, 0$, where I is the identity operator on $C(K)$. On the other hand if $U \ge T, 0$ and $f \ge 0$ then Uf is an upper bound for $\{Tg : 0 \leq g \leq f\}$ and hence for $\{Tg : 0 \leq g \leq f$ and $g(k_0) = 0\} = \{g : 0 \leq g \leq f$ and $g(k_0) = 0\}$ so that $Uf \geq f$. Thus $U \geq I$ and it follows that T^+ exists and is equal to I. If ε_{k_0} is the linear functional mapping f to $f(k_0)$ then $T^*(\varepsilon_{k_0})(f) = \varepsilon_{k_0}(Tf) = (Tf)(k_0) = 0$ for all $f \in C(K)$ so that $T^*(\varepsilon_{k_0}) = 0$. But $(T^*)^+(\varepsilon_{k_0}) = \sup\{T^*\mu : 0 \leq \mu \leq \varepsilon_{k_0}\} = 0$ as ε_{k_0} is extremal. As $(T^+)^*(\varepsilon_{k_0}) = I^*(\varepsilon_{k_0}) = \varepsilon_{k_0}$, we see that $(T^*)^+ \neq (T^+)^*$ and it follows that $|T^*| \neq |T|^*$.

We do, however, have:

Theorem 5.7. *Suppose that X and Y are Banach lattices and T is a regular operator from X into Y such that the modulus $|T|$ of T exists in $\mathfrak{L}^r(X,Y)$. Then $\mathfrak{J}_Y \circ |T| = |\mathfrak{J}_Y \circ T|$ if and only if $|T|^* = |T^*|$.*

Proof. Note first that we certainly always have $|\mathfrak{J}_Y \circ T| \leq \mathfrak{J}_Y \circ |T|$ and $|T^*| \leq |T|^*$.

If $\mathfrak{J}_Y \circ T| = \mathfrak{J}_Y \circ |T|$ then $|T^*| \geq \pm T^*$ so that $|T^*|^* \geq \pm T^{**}$ and hence $|T^*|^*{}_{|X} \geq \pm T^{**}{}_{|X} = \pm \mathfrak{J}_Y \circ T$ so that $|T^*|^*{}_{|X} \geq |\mathfrak{J}_Y \circ T| = \mathfrak{J}_Y \circ |T|$. Thus $(|T^*|^*{}_{|X})^* \geq (\mathfrak{J}_Y \circ |T|)^*$ and restricting to Y^* we see that $|T^*| \geq |T|^*$ so that $|T^*| = |T|^*$.

If $|T^*| = |T|^*$ then $|J_Y \circ T| \geq \pm \mathfrak{J}_Y \circ T$ so that $|\mathfrak{J}_Y \circ T|^* \geq \pm(\mathfrak{J}_Y \circ T)^*$ and hence $|\mathfrak{J}_Y \circ T|^*{}_{Y^*} \geq \pm(\mathfrak{J}_Y \circ T)^*{}_{|Y^*} = \pm T^*$. Thus $|\mathfrak{J}_Y \circ T|^*{}_{Y^*} \geq |T^*| = |T|^*$. Hence $(|\mathfrak{J}_Y \circ T|^*{}_{Y^*})^* \geq |T|^{**}$ and restricting to X we find that

$$|\mathfrak{J}_Y \circ T| = (|\mathfrak{J}_Y \circ T|^*{}_{Y^*})^*{}_{|}X \geq |T|^{**}{}_{|X} = \mathfrak{J}_Y \circ |T|$$

and hence $\mathfrak{J}_Y \circ |T| = |\mathfrak{J}_Y \circ T|$ as required. □

Again, it is of interest to know conditions on X and Y which guarantee that these equalities do hold.

Definition 5.8. The pair of Banach lattices (X, Y) has the *invariant modulus property* if, for every $T \in \mathfrak{L}^r(X,Y)$ for which $|T|$ exists in $\mathfrak{L}^r(X,Y)$, we have $|T|^* = |T^*|$ and, equivalently, $|\mathfrak{J} \circ T| = \mathfrak{J} \circ |T|$.

A complete characterization of the invariant modulus property again eludes us, but if we assume that Y is Dedekind σ-complete then we are very close to one. This is more useful than it might seem as without Dedekind σ-completeness there will be few operators in $\mathfrak{L}^r(X,Y)$ with a modulus. The first, and most well known, result in this area is due to Synnatzschke, [23], who established that (1)(b)⇒(2). The remainder of the proof may be found in Theorem 4.1 of [8]. We include here an alternative short proof of Synnatzschke's result.

Theorem 5.9. *Let X and Y be Banach lattices with Y Dedekind σ-complete and consider the following three conditions:*

(1) *At least one of following two conditions holds:*
   (a) *X is atomic with an order continuous norm.*
   (b) *The norm on Y is order continuous.*
(2) *The pair (X, Y) has the invariant modulus property.*

(3) *At least one of following two conditions holds:*
(a) *The lattice operations in X^* are weak* sequentially continuous.*
(b) *The norm on Y is order continuous.*

We always have (1) *implies* (2) *implies* (3) *and if X is Dedekind σ-complete or is separable or $X = C_0(\Sigma)$, where Σ is a locally compact Hausdorff space, then all three conditions are equivalent.*

Proof. If we suppose that Y has an order continuous norm, then $\mathfrak{J}_Y \circ Y$ is an ideal in Y^{**}, by Theorem 2.4.2 of [17]. It follows immediately that if $x \geq 0$ and $T \in \mathfrak{L}^r(X,Y)$ has a modulus then the set $\{\mathfrak{J}_Y(Ty) : -x \leq y \leq x\}$ has the same supremum in both $\mathfrak{J}_Y \circ Y$ and in Y^{**}. This states precisely that $\mathfrak{J}_Y \circ |T|(x) = |\mathfrak{J}_Y \circ T|(x)$ so that $\mathfrak{J}_Y \circ |T| = |\mathfrak{J} \circ T|$ as required. □

If we allow the range or domain to vary our results are much more complete, see Corollary 4.2 and Theorem 4.3 of [8].

Corollary 5.10. *A Dedekind σ-complete Banach lattice Y has an order continuous norm if and only if for every Banach lattice Y the pair (X,Y) has the invariant modulus property.*

Theorem 5.11. *A Banach lattice X is atomic with an order continuous norm if and only if for every Banach lattice Y the pair (X,Y) has the invariant modulus property.*

6. What happens for the classical Banach lattices?

The results above tell us what happens in quite a few cases, but there certainly are gaps in what the general theory tell us. If we limit our attention to some of the classical Banach lattices then we would hope to establish more. Even here our knowledge remains incomplete.

The most notable gaps in our knowledge concern operators from $C(K)$ into $C(\Omega)$ and into c_0. The latter case is relatively straightforward, at least to state. Many of the results in the first part of this section may be found in [29]. This is Xiong's Theorem 2.8.

Theorem 6.1. *If K is an infinite compact Hausdorff space then*

$$\mathfrak{L}^r(C(K), c_0) \neq \mathfrak{L}(C(K), c_0).$$

The proof of this theorem differs greatly depending on whether or not K contains a non-trivial convergent sequence. Such sequences are clearly important in this context.

Definition 6.2. If K is a compact Hausdorff space then we define X_ℓ to be the set of points $k \in K$ for which there is a *sequence* of distinct points in K which converges to k.

Definition 6.3. If $n \in \mathbb{N}$ then a compact Hausdorff space K is termed an $X(n)$ *space* if every sequence in K has a convergent subsequence and K_ℓ has precisely n members.

Theorems 2.12 and 2.13 of [29] establish the next two results.

Theorem 6.4. *If K is a compact Hausdorff space then the following are equivalent:*

(1) *For some $p \in \mathbb{N}$, K is an $X(p)$-space.*
(2) *For every compact Hausdorff space Ω, $\mathfrak{L}(C(K), C(\Omega)) = \mathfrak{L}^r(C(K), C(\Omega))$.*
(3) *For some compact Hausdorff space Ω in which there is a non-trivial convergent sequence, $\mathfrak{L}(C(K), C(\Omega)) = \mathfrak{L}^r(C(K), C(\Omega))$.*
(4) $\mathfrak{L}(C(K), C([0,1])) = \mathfrak{L}^r(C(K), C([0,1]))$.
(5) $\mathfrak{L}(C(K), c) = \mathfrak{L}(C(K), c)$.

Theorem 6.5. *If K is a compact Hausdorff space then the following are equivalent:*

(1) *K is an $X(1)$-space.*
(2) *For every compact Hausdorff space Ω, $\mathfrak{L}(C(K), C(\Omega)) = \mathfrak{L}^r(C(K), C(\Omega))$ and $\|T\| = \|T\|_r$ for all $T \in \mathfrak{L}(C(K), C(\Omega))$.*
(3) *For some compact Hausdorff space Ω in which there is a non-trivial convergent sequence, $\mathfrak{L}(C(K), C(\Omega)) = \mathfrak{L}^r(C(K), C(\Omega))$ and $\|T\| = \|T\|_r$ for all $T \in \mathfrak{L}(C(K), C(\Omega))$.*
(4) *$\mathfrak{L}(C(K), C([0,1])) = \mathfrak{L}^r(C(K), C([0,1]))$ and $\|T\| = \|T\|_r$ for all $T \in \mathfrak{L}(C(K), C([0,1]))$.*
(5) *$\mathfrak{L}(C(K), c) = \mathfrak{L}(C(K), c)$ and $\|T\| = \|T\|_r$ for all $T \in \mathfrak{L}(C(K), c)$.*

Finally, we address the question of when the space of regular operators can itself be one of the classical Banach lattices. The first result is contained in Theorem 2.1 of [28].

Theorem 6.6. *If X and Y are Banach lattices, neither of which is the zero space, then $\mathfrak{L}^r(X, Y)$ is an AL-space under the regular norm if and only if X is an AM-space and Y is an AL-space.*

Things are a little trickier when we want $\mathfrak{L}^r(X, Y)$ to be an AM-space.

Definition 6.7. If $\mathfrak{a}$ is an infinite cardinal then a Banach lattice X has an $\mathfrak{a}$-Fatou norm if whenever $A \subset X_+$ is upward directed with supremum b and the cardinality of A is at most $\mathfrak{a}$, then $\|b\| = \sup\{\|a\| : a \in A\}$.

Clearly X has a Fatou norm if and only if has a $\mathfrak{a}$-Fatou norm for very cardinal $\mathfrak{a}$. The following result is taken from [10], after a partial result was established in Theorem 2.2 of [28].

Theorem 6.8. *Let X and Y be non-zero Banach lattices, and let $\mathfrak{a}$ be the smallest cardinal that exceeds the density character of every order interval in X, then the following are equivalent:*

(1) *The following three conditions are all satisfied:*
   (a) *X is an AL-space.*
   (b) *Y is an AM-space.*
   (c) *Either X is atomic or Y is Dedekind $<\mathfrak{a}$-complete and has the $<\mathfrak{a}$-Fatou property.*
(2) *$\mathcal{L}^r(X,Y)$ is an AM-space under the operator norm.*
(3) *$\mathcal{L}^r(X,Y)$ is an AM-space under the regular norm.*

Proof. Let us start by assuming that (1) holds and prove that (2) holds. The space $\mathcal{L}^r(X,Y)$ is a Banach lattice under the regular norm by Theorem 3.16.

If X is atomic the proof is routine. Now suppose that Y has a $<\mathfrak{a}$-Fatou norm. If $S,T \in \mathcal{L}^r(X,Y)$ and $x \in X_+$ then the Riesz-Kantorovich formula tells us that

$$(S \vee T)(x) = \sup\{Sx_1 + Tx_2 : 0 \le x_1, x_2, x_1 + x_2 = x\}.$$

Consider the set $A = \{Sx_1 + Tx_2 : 0 \le x_1, x_2, x_1 + x_2 = x\} \subseteq S([0,x]) + T([0,x])$. As order intervals in X have density character strictly less than $\mathfrak{a}$ the same is true of any continuous images of them, in particular $S([0,x])$ and $T([0,x])$. It follows that their sum also has density character strictly less than $\mathfrak{a}$ and hence the same is true of A. Now let B denote the collection of all finite suprema from A, which also has density character less than $\mathfrak{a}$, is upward directed and has the same supremum as A. The $<\mathfrak{a}$-Fatou property guarantees that $\|(S \vee T)x\| = \sup\{\|b\| : b \in B\}$. If we have $0 \le x_1, x_2$ and $x_1 + x_2 = x$ then the fact that X is an AL-space shows that for any such choice of x_1 and x_2 we have

$$\begin{aligned}\|Sx_1 + Tx_2\| &\le \|Sx_1\| + T\|x_2\| \le \|S\|\|x_1\| + \|T\|\|x_2\| \\ &\le (\|S\| \vee \|T\|)(\|x_1\| + \|x_2\| = (\|S\| \vee \|T\|)\|x\|.\end{aligned}$$

Finally we use the fact that Y is an AM-space to see that if $x_1^1, x_2^1, \dots, x_1^n, x_2^n \in [0,x]$ with $x_1^k + x_2^k = x$, for $1 \le k \le n$, then

$$\left\| \bigvee_{k=1}^n (Sx_1^k + Tx_2^k) \right\| = \bigvee_{k=1}^n \|Sx_1^k + Tx_2^k)\| \le \max\{\|S\|, \|T\|\}\|x\|.$$

It follows that, for all $x \in X_+$, we have

$$\|(S \vee T)(x)\| = \|\sup(B)\| \le \max\{\|S\|, \|T\|\}\|x\|.$$

If $S \vee T \ge 0$ then this tells us that $\|(S \vee T)\| \le \max\{\|S\|, \|T\|\}$. Taking $S = -T$ we see that $\||T|\| \le \|T\|$, so that the regular and operator norms coincide. Applying the inequality to two arbitrary positive operators S and T, this suffices to show that $\mathcal{L}^r(X,Y)$ is indeed an AM-space under the operator norm.

If $\mathcal{L}^r(X,Y)$ is an AM-space under the operator norm then it is certainly a Banach lattice under the regular norm and the norm condition on positive operators tells us that it is an AM-space under the regular norm also. This establishes that (2)$\Rightarrow$(3).

Finally, let us assume (3) and establish (1). Routine considerations of rank one operators shows that X must be an AL-space and Y an AM-space. It remains only to prove that (1)(c) holds. To this end, let us recall from Theorem 3.16 that the fact that $\mathcal{L}^r(X,Y)$ is a vector lattice implies that either X is atomic (in which case there is nothing left to prove) or else that Y is Dedekind $<\mathfrak{a}$-complete. If X is not atomic choose $\mathfrak{b} < \mathfrak{a}$ such that X contains a homogenous band, B, with order intervals all having density character $\mathfrak{b}$. By Theorem 3.15 there is $e \in B_+$ and an e-integral $\phi \in B_+$ together with (ϕ, e)-Rademacher system of functionals $(\phi_\beta)_{\beta\in\mathfrak{b}}$, with $\|e\| = \|\phi\| = \|\phi_\beta\|$ for each $\beta \in \mathfrak{b}$.

In order to show that Y has a $\mathfrak{b}$-Fatou norm it suffices, by Theorem 2.1 of [2], to show that if $(y_\beta)_{\beta\in\mathfrak{b}}$ is a disjoint positive family in Y, with supremum z, then $\|z\| = \sup\{\|\sum_{\beta\in F} y_\beta\| : F \subseteq \mathfrak{b}\}$ where the supremum is taken over all finite subsets F of $\mathfrak{b}$. As Y is here known to be an AM-space, this amounts to stating that $\|z\| = \sup\{\|y_\beta\| : \beta \in \mathfrak{b}\}$. Let us denote $\sup\{\|y_\beta\| : \beta \in \mathfrak{b}\}$ by κ, then our task is to show that $\kappa \le \|z\|$, the reverse inequality being clear.

Define operators $S, T : X \to Y$ by

$$Sx = \sum_{\beta\in\mathfrak{b}} \phi_\beta^+(x) y_\beta \text{ and } Tx = \sum_{\beta\in\mathfrak{b}} \phi_\beta^-(x) y_\beta.$$

These series converges in order. In order to see that, given the Dedekind α-completeness, it suffices to note that $\phi_\beta^\pm(x)$ is bounded using, for example, the identity $\phi_\beta^+ = \frac{1}{2}(\phi + \phi_\beta)$ and condition (2) in Definition 3.14.

Our first task is to compute the norms of S and T. In fact the norms of both are at most $(\kappa + \|z\|)/2$. We show this for S, the proof for T being virtually identical. First let us note that an alternative description of S is provided by the formula

$$Sx = \sum_{\beta\in\mathfrak{b}} (\phi_\beta^+ - \frac{1}{2}\phi)(x) y_\beta + \frac{1}{2}\phi(x) z = \sum_{\beta\in\mathfrak{b}} \frac{1}{2}\phi_\beta(x) y_\beta + \frac{1}{2}\phi(x) z.$$

The series here is certainly order convergent and, for each $x \in X$, is actually norm convergent. For if $\epsilon > 0$ then there is a finite set $F \subset \mathfrak{b}$ such that $|\phi_\beta(x)| < \epsilon$ for all $\beta \in \mathfrak{b} \setminus F$. Now note that

$$\begin{aligned} |\sum_{\beta\in\mathfrak{b}} \frac{1}{2}\phi_\beta(x) y_\beta - \sum_{\beta\in F} \frac{1}{2}\phi_\beta(x) y_\beta| &= |\sum_{\beta\in\mathfrak{b}\setminus F} \frac{1}{2}\phi_\beta(x) y_\beta| = |\sum_{\beta\in\mathfrak{b}\setminus)} \frac{1}{2}\phi_\beta(x)|y_\beta| \\ &\le \sum_{\beta\in\mathfrak{b}\setminus F} |\frac{1}{2}\phi_\beta)(x)| y_\beta = \sum_{\beta\in\mathfrak{b}\setminus F} \epsilon y_\beta \le \epsilon z. \end{aligned}$$

In fact, the finite sums $\sum_{\beta\in F}(\phi_\beta^+ - \frac{1}{2}\phi)(x) y_\beta$ are actually finite suprema (because of the disjointness of the y_β) and the fact that Y is an AM-space tells us that

$$\left\| \bigvee_{\beta\in F} \frac{1}{2}\phi_\beta(x) y_\beta \right\| = \bigvee_{\beta\in F} \|\frac{1}{2}\phi_\beta(x) y_\beta\| \le \bigvee_{\beta\in F} \|\frac{1}{2}\phi_\beta\| \|x\| \|y_\beta\| \le \kappa \|x\|/2.$$

On passing to the limit, we see that

$$\left\| \sum_{\beta \in \mathfrak{b}} \frac{1}{2} \phi_\beta(x) y_n \right\| \leq \kappa \|x\|/2.$$

It is also clear that $\|\frac{1}{2}\phi(x)z\| \leq \frac{1}{2}\|x\|\|z\|$ so the triangle inequality tells us that $\|S\| \leq \frac{1}{2}(\|z\| + \kappa)$.

It is routine to verify that

$$(S \vee T)(x) = \sum_{\beta \in \mathfrak{b}} |\phi_\beta|(x) y_\beta = \sum_{\beta \in \mathfrak{b}} \phi(x) y_\beta = \phi(x) \bigvee_{\beta \in \mathfrak{b}} y_\beta = \phi(x) z,$$

so that $\|S \vee T\| = \|z\|$. If $L^r(X,Y)$ is an AM-space then $\|S \vee T\| = \|S\| \vee \|T\|$ so that

$$\|z\| = \|S \vee T\| = \|S\| \vee \|T\| \leq \frac{1}{2}(\kappa + \|z\|)$$

which shows that $\|z\| \leq \kappa$, completing the proof that Y has a $\mathfrak{b}$-Fatou norm. This will be true for all $\mathfrak{b}$ for which there is a $\mathfrak{b}$-homogenous ideal in X. The supremum of such $\mathfrak{b}$ is $\mathfrak{a}$, so that Y has a $<\mathfrak{a}$-Fatou norm, completing the proof. □

Again, it is worth recording that Elliott has given an example in [10] to show that the rather simpler version that one might hope for where "$<\mathfrak{a}$" is replaced by "$\mathfrak{a}$" is equivalent to the non-existence of weakly inaccessible cardinals. A simpler version of the preceding result may be easier to appreciate. The fact that AM-spaces with a Fatou norm are of the form $C_0(\Sigma)$ follows from a result of Nakano that may be found in [18] and [19].

Corollary 6.9. *If Y is a Banach lattice such that $\mathfrak{L}^r(X,Y)$ is an AM-space for all AL-spaces X, then Y is isometrically order isomorphic to $C_0(\Sigma)$ for some locally compact extremally disconnected space Σ.*

By way of contrast, things are not at all as interesting in the L_p case. We extract a minor technical detail from the main proof.

Lemma 6.10. *Let $1 < p, q < \infty$ then the natural embedding of ℓ_q^2 into ℓ_p^2 has norm $2^{(\frac{1}{p}-\frac{1}{q})^+}$.*

Proof. Suppose that $\theta \in [0, \pi/2]$ and consider $\cos^r(\theta) + \sin^r(\theta)$. If $r > 2$ this decreases on $[0, \pi/4]$ and increases on $[\pi/4, \pi/2]$, so has maximum value 1. If $r = 2$ it is, of course, constantly 1. If $r < 2$ then it increases on $[0, \pi/4]$ and decreases on $[\pi/4, \pi/2]$ so has a maximum value of $2^{1-\frac{r}{2}}$. To calculate the norm in question, it suffices to find the maximum value of $\|(x,y)\|_p$ subject to the constraints that $\|(x,y)\|_q = 1$ and $x, y \geq 0$. We may represent such pairs as $x = \cos^{\frac{2}{q}}(\theta)$ and $y = \sin^{\frac{2}{q}}(\theta)$ for $\theta \in [0, \pi/2]$. Then $\|(x,y)\|_p^p = x^p + y^p = \cos^{\frac{2p}{q}}(\theta) + \sin^{\frac{2p}{q}}(\theta) = \cos^r(\theta) + \sin^r(\theta)$, where $r = \frac{2p}{q}$. The maximum of this is thus 1 if $r \geq 2$, i.e., if $p \geq q$ and is $2^{1-\frac{p}{q}}$ if $p < q$. Taking pth roots gives the claimed norm. □

Theorem 6.11. *If X and Y are non-trivial Banach lattices and $\mathfrak{L}^r(X,Y)$ is isometrically isomorphic to an L_p-space, where $1 < p < \infty$, then either X or Y is one-dimensional.*

Proof. If $\mathfrak{L}^r(X,Y)$ is an L_p-space then using rank one operators shows that both X^* and Y have are L_p-spaces so that X is an L_q-space, where $\frac{1}{p}+\frac{1}{q}=1$. Suppose that both X and Y have dimension at least two. Pick disjoint positive elements of norm 1, $x_1, x_2 \in X$ and $y_1, y_2 \in Y$. By Theorem 2.7.11 of [17] there is a positive contractive projection P of X onto the two-dimensional sublattice generated by x_1 and x_2. Let Q_k denote the band projection of this two-dimensional lattice onto the linear span of x_k. The linear operator J taking x_k to y_k is a positive bijection. Define $T_k = J \circ Q_k \circ P : X \to Y$, and we see that $Q_k \circ P$ is certainly a contraction with one-dimensional image so that T_k is also a contraction. As $Tx_k = y_k$, $\|T_k\| = 1$. However, $T_1 + T_2 = J \circ (Q_1 + Q_2) \circ P = J \circ P$. Thus $\|T_1 + T_2\| \le \|J\|\|P\| = \|J\| = 2^{(\frac{1}{p}-\frac{1}{q})^+} = 2^{(\frac{2}{p}-1)^+}$, by the preceding lemma. On the other hand, if $\mathfrak{L}^r(X,Y)$ were an L_p-space then this should be $2^{\frac{1}{p}}$. If $p \ge 2$ then $(\frac{2}{p}-1)^+ = 0 \ne \frac{1}{p}$ whilst if $p < 2$ then $(\frac{2}{p}-1)^+ = \frac{2}{p} - 1$ which is only equal to $\frac{1}{p}$ if $p = 1$, which is excluded by our hypotheses. □

There are also isomorphic versions of the last three results.

References

[1] Ju.A. Abramovič, *The space of operators that act between Banach lattices*, Zap. Nauchn. Sem. Leningrad. Otdel. Mat. Inst. Steklov. (LOMI) **73** (1977), 188–192, 234 (1978), Investigations on linear operators and the theory of functions, VIII. MR513175 (80a:47069) (Russian, with English summary)

[2] Y.A. Abramovich, Z.L. Chen, and A.W. Wickstead, *Regular-norm balls can be closed in the strong operator topology*, Positivity **1** (1997), 75–96. MR1659595 (2000a:46027)

[3] Ju.A. Abramovič and V.A. Gejler, *On a question of Fremlin concerning order bounded and regular operators*, Colloq. Math. **46** (1982), 15–17. MR672357 (84b:46008)

[4] Ju.A. Abramovič and L.P. Janovskiĭ, *Application of Rademacher systems to operator characterizations of Banach lattices*, Colloq. Math. **46** (1982), 73–78. MR672365 (84h:46029)

[5] Y.A. Abramovich and A.W. Wickstead, *When each continuous operator is regular. II*, Indag. Math. (N.S.) **8** (1997), 281–294. MR1622216 (99c:47046)

[6] Y.A. Abramovich and C.D. Aliprantis, *An invitation to operator theory*, Graduate Studies in Mathematics, vol. 50, American Mathematical Society, Providence, RI, 2002, ISBN 0-8218-2146-6. MR1921782 (2003h:47072)

[7] Donald I. Cartwright and Heinrich P. Lotz, *Some characterizations of AM- and AL-spaces*, Math. Z. **142** (1975), 97–103. MR0383030 (52 #3912)

[8] Z.L. Chen and A.W. Wickstead, *Equalities involving the modulus of an operator*, Math. Proc. R. Ir. Acad. **99A** (1999), 85–92. MR1883067 (2003a:47081)

[9] Nicolae Dăneţ, *The Riesz decomposition property for the space of regular operators*, Proc. Amer. Math. Soc. **129** (2001), 539–542. MR **1707144 (2001e:**47065)

[10] M. Elliott, *Abstract Rademacher Systems in Banach Lattices*, Ph.D. thesis, Queens University Belfast, 2001.

[11] Z. Ercan and S. Onal, *A Dedekind σ-complete Banach lattice into which all bounded operators are regular*, Positivity **9** (2005), 397–399. MR **2188527 (2006f:**47047)

[12] Shizuo Kakutani, *Concrete representation of abstract (L)-spaces and the mean ergodic theorem*, Ann. of Math. (2) **42** (1941), 523–537. MR 0004095 (2,318d)

[13] ———, *Concrete representation of abstract (M)-spaces. (A characterization of the space of continuous functions.)*, Ann. of Math. (2) **42** (1941), 994–1024. MR 0005778 (3,205g)

[14] L.V. Kantorovich, *Concerning the general theory of operations in partially ordered spaces*, Dok. Akad. Nauk. SSSR **1** (1936), 271–274. (Russian)

[15] L.V. Kantorovich and B.Z. Vulikh, *Sur la représentation des opérations linéaires*, Compos. Math. **5** (1937), 119–165.

[16] J.L. Krivine, *Sous-espaces de dimension finie des espaces de Banach réticulés*, Ann. of Math. (2) **104** (1976), 1–29. MR0407568 (53 #11341)

[17] Peter Meyer-Nieberg, *Banach Lattices*, Universitext, Springer-Verlag, Berlin, 1991, ISBN 3-540-54201-9. MR1128093 (93f:46025)

[18] Hidegorô Nakano, *Über normierte teilweise geordnete Moduln*, Proc. Imp. Acad. Tokyo **17** (1941), 311–317. MR 0014174 (7,249g) (German)

[19] ———, *Über die Charakterisierung des allgemeinen C-Raumes*, Proc. Imp. Acad. Tokyo **17** (1941), 301–307. MR 0014175 (7,249h) (German)

[20] A.C.M. van Rooij, *When do the regular operators between two Riesz spaces form a Riesz space?*, Technical Report 8410, Katholieke Universiteit, Nijmegen, 1984.

[21] Helmut H. Schaefer, *Banach lattices and positive operators*, Springer-Verlag, New York, 1974, Die Grundlehren der mathematischen Wissenschaften, Band 215. MR 0423039 (54 #11023)

[22] Zbigniew Semadeni, *Banach Spaces of Continuous Functions. Vol. I*, PWN – Polish Scientific Publishers, Warsaw, 1971, Monografie Matematyczne, Tom 55. MR0296671 (45 #5730)

[23] J. Synnatzschke (Ju. Synnačke), *The operator that is conjugate to a regular one, and certain applications of it to the question of the complete continuity and weak complete continuity of regular operators*, Vestnik Leningrad. Univ. (1972), 60–69. MR 0312314 (47 #876) (Russian, with English summary)

[24] A.I. Veksler and V.A. Geĭler, *Order completeness and disjoint completeness of linear partially ordered spaces*, Sibirsk. Mat. Ž. **13** (1972), 43–51. MR0296654 (45 #5713) (Russian)

[25] A.W. Wickstead, *The regularity of order bounded operators into $C(K)$*, Quart. J. Math. Oxford Ser. (2) **41** (1990), 359–368. MR1067490 (91k:47086)

[26] ———, *Spaces of operators with the Riesz separation property*, Indag. Math. (N.S.) **6** (1995), 235–245. MR1338329 (96g:47032)

[27] ———, *Order bounded operators may be far from regular*, Functional Analysis and Economic Theory (Samos, 1996), Springer, Berlin, 1998, pp. 109–118. MR1730124 (2000i:47071)

[28] ———, *AL-spaces and AM-spaces of operators*, Positivity **4** (2000), 303–311, Positivity and its applications (Ankara, 1998). MR1797132 (2001j:47038)

[29] Hong Yun Xiong, *On whether or not* $\mathcal{L}(E,F) = \mathcal{L}^r(E,F)$ *for some classical Banach lattices* E *and* F, Nederl. Akad. Wetensch. Indag. Math. **46** (1984), 267–282. MR **763464 (86a:**47042)

A.W. Wickstead
Pure Mathematics Research Centre
Queen's University Belfast
University Road
Belfast BT7 1NN
Northern Ireland, UK

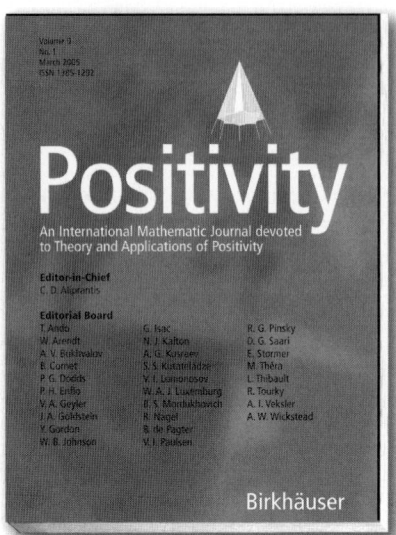